A Chanticleer Press Edition

National Audubon Society®
Field Guide to
North American Trees

National Audubon Society® Field Guide to North American Trees

Eastern Region

Elbert L. Little
former Chief Dendrologist,
U.S. Forest Service

Photographs by
Sonja Bullaty and Angelo Lomeo
and others

Visual Key by
Susan Rayfield and Olivia Buehl

Alfred A. Knopf, New York

This is a Borzoi Book
Published by Alfred A. Knopf

 Published in the United States by Alfred A. Knopf, a division of Penguin Random House LLC, New York, and distributed in Canada by Random House of Canada, a division of Penguin Random House Ltd., Toronto.

www.aaknopf.com

Prepared and produced by Chanticleer Press, Inc., New York.

Color reproductions by Nievergelt Repro AG, Zurich, Switzerland. Typeset in Garamond by Dix Type Inc., Syracuse, New York. Printed and bound by Toppan Leefung Printing Limited, China.

Published July 2, 1980
Thirty-second printing, October 2015

Library of Congress Cataloging-in-Publication Number: 79-3474
ISBN: 978-0-394-50760-6

CONTENTS

NATIONAL AUDUBON SOCIETY

The mission of NATIONAL AUDUBON SOCIETY, *founded in 1905, is to conserve and restore natural ecosystems, focusing on birds, other wildlife, and their habitats for the benefit of humanity and the earth's biological diversity.*

One of the largest, most effective environmental organizations, AUDUBON has nearly 550,000 members, numerous state offices and nature centers, and 500 + chapters in the United States and Latin America, plus a professional staff of scientists, educators, and policy analysts. Through its nationwide sanctuary system AUDUBON manages 160,000 acres of critical wildlife habitat and unique natural areas for birds, wild animals, and rare plant life.

The award-winning *Audubon* magazine, which is sent to all members, carries outstanding articles and color photography on wildlife, nature, environmental issues, and conservation news. AUDUBON also publishes *Audubon Adventures,* a children's newspaper reaching 450,000 students. Through its ecology camps and workshops in Maine, Connecticut, and Wyoming, AUDUBON offers nature education for teachers, families, and children; through *Audubon Expedition Institute* in Belfast, Maine, AUDUBON offers unique, traveling undergraduate and graduate degree programs in Environmental Education.

AUDUBON sponsors books and on-line nature activities, plus travel programs to exotic places like Antarctica, Africa, Baja California, the Galápagos Islands, and Patagonia. For information about how to become an AUDUBON member, subscribe to *Audubon Adventures,* or to learn more about any of its programs, please contact:

AUDUBON
225 Varick Street, 7th Floor
New York, NY 10014
(212) 979-3000
(800) 274-4201
www.audubon.org

ACKNOWLEDGMENTS

Many people have cooperated in the preparation of this field guide. Sonja Bullaty and Angelo Lomeo, who spent months photographing more than half the species in the book, deserve special mention. Their work and that of other photographers make the collection of color photographs in this book unrivaled by any other. I also wish to thank Richard Weaver and the staff of the Arnold Arboretum, Harvard University, for making their large collection of trees available to our photographers.

A special note of appreciation is due Thomas H. Everett, New York Botanical Garden, for his review of the manuscript and photographs. Marshall C. Johnston of the University of Texas, Austin, and his associate Scooter Cheatham tracked down and photographed many uncommon and local species. Thomas L. Gates, Director of the Plant Sciences Data Center, Mt. Vernon, Virginia, was of great assistance in locating many trees. The following individuals have also given graciously of their time: James L. Kelly and Bob Hopeful, Monroe County (N.Y.) Park System; Clair A. Brown, Louisiana State University; George W. Argus, National Museum of Canada; Benjamin Blackburn,

Willowwood Arboretum; Robert Hebb and Michael Ruggerio, Cary Arboretum; Bill Jordan, University of Wisconsin Arboretum; John E. Ford, Secrest Arboretum; Peter M. Mazzeo, U.S. National Arboretum; Peter W. Bristol, Holden Arboretum; Donald R. Hendricks, Hayes Regional Arboretum; Daniel Tompkins, Bayard Cutting Arboretum; Gurdon L. Tarbox, Jr., and William McBee, Brookgreen Gardens; Walter O. Petroll and Theresa George, Winterthur Gardens; Gregory Armstrong, Botanic Garden, Smith College; John M. Fogg, Jr., Barnes Foundation; Wilbur H. Duncan, University of Georgia; Fred C. Galle, Calloway Gardens; Mitchell Flinchum, School of Forestry, University of Florida; Fred Lape, Susan Shafner, and John Bak, George Landis Arboretum; David Patterson, Longwood Gardens; Marion T. Hall and George Ware, Morton Arboretum; Benny J. Simpson, Texas A & M University.

My wife, Ruby Rice Little, typed the manuscript. The dedicated staff of Chanticleer Press has contributed in many ways. Olivia Buehl located nearly every native tree species, assembled the photographs, and edited the manuscript. Susan Rayfield created the visual keys. Mary Beth Brewer coordinated the numerous maps and marginal artwork. My sincere thanks go to Paul Steiner, Gudrun Buettner, and Milton Rugoff for their support and encouragement, and to Carol Nehring, Helga Lose, Ray Patient, and Dean Gibson for their special efforts in the production of the book.

INTRODUCTION

Among the most beautiful and useful products of nature, trees have been cherished since ancient times. The oxygen we breathe is released by trees and other green plants; trees prevent soil erosion and provide food and cover for animals. They also supply almost countless products ranging from coal and timber to paper and plastics. With urban sprawl overtaking more and more of the landscape, the importance of national and state parks and forests for recreation is enormous.

Worldwide, the number of tree species may exceed 50,000; of these, about 680 are native to the United States and Canada. Yet even these species exhibit great variation. Some trees, like most conifers and hollies, are evergreen year-round; others, like many maples, put on a blaze of fall color before losing their leaves in winter. Even the shapes of trees vary greatly, with crowns that may be pyramidal, conical, columnar, spreading, vase-shaped, or round.

This book is designed for everyone who wants to know how to identify the trees in backyards, along streets, and in city parks as well as in forests. The combination of vivid color photographs and clear, non-technical descriptions in this guide should make tree identification easy and enjoyable for all.

Tree Shapes

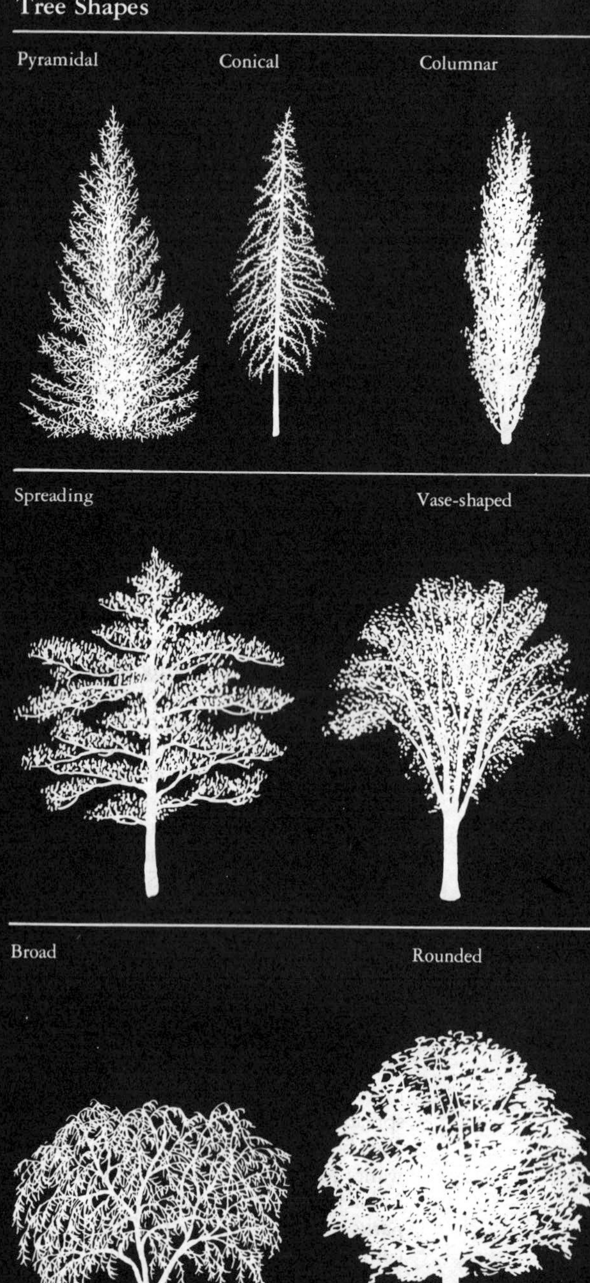
Pyramidal
Conical
Columnar
Spreading
Vase-shaped
Broad
Rounded

Definition of a Tree: A tree is a woody plant with an erect perennial trunk at least 3 inches (7.5 centimeters) in diameter at breast height (4½ feet or 1.3 meters), a definitely formed crown of foliage, and a height of at least 13 feet (4 meters). In contrast, a shrub is a small woody plant usually with several perennial stems branching from the base. Some species that are normally trees remain shrubs under severe conditions, such as particularly cold or dry climates; these are included in this guide. Other species, which are usually shrubs and rarely attain tree size, are excluded. The trunk of a tree includes the fibrous inner wood and the outer covering, or bark. Leaves manufacture food by photosynthesis and may be either evergreen (remaining on the tree all year) or deciduous (falling off in winter or during a dry period).

Leaves: Each leaf is made up of a broad, flat blade and a leafstalk, which is usually short and narrow (or sometimes absent) and is attached to the twig. Most leaves have one blade and are known as simple leaves. Compound leaves have the blade divided into three or more leaflets, which may be arranged in two rows along an axis (pinnately compound), on side branches off an axis (bipinnately compound), or radiating like a fan (palmately compound). Whether simple or compound, leaves are attached to the stem in three different ways. Those attached singly at different levels (nodes) on the stem are called alternate, those attached at the same level in pairs are opposite, while three or more attached at one level in a ring are whorled. Alternate arrangement is by far the most common.

Alternate

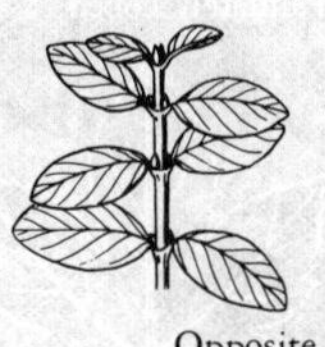
Opposite

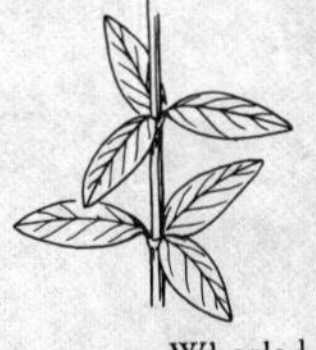
Whorled

Leaves can be grasslike (linear); shaped like a lance, with a pointed tip and a broader base (lanceolate); reverse lance-shaped, broadest near the tip and pointed at the base (oblanceolate);

Leaf Types

Scales

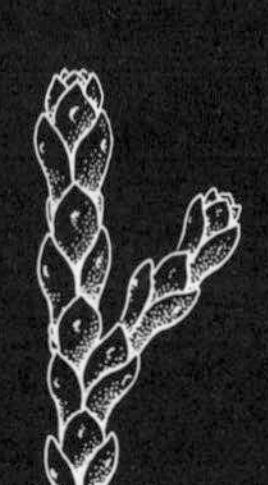

Needles in Cluster

Needles in Bundle

Oblanceolate

Spatulate

Obovate

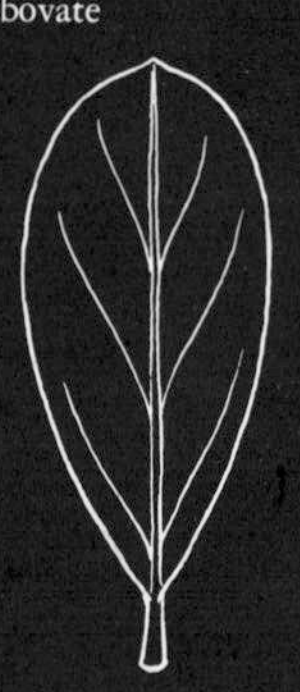

Elliptical

Pinnately Lobed

Palmately Lobed

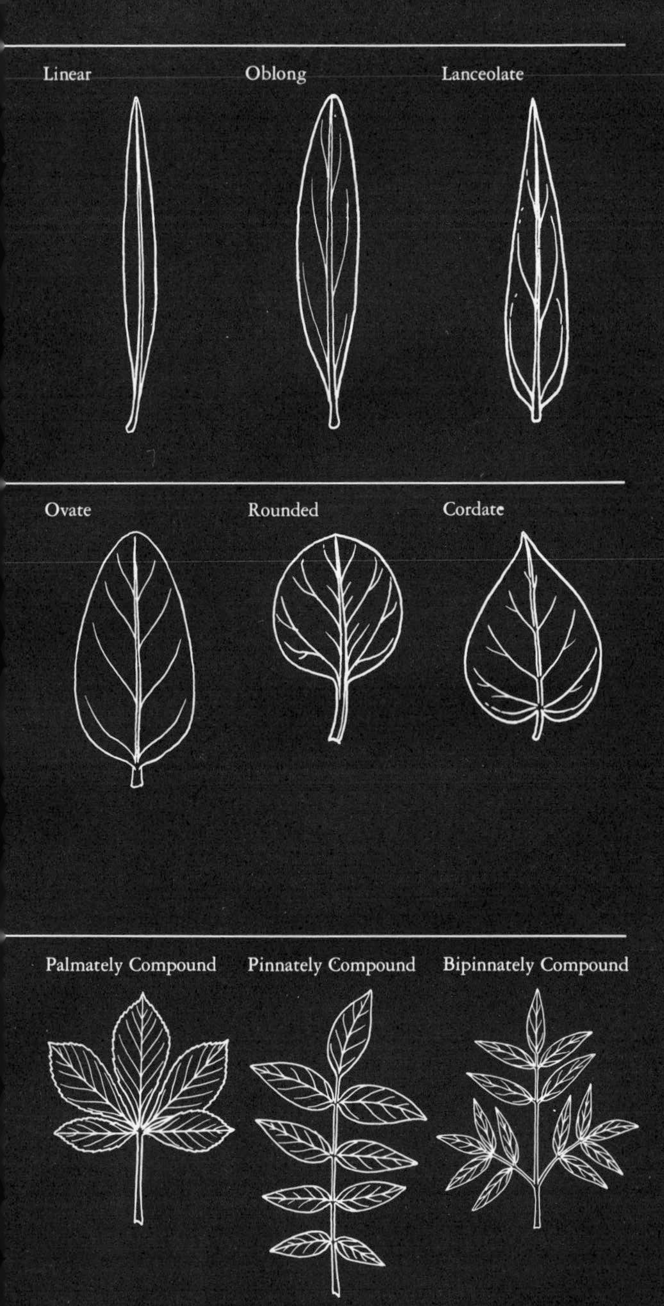

Linear
Oblong
Lanceolate
Ovate
Rounded
Cordate
Palmately Compound
Pinnately Compound
Bipinnately Compound

shaped like an egg (ovate); reverse egg-shaped, broadest toward the tip (obovate); oval, broadest in the middle (elliptical); with nearly parallel edges (oblong); rounded, in the shape of a circle; or spoon-shaped (spatulate).
In addition to overall shape, leaves may be further distinguished by having edges without teeth (entire) or toothed; the teeth may point forward as in a saw (saw-toothed). The edges may also be wavy, turned under, or deeply divided into parts (lobed). In some species the leaves of young plants or of vigorous twigs are of a different shape and are known as juvenile leaves. Leaves of conifers are either long and narrow (needlelike) or short, pointed, and often overlapping (scalelike).

Flowers: Flowers usually consist of four parts. The outermost calyx is composed of leaflike parts called sepals. The corolla consists of the colored petals. (The corolla may be regular, with equal petals like the spokes of a wheel; irregular, with petals of unequal size; or tubular, with a tube and lobes formed by united petals.) The stamens, or male organs of the flower, have a filament, or stalk, and an anther, or enlarged part (usually yellow), composed of four or fewer pollen-sacs containing tiny pollen grains. In the center of the flower is the pistil(s), or female organ, which has three parts: the stigma at the tip which receives the pollen, the style or stalk (sometimes absent), and the enlarged ovary at the base. The ovary has one to several cells, each containing one to many ovules, the rounded, whitish egg-bearing units. The mature ovary is the fruit, while the mature ovules become seeds after fertilization, each capable of germinating into a tiny plant.
Most tree flowers are bisexual, possessing both pollen-bearing male stamens and female pistil(s). Others are of one sex only, and depending upon

Leaf and Flower Parts

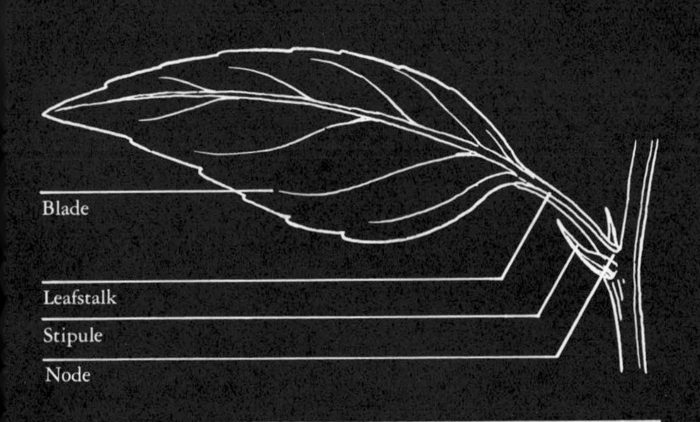

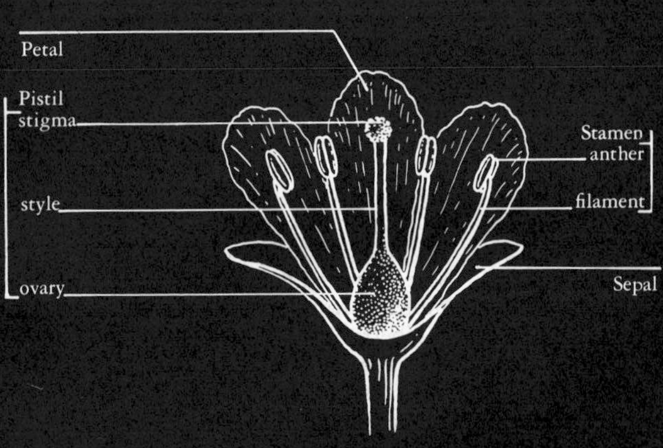

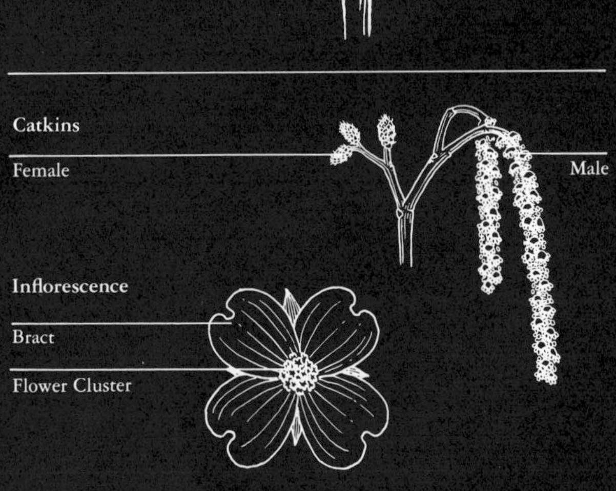

the species, the two sexes appear either on the same or on separate trees. A few species have both unisexual and bisexual flowers on the same plant. Some flowers are borne singly; more commonly, they are clustered. Catkins, common in willows and poplars, are unbranched clusters, usually with a drooping axis, scales, and stalkless flowers of one sex. Dogwood flowers are characterized by inflorescences made up of a flower cluster and petal-like bracts.

Fruit: The fruit may be a berry, drupe, pome, multiple fruit, aggregate fruit, acorn or other nut, key (samara), achene, pod, capsule, or follicle. Commonly, the fruit develops from the ovary of a single pistil and is known as a simple fruit. Simple fruits may be dry (hard), as in the hickories, or fleshy with seeds in mealy or juicy pulp, as in most hawthorns. A fruit that develops from several pistils within a flower is known as an aggregate fruit, such as that of a magnolia. A fruit that develops from several united flowers is known as a multiple fruit; an example is a mulberry. Fleshy fruits, which do not open, include the berry, which has several seeds as in persimmons; the drupe which has a central hard stone, as in cherries; and the pome which has a fleshy outer part and hard or papery core containing seeds, as in apples. A nut is a one-seeded fruit with a thick, hard shell (a nutlet is similar but smaller); an acorn is a distinctive, pointed nut that has a scaly cup at the base. A key (samara)—characteristic of maples, elms, and ashes—is a one-seeded fruit with a thin, flat, dry wing, while an achene is a small, dry seedlike fruit with a thin wall. The Legume Family is characterized by dry, one-celled pods that split open, usually along two grooved lines; in contrast, a follicle opens along one line. Capsules are dry fruits with two or more cells,

Types of Cones and Fruit

Cone

Nut

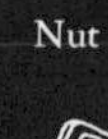

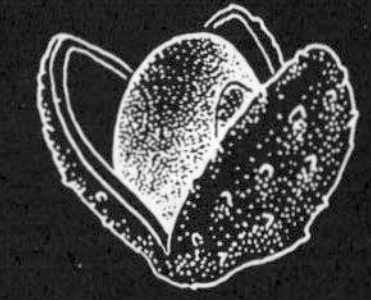

Acorn

Pod

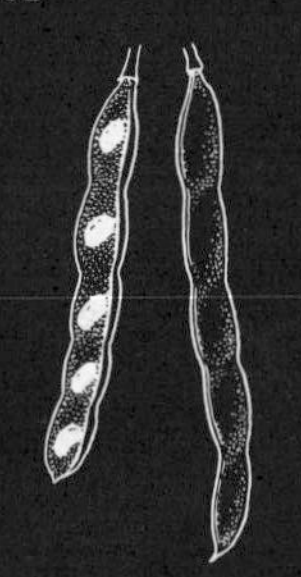

Achene

Key

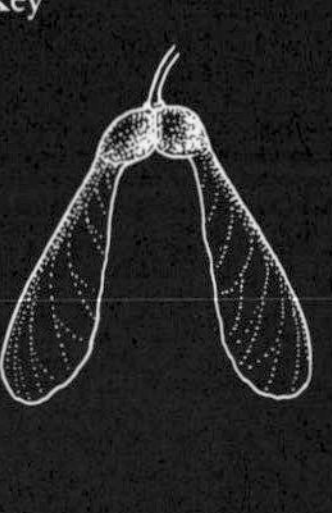

Drupe

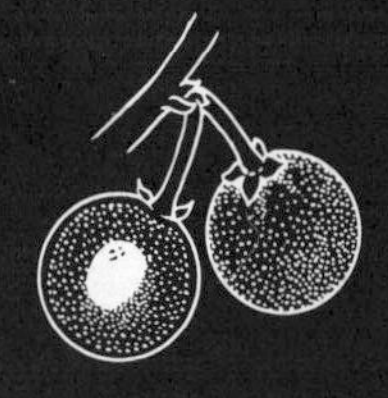

Berry

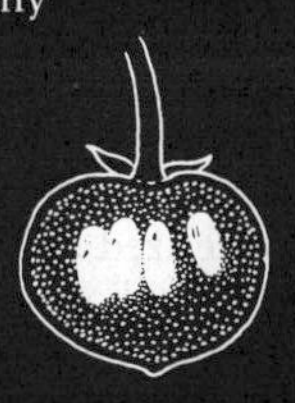

Pome

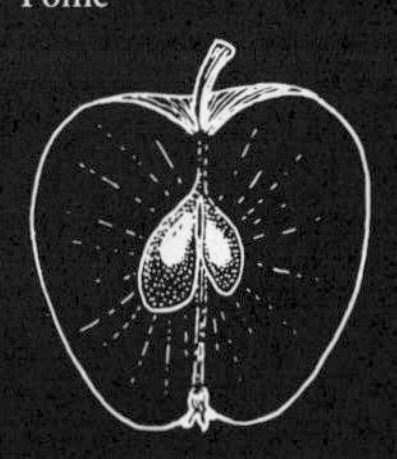

Capsule

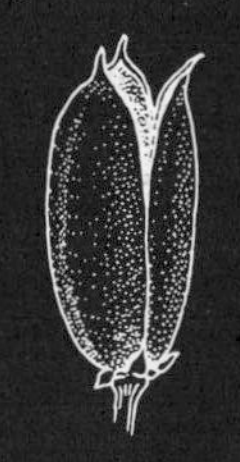

Aggregate of Capsules

Multiple of Nutlets

opening along as many lines as there are cells, as in some yuccas. While conifers technically lack flowers and fruit, they have separate male and female parts known as cones, which are usually on the same tree. The smaller, non-woody, male cones, which resemble flowers, produce the pollen; the larger, more familiar, female cones are composed of hard cone-scales, each commonly above a smaller scale, with exposed seeds at the base of the cone-scale.

Major Eastern Habitats: Tree species are not distributed at random but are associated with special habitats. The northern coniferous forests of spruces and firs extend across Canada and into the northeastern United States, where pines are also characteristic, and southward at high altitudes in the Appalachian Mountains. In the eastern United States there are extensive deciduous forests of hardwoods, such as maple–beech in the North and oak–hickory in the South. The Southeast has vast pine forests, mostly at low altitudes on the Coastal Plain. High mountains, such as the southern Appalachians, with climates that are colder than nearby lowlands are characterized by spruce–fir forests similar to those of eastern Canada. Some tree species are found commonly in wet soils of bogs, river banks, and flood plains. Other trees are adapted to dry uplands. Jack Pine is common on sandy upland soils in the Great Lakes region. Eastern Redcedar thrives on rock outcrops and cliffs forming cedar glades. Loblolly Pine is called "Oldfield Pine" because it invades abandoned farm lands; Pin Cherry, or "Fire Cherry," characteristically grows in areas recently razed by forest fire.

Geographic Scope: This volume covers the 315 native, or indigenous, trees of the eastern two thirds of North America, with the exception of roughly 100 tropical species confined to southern Florida. Also included are many common naturalized trees, species which have been introduced from another area and have become common and established as though wild, reproducing naturally and spreading. Additionally, a number of cultivated, or planted, tree species often seen in parks and along streets are included. Our range is bounded by the Arctic tree line across northern Canada, the Atlantic Ocean in the east, the Gulf of Mexico to the south, and the Rocky Mountains in the west. This division follows the natural boundary between eastern and western North America for most plants and animals, running from the eastern base of the Rocky Mountains in Alberta and eastern Montana south to the Big Bend region of southwestern Texas (see map). Except along streams, eastern trees gradually thin out westward in the generally treeless, prairie-plains grasslands of the interior. Western tree species are treated in the companion volume by the same author.

Use of Color Photographs: We have chosen to use color photographs rather than conventional drawings because they show bark details and leaves, flowers, and fruit growing on the trees as you would find them in their natural setting, making identification much easier. In general, we have selected photographs that show the most typical examples of a species. Where a species is variable, we have tried to indicate its range, sometimes showing more than one example, such as the White Mulberry, which has leaves both toothed and lobed.

Organization of the Color Plates: Unlike birds, mammals, reptiles, insects, and wildflowers, whose overall color and shape can be taken in quickly at a glance, trees are too large and have too many components to be easily identified from just one photograph.

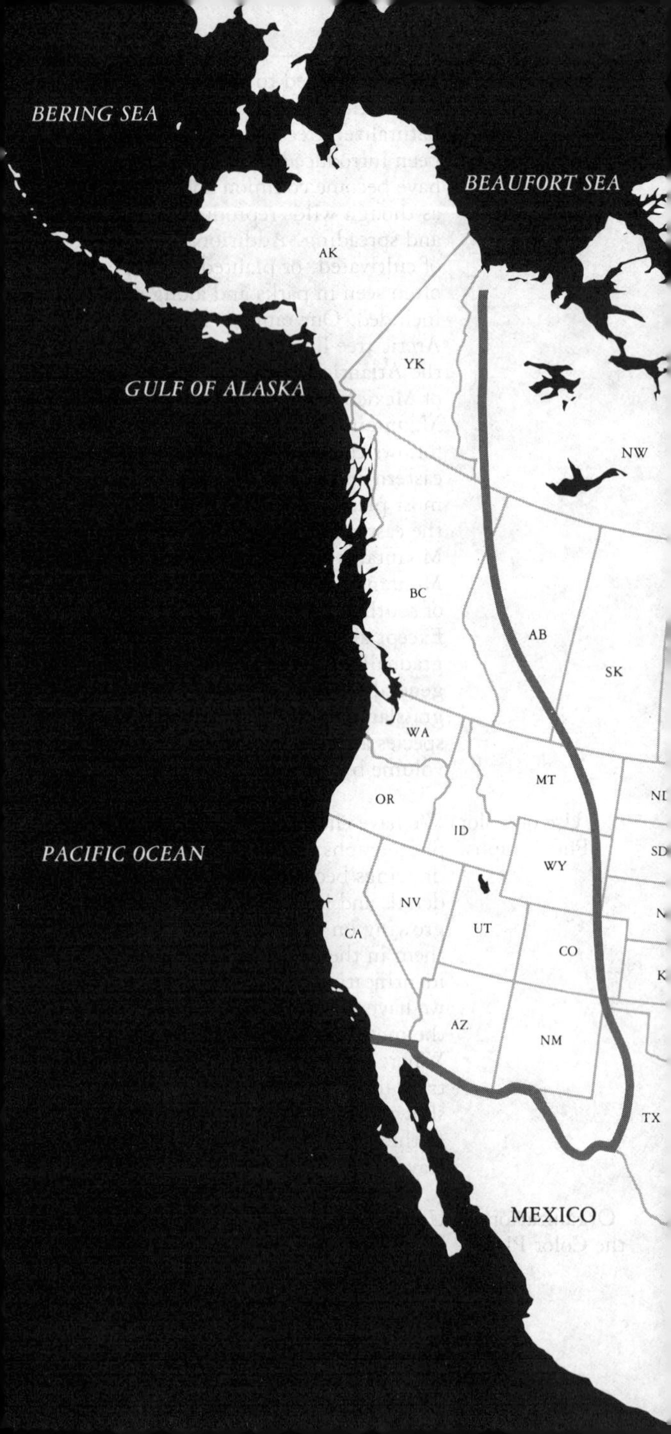

BERING SEA

BEAUFORT SEA

AK

YK

GULF OF ALASKA

NW

BC

AB

SK

WA

MT

OR

ID

PACIFIC OCEAN

SD

WY

NV

UT

CA

CO

AZ

NM

TX

MEXICO

GREENLAND

BAFFIN BAY

HUDSON BAY

MB

ON

QU

NF

PEI

NB

NS

ME

VT

NH

MA

CT

RI

NY

MN

WI

MI

IA

PA

NJ

IL

IN

OH

MD

DE

MO

WV

VA

KY

ATLANTIC OCEAN

OK

NC

TN

AR

SC

MS

AL

GA

LA

FL

GULF OF MEXICO

CARIBBEAN SEA

Therefore, we have created four separate visual keys—leaf, flower, cones and fruit, and autumn foliage—each corresponding to a major identification feature of the tree. The four keys are explained and the symbols introduced in the thumb print guide which appears before the color plates.

Leaf Visual Key: Since leaves are the most noticeable feature of an unfamiliar tree, and since they remain on the tree a relatively long time, our primary visual key to identification is by leaf photographs, paired with characteristic bark photographs. These leaf photographs are grouped by similar shape, regardless of their family or genus relationships. Many related trees, such as those in the Maple Family, have similarly shaped leaves, and therefore will appear near each other in the photo section. On the other hand, various groups of trees such as oaks or hollies, whose species have widely different-looking leaves, will be found in two sections. The leaf photographs are arranged in the following categories:

Needle-leaf Conifers
Scale-leaf Conifers
Untoothed Simple Leaves
Toothed Simple Leaves
Lobed Simple Leaves
Compound Leaves
Yuccas and Palmettos

Within each group, we have further loosely organized the photographs according to surface texture and vein patterns, keeping in mind that leaves can vary from one individual tree of a species to another and even within one tree. The needle-leaf conifers have been arranged by needle length, from longest to shortest.

The bark photograph of each tree usually shows the bark of a mature trunk of average size. The color, texture, and surface pattern are often important in distinguishing different species in the same family, such as birches, and are a secondary characteristic in others. It should be noted, however, that many related species have similar bark. Also, the bark of a tree often changes with age, as outer layers are cracked and shed. Therefore, the bark photographs usually serve only as a secondary tool for identification.

Flower Visual Key: Flowers are also important in the identification of trees. Our flower visual key includes many photographs of the most common and interesting flowers, ranging from the tiny Red Maple, just ⅛″ long, to the Horsechestnut, which has showy blooms in a cluster 10″ long. Since most people notice color first, we have arranged flowers by color in the following groups:

Red
Green
Yellow
Cream-colored
White
Pink
Purple

Other important features to note in the flower photographs are shape, number of petals, and whether a flower is clustered or solitary.

Cones and Fruit Visual Key: Cones and fruit are also extremely important in identification, particularly since some remain on the tree or the ground long after they have matured. We have grouped them by shape in the following easily recognizable categories:

Cones
Keys
Pods
Balls, Capsules, and Tufted Fruit
Acorns

Nuts
Berrylike Fruit
Fleshy Fruit

Within these groups the photographs are further arranged by color. Conelike fruits are placed with the true conifer cones. Some fruits that appear berrylike are grouped with the berries.

Autumn Leaves Visual Key: The fourth major visual key illustrates some of the most spectacular examples of fall foliage, arranged according to color:

Red
Orange and Brown
Yellow

Certain trees, such as Black Tupelo and Flowering Dogwood, are characteristically scarlet in the fall. Others, such as birches, aspens, and alders, are brilliant yellow. Still others, such as American Beech, may be either yellow or brownish.

How to Read the Captions: The caption beneath each photograph gives the plate number, common name, measurement, and text page on which the species is described. Where helpful, other pertinent information is included, such as the number of pine needles in a bundle; if the tree is evergreen; whether the leaves are opposite instead of in the usual alternate arrangement; and if the leaves are compound, whether they are pinnate, bipinnate, or palmate. In the flower section, where important, the sex of the tree flower is given: either male (♂) or female (♀). Measurements include cluster width (cw.) and cluster length (cl.) as well as the width (w.) and length (l.) of a single flower.

Common Names: Each tree has at least two names, one common and one scientific. The English common names in this guide include some that have been adapted from other languages, particularly Spanish names of trees in regions bordering Mexico. The main common names used here follow the latest U.S. Forest Service *Checklist of United States Trees (Native and Naturalized)* by the author (U.S. Department of Agriculture, Agriculture Handbook 541; 1979). Other English names in use locally are also included, designated by quotation marks. Since a tree may have several common names, botanists have also given it a scientific name, which is used throughout the world.

Scientific Names: Scientific names are usually from Latin or, if derived from Greek or other languages, are in Latin form. Each scientific name consists of two words: the genus name (always capitalized), followed by the species name (lower case). For example, all oaks are in the genus *Quercus*; the species known as White Oak is designated *Quercus alba* from the genus *Quercus* and the species *alba*. A related species in the same genus, Chestnut Oak, is known as *Quercus prinus*. The scientific name of a tree may be followed by the name of the author (or authors) who named it, usually abbreviated. For example, *Picea rubens* (Red Spruce) is followed by Sarg., the abbreviation for Charles Sprague Sargent, author of the 14-volume *Silva of North America*. The author's name helps in tracing the history of the scientific name and the discovery of the species. Varieties of trees are expressed with the variety name preceded by "var."—for example, *Populus deltoides* var. *occidentalis,* a variety of Eastern Cottonwood.

Classification: Trees are classified in two major plant groups: Gymnosperms, which include the conifers, and Angiosperms, which include all the flowering plants. The conifers, or softwoods, are usually evergreen, having narrow, needlelike or scalelike leaves, and bear exposed seeds, usually in cones. The flowering plants bear seeds enclosed in fruits and are further divided into two groups. Among the Monocotyledons, only yuccas and palms reach tree size; they have large, evergreen, bayonetlike or fanlike leaves that are usually clustered at the top of the trunk or on a few large branches. Dicotyledons are the largest group and contain all the hardwoods, or broadleaf trees, which are mostly deciduous and have net-veined leaves.

Organization of the Text: The text contains a description of each native tree species, as well as common introduced species, found within the geographical range of this guide. Arrangement of families is botanical; within each family the listing is alphabetical by scientific name.

Text Descriptions: Each species description begins with a general statement about the shape and appearance of the tree as well as any key features that are useful in immediate identification.

Height and Diameter: Also given are measurements (in both the English and metric systems) for the height and diameter (at breast level) of mature specimens under favorable growing conditions. Naturally, height and trunk diameter vary greatly with age, location, and climate.

The description of the tree is subdivided into sections on leaves, bark, twigs, flowers, and fruit. Where flowers do not vary greatly from species to species, as among the oaks, they are discussed in the family description. Wherever possible, familiar terms are used in the description of trees; where technical terms are necessary, they are defined in the Glossary (page 681). Important details are in italics.

Leaves: The leaf description includes information on size, shape, edges, texture, and color. Since most leaves are arranged in alternate positions on the stem, leaf arrangement is usually mentioned only when it is opposite or whorled. Similarly, since most trees are deciduous, only the less common evergreen characteristic is specified. The color of autumn foliage is also described.

Bark: The bark description gives details of color, texture, thickness, shagginess, and other relevant details. The barks of some species change significantly with age, and in these cases, both young and mature trees are described.

Twigs: Twig descriptions include color, stoutness, type of branching, hairiness, fragrance, and stickiness, if applicable.

Flowers: The flower description covers size, shape, and color as well as time of flowering and fragrance. If the tree has both male and female flowers, each is described. While many flowers are solitary, others are grouped in clusters; in these cases the description includes the size of the cluster and where it is placed on the twig.

Fruit: Fruit descriptions indicate size, shape, color, surface texture, dryness or pulpiness, shape and number of seeds, and time of maturity.

Habitat: The habitat, or place where a tree grows naturally, is given for each species and is often important in identification. Commonly associated vegetation is also mentioned.

Range: The boundaries of the natural range, or geographic distribution, of each tree species are given, generally in a clockwise direction from northwest to east, then south, west, and north. Additionally, the approximate range in altitude is cited. For introduced species the native continent or region is given, followed by information on the distribution in North America. Range maps for each native species supplement the range description. Condensed from the author's 5-volume *Atlas of United States Trees,* these maps reflect the most accurate and recent information on distribution available.

Comments: The description of each species concludes with notes on the uses of the tree and its products, cultivation, history and lore, and frequently explains the origin of names.

Drawings: Typical winter silhouettes of many species appear in the margins as a further aid to identification. The drawings show the characteristic shape and manner of branching of trees growing under ideal conditions. Since evergreen trees have foliage year round, they are shown with their leaves or needles. Fruits not illustrated in color photographs are shown in small drawings beside the text. Additionally, we have included in the margins next to the descriptions, drawings of the leaves of a few rare or inaccessible species for which photographs were not available.

Poisonous Trees: A few tree species are poisonous, particularly Poison-sumac (*Toxicodendron vernix*), which may cause a skin rash. Various parts of certain trees, such as seeds of buckeyes (*Aesculus*), are also toxic. A safe rule is not to eat unknown fruits and seeds.

Rare Species: Relatively few species of native trees are rare or local in distribution. According to lists proposed under the Endangered Species Act of 1973, very few tree species are threatened with extinction. Rare tree species get some protection in public parks, forests, and other preserves. However, the destruction and disturbance of habitats make the establishment of additional preserves desirable. In the wild, one may usually obtain small botanical specimens without damage; of course, collecting should never be done in a park or arboretum.

HOW TO USE THIS GUIDE

Example 1
A Broadleaf Tree

You are in a lowland swamp surrounded by trees with broad, saw-toothed leaves that have 3 pointed lobes and 2 smaller lobes near the base. The leaves are in opposite arrangement on the twig and have red leafstalks. The bark is grayish brown with scaly ridges.

1. In the Thumb Tab Guide preceding the color plates, look for the section called *Lobed Simple Leaves*. The only ones with a shape resembling your specimen are *Maples* and *Sycamores* (plates 253–264).
2. Turn to the color plates for these 2 groups. Reading the captions eliminates the sycamores, which have alternate leaves; you now know your leaf is a maple. In studying the photographs you discover that only Planetree Maple, Red Maple, and Chalk Maple have red leafstalks. Looking more closely at your leaf specimen reveals that it has saw-toothed edges like the Red and Planetree maples but unlike Chalk Maple. The captions indicate that the size of your specimen is closer to that of Red Maple.
3. The caption under the color plates refers you to the species description of Red Maple. Reading the text confirms your identification of Red Maple, a widespread tree in the East, typically occupying damp habitats.

Example 2
A Bell-shaped White Flower

Next to a stream in a forest you see a small tree with tiny, bell-shaped, white flowers in drooping clusters.

1. In the Thumb Tab Guide you find that the flower section is divided into several color groupings. White flowers appear in plates 412–453.
2. Turn to the color plates of white flowers, where you find that 6 are bell-shaped, like your specimen. 2 of these are yuccas with much larger flowers, and 2 are silverbells which are attached to the twig instead of in clusters. The photographs of Tree Sparkleberry and American Bladdernut look most like the flower before you.
3. The text section indicated in the captions for these 2 species confirms that both trees have bell-shaped flowers. Your flower has 5 separate petals, as in American Bladdernut, and not the slightly 5-lobed corolla of Tree Sparkleberry. Moreover, the habitat description—dry upland areas—does not fit the tree you have found. The leaf description of palmately compound leaves with saw-toothed leaflets indicates this is American Bladdernut, which inhabits moist soils.

Example 3
A Beanlike Fruit

In a wooded area you find a spiny tree. The leaves have fallen, and many dark brown pods are lying on the ground. Many of the pods are over a foot long.

1. In the Thumb Tab Guide the *Cones and Fruit* section is divided into several groups including *Pods* (plates 502–507), which most closely resemble the example you have found.
2. Turn to the color plates. Of the 6 pods only 2 are similar to the ones you have found, Black Locust and Honeylocust. The caption for Black Locust eliminates this tree, since the pods are only 2–4″ long.
3. Turn to the text description of Honeylocust which confirms your identification of this spiny member of the Legume Family.

Keys to the Color Plates

The color plates are arranged according to four keys: leaves, flowers, fruit and cones, and autumn leaves. Within each key the photographs are organized in the following groups:

Leaf Key
Conifers
Untoothed Simple
Toothed Simple
Lobed Simple
Compound
Yuccas and Palmettos

Flower Key
Red
Green
Yellow
Cream-colored
White
Pink
Purple

Fruit and Cones Key
Cones
Keys
Pods
Balls, Capsules, and Tufted Fruit
Acorns
Nuts
Berrylike Fruit
Fleshy Fruit

Autumn Leaves Key
Red
Orange and Brown
Yellow

Thumb Tabs Each group of photographs within a visual key is represented by a silhouette which appears as a thumb tab at the left edge of each double-page of plates. A chart of the thumb tab organization appears on the pages preceding the color plates. The flower and autumn leaf groups are each indicated by one symbol, shown in various colors in the thumb tab.

Captions The caption under each photograph contains the plate number, common name of the tree, measurement, and the page on which it is described. When pertinent, additional information is given, such as whether a tree is evergreen, if the leaves are opposite each other on the stem (rather than the usual alternate arrangement), or if the leaf is pinnately, bipinnately, or palmately compound. The flower captions include length (l.) or width (w.) of an individual flower, cluster length (cl.) or width (cw.), and, if a flower is not bisexual, the male (♂) or female (♀) symbol. The fruit captions indicate the width, unless the fruit is oblong, when length is given.

Part 1
Color Keys

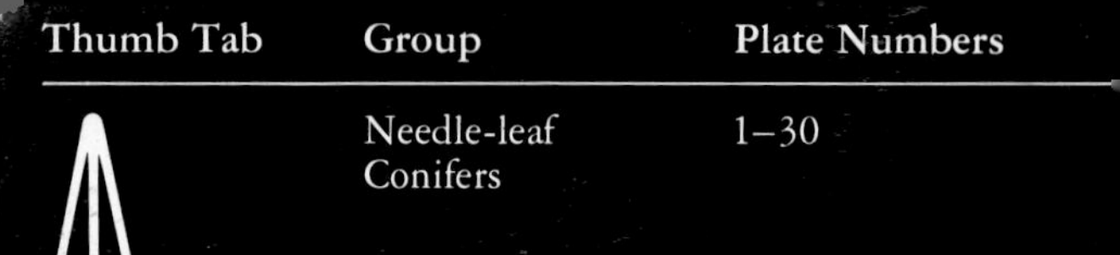

Thumb Tab	Group	Plate Numbers
	Needle-leaf Conifers	1–30
	Scale-leaf Conifers	31–39

Typical Leaf Shapes		Plate Numbers
	pines	1–15
	larches	17, 18
	hemlocks	19, 20
	firs	21, 22
	spruces	23–26
	baldcypresses	27, 28
	yews	29, 30
	cedars	31, 32, 34–36
	cypress	33
	junipers	37–39

Thumb Tab	Group	Plate Numbers
	Untoothed Simple Leaves	40–105

Typical Leaf Shapes		Plate Numbers
	persimmons	40, 83
	bumelias	42–44
	oaks	46–51, 101
	hollies	52, 54
	viburnums	53, 80
	mountain-laurel and rhododendrons	57, 62, 63
	magnolias	58, 85, 86, 88–93
	tupelos	66, 72
	dogwoods	76–79
	catalpas	94, 97

Thumb Tab	Group	Plate Numbers
	Toothed Simple Leaves	106–234

Typical Leaf Shapes		Plate Numbers
	willows	106–111, 113, 114, 116, 117
	hollies	115, 118, 143, 197, 207, 208, 212, 213
	apple and crab apples	121, 130, 133, 135
	plums	122, 123, 125, 127, 128, 140, 142, 144, 146
	cherries	124, 126, 129, 132, 134, 136, 138
	hackberries	139, 169–171
	buckthorns	145, 201
	hornbeam and hophornbeam	147, 222
	chestnut and chinkapins	148–151
	beeches	152, 174

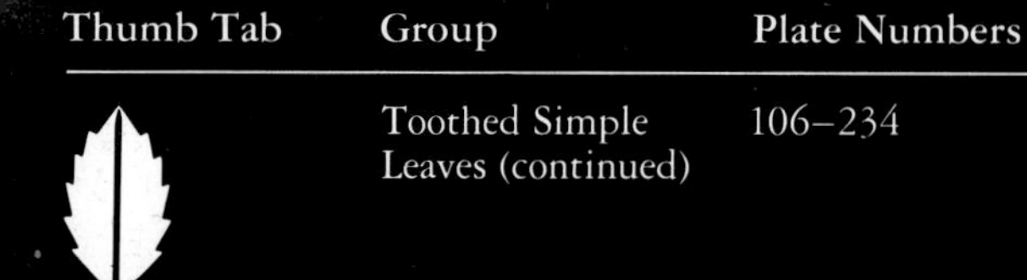

Thumb Tab	Group	Plate Numbers
	Toothed Simple Leaves (continued)	106–234

Typical Leaf Shapes		Plate Numbers
	mulberries	154, 157, 183
	basswoods and linden	155, 156, 158, 159
	elms	160–168, 205
	viburnums	173, 203, 204, 223
	serviceberries	175, 199, 200
	birches	176–182, 202
	aspens, poplars and cottonwoods	184–188, 190, 191
	silverbells	192, 195, 215
	hawthorns	210, 216–221, 224–226, 231, 232, 234
	alders	227–229, 233

Thumb Tab	Group	Plate Numbers
	Lobed Simple Leaves	235–294

Typical Leaf Shapes		Plate Numbers
	ginkgo	235
	mulberries	236, 237
	poplar	238
	sassafras	239
	hawthorns	240–252
	maples	253–255, 258–264
	sycamores	256, 257
	sweetgum	266
	yellow-poplar	267
	oaks	268–294

Thumb Tab	Group	Plate Numbers
	Compound Leaves	295–360
	Yuccas and Palmettos	361–363

Typical Leaf Shapes		Plate Numbers
	acacias	295–297, 300
	prickly-ashes	301, 339
	locusts	304–306, 309
	sumacs	307, 314, 315, 342, 349
	walnuts and butternut	316, 317, 330
	buckeyes	318, 355–359
	mountain-ashes	319–321
	hickories and pecan	324, 327, 332, 333, 337, 338, 343–345, 347, 348
	ashes	325, 326, 328, 329, 331, 334, 351
	yuccas	361, 362
	palmetto	363

Thumb Tab	Flower Key	Plate Numbers
	Red Flowers	364–373
	Green Flowers	374–390
	Yellow Flowers	391–399
	Cream-colored Flowers	400–411
	White Flowers	412–455
	Pink Flowers	456–468
	Purple Flowers	469–471

Thumb Tab	Cones/Fruit Key	Plate Numbers
	Cones	472–488
	Keys	490–501
	Pods	502–507, 510
	Balls, Capsules, and Tufted Fruit	508, 509, 511–513
	Acorns	514–525
	Nuts	526–537
	Berrylike Fruit	538–567
	Fleshy Fruit	568–576

Thumb Tab	Autumn Leaves	Plate Numbers
	Red Leaves	577–594
	Orange and Brown Leaves	595–606
	Yellow Leaves	607–630

The color plates on the following pages are numbered to correspond with the numbers preceding the text descriptions.

Needle-leaf and Scale-leaf Conifers

Most conifers, or softwoods, have narrow, needlelike leaves. Pines, the largest group, are recognized by relatively long needles in bundles of 2–5. The number of needles per bundle is indicated in parentheses in the captions. Larches have many crowded needles that are shed in autumn. Spruces have stiff, 4-angled, sharp-pointed needles, and hemlocks have flat, blunt-pointed needles; both have rough twigs with peglike bases left after the needles have fallen. Firs have softer needles and a distinct fragrance of balsam. The short needles of baldcypresses are shed with their slender twigs in autumn. Although yews are not conifers, they have been included here because they have flat, evergreen needles.

Some conifers have small, short, evergreen scalelike leaves arranged on branchlets in pairs or in 3's. Junipers have slender, angled, scaled branchlets, while twigs of white-cedars are usually flattened or 4-angled.

Oriental Arborvitae and Sawara False-cypress are cypresses, also known as "cedars," with flattened, scalelike leaves and twigs.

1 Eastern White Pine, (5) 2½–5″, *p. 296*

2 Longleaf Pine, (3) 10–15″, *p. 291*

3 Loblolly Pine, (3) 5–9″, *p. 297*

4 Pond Pine, (3) 5–8″, *p. 295*

5 Pitch Pine, (3) 3–5″, *p. 294*

6 Slash Pine, (2–3) 7–10″, *p. 288*

7 Red Pine, (2) 4¼–6½″, *p. 293*

8 Austrian Pine, (2) 3½–6″, *p. 290*

9 Shortleaf Pine, (2–3) 2¾–4½″, *p. 287*

10 Table Mountain Pine, (2–3) 1¼–2½″, *p. 292*

11 Spruce Pine, (2) 1½–4″, *p. 289*

12 Scotch Pine, (2) 1½–2¾″, *p. 297*

13 Sand Pine, (2) 2–3½″, *p. 287*

14 Virginia Pine, (2) 1½–3″, *p. 298*

15 Jack Pine, (2) ¾–1½″, *p. 286*

16 Cedar-of-Lebanon, 1–1¼″, *p. 279*

17 European Larch, *deciduous,* ¾–1¼″, *p. 280*

19 Carolina Hemlock, ⅜–¾″, *p. 300*

20 Eastern Hemlock, ⅜–⅝″, *p. 299*

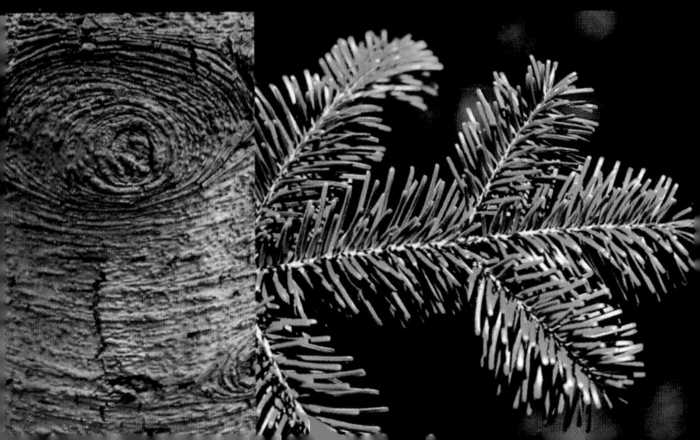

22 Balsam Fir, ½–1″, *p. 277*

23 White Spruce, ½–¾″, *p. 283*

25 Red Spruce, ½–⅝″, *p. 285*

26 Norway Spruce, ½–1″, *p. 282*

28 Montezuma Baldcypress, ¼–½″, *p. 303*

29 Florida Yew, ¾–1″, *p. 273*

30 Florida Torreya, 1–1½″, *p. 274*

31 Oriental Arborvitae, *opp.*, 1/16–1/8", *p. 313*

32 Northern White-cedar, *opp.*, 1/16–1/8", *p. 312*

33 Sawara False-cypress, *opp.*, 1/8", *p. 305*

34 Atlantic White-cedar, *opp.*, 1⁄16–1⁄8″, *p. 306*

35 Eastern Redcedar, *opp.*, 1⁄16″, *p. 310*

37 Ashe Juniper, *opp.*, ⅟16″, *p. 307*

38 Pinchot Juniper, ⅟16″, *p. 309*

Untoothed Simple Leaves

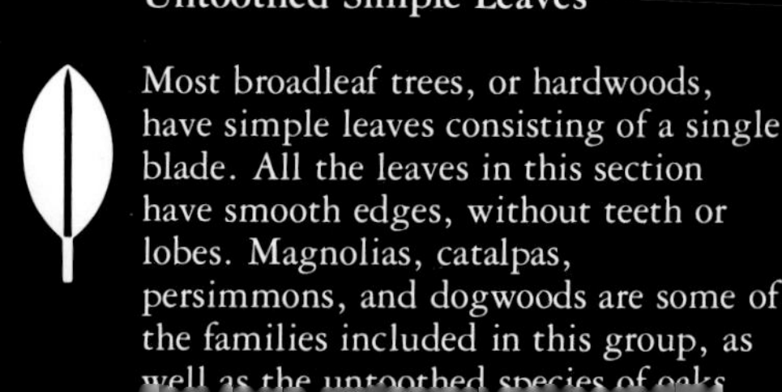

Most broadleaf trees, or hardwoods, have simple leaves consisting of a single blade. All the leaves in this section have smooth edges, without teeth or lobes. Magnolias, catalpas, persimmons, and dogwoods are some of the families included in this group, as well as the untoothed species of oaks

40 Texas Persimmon, ¾–1½″, *p. 634*

41 Bluewood, ⅝–1″, *p. 592*

42 Tough Bumelia, *evergreen,* 1-2½″, *p. 632*

43 Buckthorn Bumelia, 2-4″, *p. 631*

44 Gum Bumelia, 1-3″, *p. 630*

45 Buckwheat-tree, *evergreen*, 1–2″, *p. 554*

46 Bluejack Oak, 2–4″, *p. 392*

47 Live Oak, *evergreen*, 1½–4″, *p. 410*

48 Myrtle Oak, *evergreen*, ¾–2″, *p. 400*

49 Oglethorpe Oak, 2–5″, *p. 403*

50 Shingle Oak, 3–6″, *p. 391*

52 Dahoon, *evergreen*, 1½–3½", *p. 560*

53 Walter Viburnum, *evergreen*, *opp.*, 1-2½", *p. 673*

55 Sweetleaf, 3-5″, *p. 642*

56 Black-mangrove, *evergreen, opp.*, 2-4″, *p. 659*

58 Southern Magnolia, *evergreen*, 5–8″, *p. 440*

59 Florida Anise-tree, *evergreen*, 2½–6″, *p. 434*

61 Odorless Bayberry, *evergreen,* 2–4″, *p. 340*

62 Catawba Rhododendron, *evergreen,* 3–6″, *p. 626*

63 Rosebay Rhododendron, *evergreen,* 4–10″, *p. 627*

64 American Smoketree, 2–6″, *p. 545*

65 Elliottia, 2¾–4″, *p. 622*

66 Ogeechee Tupelo, 4–5½″, *p. 619*

67 Swamp Cyrilla, 1½–3″, *p.* 555

68 Devilwood, *evergreen, opp.,* 3½–5″, *p.* 656

69 Redbay, *evergreen,* 3–6″, *p.* 449

70 Camphor-tree, *evergreen, 2½–4", p. 448*

71 Osage-orange, *2½–5", p. 429*

72 Black Tupelo, *2–5", p. 620*

73 Buttonbush, *opp.*, 2½–6″, *p. 666*

74 Anacua, *evergreen*, 1¼–3¼″, *p. 657*

76 Flowering Dogwood, *opp.*, 2½–5″, *p. 615*

77 Alternate-leaf Dogwood, 2½–4½″, *p. 613*

79 Roughleaf Dogwood, *opp.,* 1½–3½″, *p. 614*

80 Possumhaw Viburnum, *opp.,* 2–5″, *p. 672*

81 Pinckneya, *opp.,* 4–8″, *p. 667*

82 Corkwood, 3½–6″, *p. 342*

83 Common Persimmon, 2½–6″, *p. 635*

84 Fringetree, *opp.*, 4–8″, *p. 644*

85 Sweetbay, 3–6″, *p. 444*

86 Saucer Magnolia, 5–8″, *p. 443*

88 Umbrella Magnolia, 10–20″, *p. 444*

89 Ashe Magnolia, 8–24″, *p. 438*

91 Fraser Magnolia, 8–18″, *p. 439*

92 Pyramid Magnolia, 3½–4½″, *p. 442*

93 Cucumbertree, 5–10″, *p. 437*

94 Southern Catalpa, *opp.*, 5–10″, *p. 663*

95 Tallowtree, 1½–3″, *p. 543*

96 Lindheimer Hackberry, 1½–2¾″, *p. 413*

97 Northern Catalpa, *opp.*, 6–12″, *p. 664*

98 Royal Paulownia, *opp.*, 6–16″, *p. 661*

99 Eastern Redbud, 2½–4½″, *p. 518*

100 Russian-olive, 1½–3¼″, *p. 609*

101 Willow Oak, 2–4½″, *p. 404*

102 Tree Lyonia, *evergreen*, 1–3½″, *p. 624*

103 Chinese Privet, *opp.*, 1–2½", *p. 655*

104 Florida-privet, *evergreen*, *opp.*, ¾–2¼", *p. 646*

Toothed Simple Leaves

The largest group of broadleaf trees has simple leaves with teeth along the edges. Often the teeth are saw-toothed —uniform and pointed forward—as in most willows. Some trees, such as elms, have saw-toothed leaves with large and small teeth alternating. In some hollies the teeth are bristle-tipped, while in others they are blunt and wavy, as in cottonwoods. Other typical trees with toothed leaves are hawthorns, cherries, plums, hollies, basswoods, chestnuts, and chinkapins.

106 Black Willow, 3–5″, *p. 335*

107 Sandbar Willow, 1½–4″, *p. 333*

108 Weeping Willow, 2½–5″, *p. 329*

109 Coastal Plain Willow, 2–4″, *p. 331*

110 White Willow, 2–4½″, *p. 327*

111 Crack Willow, 4–6″, *p. 334*

112 Swamp-privet, *opp.*, 1½–4″, *p. 645*

113 Balsam Willow, 2–3½″, *p. 336*

114 Florida Willow, 2–5″, *p. 334*

115 Sarvis Holly, 1½–3½", *p. 558*

116 Pussy Willow, 1½–4¼", *p. 332*

117 Bebb Willow, 1–3½", *p. 330*

118 Balsam Poplar, 3–5″, *p. 321*

119 Loblolly-bay, *evergreen,* 4–6″, *p. 605*

120 Texas Madrone, *evergreen,* 1-3½″, *p. 621*

121 Southern Crab Apple, 1–2¾″, *p. 489*

122 Allegheny Plum, 2–3½″, *p. 492*

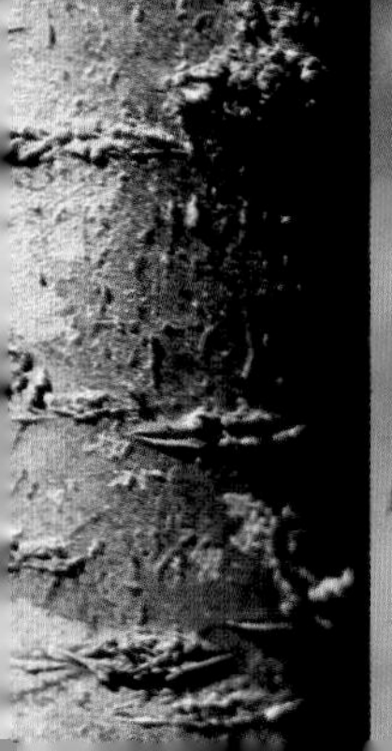

124 Pin Cherry, 2½–4½″, *p. 504*

125 American Plum, 2½–4″, *p. 493*

127 Flatwoods Plum, 1½–2¾″, *p. 507*

128 Garden Plum, 2–4″, *p. 498*

129 Sweet Cherry, 3–6″, *p. 495*

130 Prairie Crab Apple, 2½–4″, *p. 490*

131 Pear, 1½–3″, *p. 509*

132 Mahaleb Cherry, 1¼–2¾″, *p. 500*

133 Apple, 2–3½″, *p. 491*

134 Common Chokecherry, 1½–3¼″, *p. 508*

136 Black Cherry, 2–5″, *p. 506*

137 Sourwood, 4–7″, *p. 625*

139 Sugarberry, 2½–4″, *p. 412*

140 Hortulan Plum, 2¾–4½″, *p. 499*

142 Wildgoose Plum, 2½–4″, *p. 502*

143 Mountain Winterberry, 2½–5″, *p. 562*

145 Carolina Buckthorn, 2–5″, *p. 593*

146 Canada Plum, 2½–4″, *p. 503*

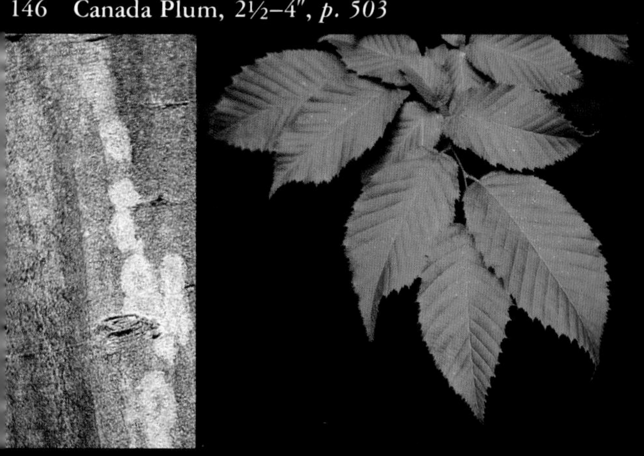

147 American Hornbeam, 2–4½″, *p. 372*

148 Allegheny Chinkapin, 3–6″, *p. 379*

149 Ozark Chinkapin, 5–8″, *p. 378*

150 American Chestnut, 5–9″, *p. 377*

151 Florida Chinkapin, 3–4″, *p. 376*

152 American Beech, 2½–5″, *p. 380*

153 Japanese Zelkova, 1–3½″, *p. 426*

154 White Mulberry, 2½–7″, *p. 430*

155 White Basswood, 3–7″, *p. 600*

156 European Linden, 2–4″, *p. 599*

157 Red Mulberry, 4–7″, *p. 432*

158 American Basswood, 3–6″, *p. 597*

160 Slippery Elm, 4–7″, *p. 423*

161 Rock Elm, 2–3½″, *p. 425*

163 English Elm, 2–3¼″, *p. 421*

164 September Elm, 2–3½″, *p. 424*

166 Siberian Elm, ¾–2″, *p. 422*

167 Water-elm, 2–2½″, *p. 417*

169 Hackberry, 2–5″, *p. 414*

170 Netleaf Hackberry, 1–2½″, *p. 415*

172 Eastern Burningbush, *opp.,* 2–4½″, *p. 566*

173 Nannyberry, *opp.,* 2½–4″, *p. 671*

175 Downy Serviceberry, 1½–4″, *p. 460*

176 Sweet Birch, 2½–5″, *p. 365*

178 River Birch, 1½–3″, *p. 366*

179 Paper Birch, 2–4″, *p. 368*

181 European White Birch, 1¼–2¾″, *p. 369*

182 Gray Birch, 2–3″, *p. 370*

183 Texas Mulberry, 1–2½″, *p. 431*

184 Quaking Aspen, 1¼–3″, *p. 326*

185 Eastern Cottonwood, 3–7″, *p. 322*

186 Eastern Cottonwood, 3–7″, *p. 322*

187 Lombardy Poplar, 1½–3″, *p. 325*

188 White Poplar, 2½–4″, *p. 320*

190 Swamp Cottonwood, 4–7″, *p. 324*

191 Bigtooth Aspen, 2½–4″, *p. 323*

193 Virginia Stewartia, 2½–4", *p. 606*

194 Mountain Stewartia, 2½–4½", *p. 607*

195 Two-wing Silverbell, 2½–4½", *p. 638*

196 Bigleaf Snowbell, 2½–5½″, *p. 640*

197 Carolina Holly, 1–2″, *p. 557*

198 Water Tupelo, 5–8″, *p. 618*

199 Roundleaf Serviceberry, 1¼–2¾", *p. 461*

200 Western Serviceberry, ¾–2", *p. 459*

202 Virginia Roundleaf Birch, 1½–2½", *p. 371*

203 Blackhaw, *opp.*, 1½–3", *p. 674*

205 Cedar Elm, 1–2″, *p. 420*

206 Tree Sparkleberry, ½–1¾″, *p. 629*

207 Possumhaw, 1–3″, *p. 561*

208 Yaupon, *evergreen*, ¾–1¼", *p. 565*

209 Southern Bayberry, *evergreen*, 1½–3½", *p. 339*

210 Riverflat Hawthorn, 2–2½", *p. 478*

211 Eastern Baccharis, 1–2½", *p. 677*

212 American Holly, *evergreen,* 2–4", *p. 564*

214 Franklinia, 5–7″, *p. 604*

215 Little Silverbell, 2½–4″, *p. 639*

217 Pensacola Hawthorn, ⅜–1″, *p. 475*

218 Yellow Hawthorn, 2–2½″, *p. 471*

219 Cockspur Hawthorn, 1–4″, *p. 468*

220 Downy Hawthorn, 3–4″, *p. 477*

221 Green Hawthorn, 1–2½″, *p. 488*

222 Eastern Hophornbeam, 2–5″, *p. 374*

223 Arrowwood, *opp.*, 1½–4″, *p. 670*

224 Brainerd Hawthorn, 2–3″, *p. 464*

226 Fleshy Hawthorn, 2–2½", *p. 484*

227 Hazel Alder, 2–4½", *p. 363*

229 Speckled Alder, 2–4″, *p. 362*

230 Witch-hazel, 3–5″, *p. 452*

231 Harbison Hawthorn, 1½–3″, *p. 473*

232 Dotted Hawthorn, 2–3″, *p. 481*

233 European Alder, 1½–4″, *p. 360*

234 Oneflower Hawthorn, ¾–1½″, *p. 487*

Lobed Simple Leaves

Some broadleaf trees have leaves with edges that are shallowly or deeply cut into narrow or broad lobes. Included here are most of the oaks, maples, sycamores, and many hawthorns, as well as the odd-shaped leaves of the Sassafras, Ginkgo, Sweetgum, and Yellow-poplar.

235 Ginkgo, 1–2″, *p. 271*

236 Paper-mulberry, 3–8″, *p. 428*

238 White Poplar, 2½–4″, *p. 320*

239 Sassafras, 3–5″, *p. 450*

241 May Hawthorn, 1–2″, *p. 461*

242 Parsley Hawthorn, ¾–2″, *p. 476*

244 Frosted Hawthorn, 1–2″, *p. 480*

245 Fanleaf Hawthorn, 1½–3″, *p. 470*

247 Fireberry Hawthorn, 1½–2″, *p. 466*

248 Texas Hawthorn, 3–4″, *p. 485*

250 Biltmore Hawthorn, 1–2½″, *p. 474*

251 Scarlet Hawthorn, 2–4″, *p. 467*

253 Planetree Maple, *opp.*, 3½–6″, *p. 576*

254 Mountain Maple, *opp.*, 2½–4½″, *p. 580*

256 Sycamore, 4–8″, *p. 456*

257 London Planetree, 5–10″, *p. 455*

258 Sugar Maple, *opp.*, 3½–5½″, *p. 579*

259 Norway Maple, *opp.*, 4–7″, *p.* 575

260 Black Maple, *opp.*, 4–5½″, *p.* 573

261 Red Maple, *opp.*, 2½–4″, *p.* 577

262 Florida Maple, *opp.*, 1½–3″, *p. 570*

263 Chalk Maple, *opp.*, 2–3½″, *p. 571*

264 Silver Maple, *opp.*, 4–6″, *p. 578*

265 Chinese Parasoltree, 6–12″, *p. 602*

266 Sweetgum, 3–6″, *p. 453*

267 Yellow-poplar, 3–6″, *p. 436*

268 Arkansas Oak, 2–5″, *p. 383*

269 Chapman Oak, 1½–3½″, *p. 385*

271 Water Oak, 1½–5″, *p. 401*

272 Lacey Oak, 2–4½″, *p. 390*

274 Mohr Oak, 1–3″, *p. 398*

275 Chinkapin Oak, 4–6″, *p. 399*

276 Chestnut Oak, 4–8″, *p. 405*

277 Swamp Chestnut Oak, 4–9″, *p. 398*

278 Overcup Oak, 5–8″, *p. 395*

279 English Oak, 2–5″, *p. 406*

280 Swamp White Oak, 4–7″, *p. 384*

281 Bur Oak, 4–10″, *p. 395*

283 Southern Red Oak, 4–8″, *p. 388*

284 Turkey Oak, 4–8″, *p. 393*

286 Northern Pin Oak, 3–5″, *p. 387*

287 Scarlet Oak, 3–7″, *p. 385*

288 Pin Oak, 3–5″, *p. 403*

289 Bear Oak, 2–4″, *p. 390*

290 Georgia Oak, 2–4″, *p. 389*

291 Blackjack Oak, 2½–5″, *p. 397*

292 Northern Red Oak, 4–9″, *p. 407*

293 Shumard Oak, 3–7″, *p. 408*

Compound Leaves

Compound leaves are composed of 3 or more small leaflets. When the leaflets are arranged along a central stalk, the leaf is pinnately compound, as in hickories, pecans, sumacs, and ashes. When the central stalk has side branches, the leaf is bipinnately compound; examples include acacias, Kentucky Coffeetree, and Devils-walkingstick. When the leaflets are attached at the end of the leafstalk and spread like the fingers of a hand, the leaf is palmately compound, as in the buckeyes, the Common Hoptree, and the American Bladdernut.

295 Gregg Catclaw, *bipinnate,* 1–3″, *p.* 514

296 Wright Catclaw, *bipinnate,* 1–2″, *p.* 517

298 Texas Lignumvitae, *opp., pinnate,* 1–3″, *p. 532*

299 Honey Mesquite, *bipinnate,* 3–8″, *p. 525*

300 Roemer Catclaw, *bipinnate,* 1½–4″, *p. 516*

301 Lime Prickly-ash, *evg.*, *pinnate*, 3–4″, *p.* 537

302 Texas Sophora, *pinnate*, 6–9″, *p.* 528

303 Texas Pistache, *evergreen*, *pinnate*, 2–4″, *p.* 546

304 Black Locust, *pinnate,* 6–12″, *p. 526*

305 Clammy Locust, *pinnate,* 5–10″, *p. 527*

306 Honeylocust, *pinnate, bipinnate,* 4–8″, *p. 523*

307 Prairie Sumac, *pinnate,* 9″, *p. 550*

308 Western Soapberry, *pinnate,* 5–8″, *p. 589*

309 Waterlocust, *pinnate, bipinnate,* 4–8″, *p. 522*

310 Silktree, *bipinnate,* 6–15″, *p. 517*

311 Kentucky Coffeetree, *bipinnate,* 12–30″, *p. 524*

312 Chinese Scholartree, *pinnate,* 6–10″, *p. 529*

313 Ailanthus, *pinnate,* 12–24", *p.* 539

314 Staghorn Sumac, *pinnate,* 12–24", *p.* 551

315 Smooth Sumac, *pinnate,* 12", *p.* 548

316 Little Walnut, *pinnate,* 8–13″, *p. 357*

317 Black Walnut, *pinnate,* 12–24″, *p. 358*

319 Showy Mountain-ash, *pinnate*, 4–6″, *p. 512*

320 American Mountain-ash, *pinnate*, 6–8″, *p. 510*

322 Devils-walkingstick, *bipinnate,* 15–30″, *p. 611*

323 Hercules-club, *pinnate,* 5–10″, *p. 536*

325 Carolina Ash, *opp.*, *pinnate*, 6–12″, *p. 649*

326 Blue Ash, *opp.*, *pinnate*, 8–12″, *p. 653*

328 Green Ash, *opp., pinnate,* 6–10″, *p. 651*

329 White Ash, *opp., pinnate,* 8–12″, *p. 647*

331 Berlandier Ash, *opp., pinnate,* 3–7″, *p. 648*

332 Bitternut Hickory, *pinnate,* 6–10″, *p. 345*

334 Black Ash, *opp., pinnate,* 12–16″, *p. 650*

335 Boxelder, *opp., pinnate,* 6″, *p. 572*

336 American Elder, *opp., pinnate,* 5–9″, *p. 669*

337 Sand Hickory, *pinnate,* 7–15″, *p. 353*

338 Scrub Hickory, *pinnate,* 4–8″, *p. 346*

339 Common Prickly-ash, *pinnate,* 5–10″, *p. 535*

340 Mescalbean, *evergreen, pinnate,* 3–6″, *p. 530*

341 Chinaberry, *bipinnate,* 8–18″, *p. 541*

343 Pignut Hickory, *pinnate,* 6–10″, *p. 347*

344 Shagbark Hickory, *pinnate,* 8–14″, *p. 352*

346 Pumpkin Ash, *opp.*, *pinnate*, 8–16″, *p. 652*

347 Nutmeg Hickory, *pinnate*, 7–14″, *p. 351*

349 Poison-sumac, *pinnate,* 7–12″, *p. 552*

350 Yellowwood, *pinnate,* 8–12″, *p. 519*

352 Common Hoptree, *palmate*, 4–7″, *p.* 534

353 American Bladdernut, *opp.*, *palmate*, 6–9″, *p.* 568

355 Texas Buckeye, *opp., palmate,* 2–5″, *p. 582*

356 Painted Buckeye, *opp., palmate,* 4–6″, *p. 588*

358 Yellow Buckeye, *opp., palmate,* 3½–7″, *p. 586*

359 Ohio Buckeye, *opp., palmate,* 2–6″, *p. 583*

Yuccas and Palmettos

Yuccas and palmettos differ from other trees in having large evergreen leaves clustered at the top of a stout trunk or at the end of a few large branches. Yuccas have long, narrow, bayonetlike leaves, usually ending in a sharp point. The Cabbage Palmetto has enormous fanlike leaves.

361 Aloe Yucca, 12–20″, *p. 316*

362 Moundlily Yucca, 16–24″, *p. 317*

363 Cabbage Palmetto, 4–7′, *p. 314*

Flowers

Tree flowers vary greatly in color, shape, size, and arrangement. Some, such as elm flowers, are a fraction of an inch wide; others, like those of the yuccas, grow in showy clusters several feet long. Some flowers are borne singly; others form branched clusters or drooping catkins. The flowers shown here are grouped by color in the following order: red, green, yellow, cream-colored, white, pink, and purple. Most flowers are bisexual; those that are only male or only female are indicated in the captions.

364 Pawpaw, *w.* 1½", *p. 446*

365 Florida Anise-tree, *w.* 2", *p. 434*

367 Red Buckeye, *cl.* 4–8″, *p.* 587

368 Southeastern Coralbean, *cl.* 8–12″, *p.* 521

369 Red Maple ♂, *cw.* ½–¾″, *p.* 577

370 Northern Red Oak ♂, *cl.* 4–5″, *p. 407*

371 Eastern Hophornbeam ♂, *cl.* 1½–2½″, *p. 374*

372 Speckled Alder ♂, *cl.* 1½–3″, *p. 362*

373 American Elm, *cl.* ¾–1″, *p. 419*

374 Sugar Maple ♂, *cl.* 1–3″, *p. 579*

375 English Oak ♂, *cl.* 2–5″, *p. 406*

376 Little Walnut ♂, *cl.* 3–4", *p.* 357

377 Pignut Hickory ♂, *cl.* 3–6", *p.* 347

378 Crack Willow ♀, *cl.* 1–2¼", *p.* 334

379 Sweet Birch ♂, *cl.* 3–4″, *p. 365*

380 White Oak ♂, *cl.* 2½–3″, *p. 382*

382 Ohio Buckeye, *cl.* 4–6″, *p.* 583

383 Painted Buckeye, *cl.* 4–6″, *p.* 588

385 Ailanthus ♂, *cl.* 6–10″, *p.* 539

386 Smooth Sumac, *cl.* 8″, *p.* 548

387 Shining Sumac, *cw.* 3″, *p.* 547

388 Norway Maple ♂, *cw.* 3″, *p.* 575

389 Cucumbertree, *w.* 2½–3½″, *p.* 437

390 Yellow-poplar, *w.* 1½–2″, *p.* 436

391 Osage-orange ♂, *cw.* ¾–1″, *p. 429*

392 Russian-olive, *l.* ⅜″, *p. 609*

394 Sassafras ♂, *cw.* 1″, *p. 450*

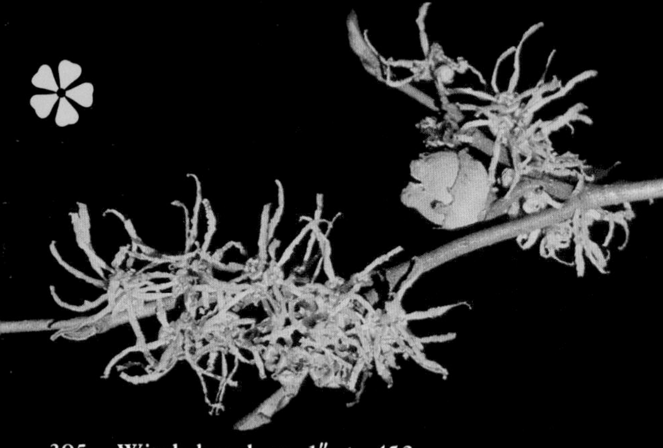

395 Witch-hazel, *w.* 1″, *p. 452*

397 Gregg Catclaw, *cl.* 1–2″, *p. 514*

398 Honey Mesquite, *cl.* 2–3″, *p. 525*

399 Pussy Willow ♂, *cl.* 1–2½″, *p. 332*

400 Lime Prickly-ash ♂, *cw.* ½", *p. 537*

401 Sweetleaf, *cw.* ¾", *p. 642*

402 Buttonbush, *cw.* 1–1½", *p. 666*

403 Allegheny Chinkapin ♂, *cl.* 4–6″, *p.* 379

404 Swamp Cyrilla, *cl.* 4–6″, *p.* 555

405 Sourwood, *cl.* 4–10″, *p.* 625

406 American Basswood, *cw.* 2″, *p. 597*

407 Horsechestnut, *cl.* 10″, *p. 585*

408 Carolina Laurelcherry, *cw.* ¾″, *p. 496*

409 Alternate-leaf Dogwood, *cw.* 2″, *p. 613*

410 Possumhaw Viburnum, *cw.* 2½–5″, *p. 672*

411 Rusty Blackhaw, *cw.* 3–6″, *p. 675*

412 Cockspur Hawthorn, *w.* ½–⅝", *p. 468*

413 American Elder, *cw.* 4–8", *p. 669*

414 Nannyberry, *cw.* 3–5", *p. 671*

415 Showy Mountain-ash, *cw.* 2–4″, *p. 512*

416 American Mountain-ash, *cw.* 3–5″, *p. 510*

418 Downy Serviceberry, *w.* 1¼″, *p. 460*

419 Fringetree ♀, *cl.* 4–8″, *p. 644*

421 Garden Plum, *w.* ¾–1″, *p. 498*

422 Mahaleb Cherry, *w.* ⅝″, *p. 500*

424 Tree Sparkleberry, *l.* ¼", *p. 629*

425 American Bladdernut, *l.* ½", *p. 568*

427 Carolina Silverbell, *l.* ½–1″, *p. 637*

428 Aloe Yucca, *l.* 1½–2¼″, *p. 316*

429 Moundlily Yucca, *l.* 1½–2″, *p. 317*

430 Bigleaf Magnolia, *w.* 10–12″, *p. 441*

431 Umbrella Magnolia, *w.* 7–10″, *p. 444*

432 Southern Magnolia, *w.* 6–8″, *p. 440*

433 Sweetbay, *w.* 2–2½", *p.* 444

434 Fraser Magnolia, *w.* 8–10", *p.* 439

436 Virginia Stewartia, *w.* 4″, *p. 606*

437 Franklinia, *w.* 3″, *p. 604*

439 Rosebay Rhododendron, *w.* 1½″, *p. 627*

440 Mountain-laurel, *w.* ¾–1″, *p. 623*

441 Northern Catalpa, *l.* 2–2¼″, *p. 664*

442 Black Locust, *cl.* 4–8″, *p. 526*

443 Elliottia, *cl.* 4–10″, *p. 622*

444 Chinese Privet, *cl.* 1½–4″, *p. 655*

445 Allegheny Plum, *w.* ½″, *p. 492*

446 Common Chokecherry, *cl.* 4″, *p. 508*

448 Washington Hawthorn, *w.* ½″, *p. 479*

449 Broadleaf Hawthorn, *w.* 1″, *p. 470*

451 Parsley Hawthorn, *w.* ⅝″, *p.* 476

452 Downy Hawthorn, *w.* 1″, *p.* 477

453 Green Hawthorn, *w.* ⅝″, *p.* 488

454 Flowering Dogwood, *cw.* 3–4", *p. 615*

455 Pear, *w.* 1¼", *p. 509*

456 Apple, *w.* 1¼", *p. 491*

457 Prairie Crab Apple, *w.* 1½–2″, *p. 490*

458 Sweet Crab Apple, *w.* 1½″, *p. 490*

459 Peach, *w.* 1–1¼″, *p. 505*

460 American Smoketree ♀, *cl.* 6″, *p.* 545

461 Royal Paulownia, *cl.* 6–12″, *p.* 661

462 Mountain-laurel, *w.* ¾–1″, *p.* 623

463 Silktree, *cw.* 1½–2″, *p. 517*

464 Pinckneya, *cw.* 6″, *p. 667*

465 Rosebay Rhododendron, *w.* 1½″, *p. 627*

466 Eastern Redbud, *l.* ½", *p. 518*

467 Clammy Locust, *cl.* 3", *p. 527*

468 Catawba Rhododendron, *w.* 2¼", *p. 626*

469 Chinaberry, *w.* ¾″, *p.* 541

470 Texas Lignumvitae, *w.* ½–¾″, *p.* 532

Cones and Fruit

This group of photographs is divided into several categories. First are the cones of pines, spruces, and other conifers as well as the conelike fruit of birches and alders. The papery winged keys are represented by maples, elms, ashes, Ailanthus, and the Common Hoptree. Pods typify locusts and other members of the legume family, while oaks are characterized by their acorns. Nuts include the fruits of chestnuts, beeches, buckeyes, walnuts, and hickories. Balls, capsules, and tufted fruit encompass several oddities such as Sycamore, Sweetgum, Sourwood, cottonwoods, and Eastern Baccharis. Berrylike fruit includes true berries like hollies as well as cherries and other small fleshy fruit. The larger plums, crab apples, Apple, Pear, and Peach are in the fleshy fruit category.

472 Pitch Pine, 1¼–2¾", *p. 294*

473 Virginia Pine, 1½–2¾", *p. 298*

474 Shortleaf Pine, 1½–2½", *p. 287*

475 Eastern White Pine, 4–8″, *p. 296*

476 Jack Pine, 1¼–2″, *p. 286*

478 Red Spruce, 1¼–1½″, *p. 285*

479 White Spruce, 1¼–2½″, *p. 283*

81 Black Spruce, ⅝–1¼", *p. 284*

82 Tamarack, ½–¾", *p. 281*

83 Eastern Hemlock, ⅝–¾", *p. 299*

484 Seaside Alder, ⅝–¾", *p. 361*

485 Northern White-cedar, ⅜", *p. 312*

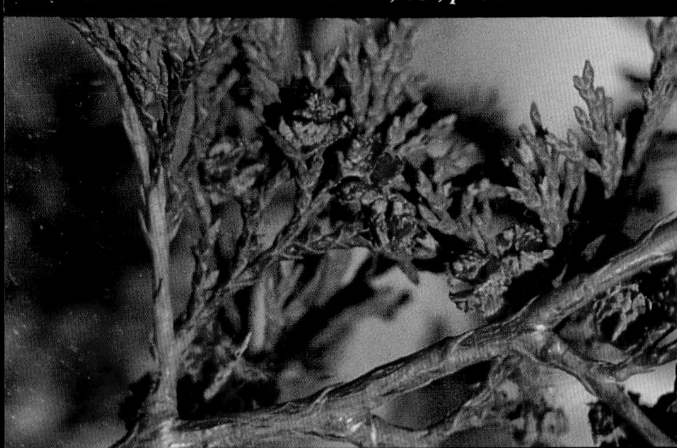

486 Atlantic White-cedar, ¼", *p. 306*

487 Yellow Birch, ¾–1¼", *p. 364*

488 Sweet Birch, ¾–1½", *p. 365*

490 Yellow-poplar, 2½–3″, *p. 436*

491 Eastern Hophornbeam, 1½–2″, *p. 374*

493 Silver Maple, 1½–2½", *p. 578*

494 Boxelder, 1–1½", *p. 572*

496 Ailanthus, 1½", *p. 539*

497 Winged Elm, ⅜", *p. 418*

499 Rock Elm, ⅜–¾″, *p. 425*

500 Slippery Elm, ½–¾″, *p. 423*

501 Common Hoptree, ⅞″, *p. 534*

502 Wright Catclaw, 2–6″, *p. 517*

503 Kentucky Coffeetree, 4–7″, *p. 524*

504 Mescalbean, 1–5″, *p. 530*

505 Honey Mesquite, 3½–8″, *p. 525*

506 Honeylocust, 6–16″, *p. 523*

507 Black Locust, 2–4″, *p. 526*

508 Sycamore, 1″, *p. 456*

509 Sweetgum, 1–1¼″, *p. 453*

510 Texas Sophora, 1–5″, *p. 528*

511 Sourwood, ⅜", *p. 625*

512 Eastern Cottonwood, ⅜", *p. 322*

513 Eastern Baccharis, ½", *p. 677*

514 Post Oak, ½–1″, *p. 409*

515 Swamp White Oak, ¾–1¼″, *p. 384*

516 Black Oak, ⅝–¾″, *p. 410*

517 Pin Oak, ½″, *p. 403*

518 Scarlet Oak, ½–1″, *p. 385*

520 Myrtle Oak, ⅜–½″, *p. 400*

521 Live Oak, ⅝–1″, *p. 410*

523 Northern Red Oak, ⅝–1⅛″, *p. 407*

524 White Oak, ⅜–1¼″, *p. 382*

525 Bur Oak, ¾–2″, *p. 395*

526 American Chestnut, 2–2½″, *p. 377*

527 Florida Chinkapin, ¾–1″, *p. 376*

528 American Beech, ½–¾″, *p. 380*

529 Witch-hazel, ½″, *p. 452*

530 Ohio Buckeye, 1–2″, *p. 583*

532 Black Walnut, 1½–2½", *p. 358*

533 Butternut, 1½–2½", *p. 356*

535 Pecan, 1¼–2″, *p. 348*

536 Shellbark Hickory, 1¾–2½″, *p. 350*

538 Devils-walkingstick, ¼", *p. 611*

539 American Elder, ¼", *p. 669*

541 Arrowwood, ¼–¾″, *p. 670*

542 Arrowwood, ¼–⅜″, *p. 670*

544 Sassafras, 3/8", *p. 450*

545 Glossy Buckthorn, 5/16", *p. 595*

547 Sweet Cherry, ¾–1″, *p. 495*

548 Black Cherry, ⅜″, *p. 506*

550 Pin Cherry, ¼″, *p. 504*

551 Hackberry, ¼–⅜″, *p. 414*

552 Washington Hawthorn, ¼″, *p. 479*

553 Texas Lignumvitae, ½″, *p. 532*

554 Flowering Dogwood, ⅜–⅝″, *p. 615*

556 Dahoon, ¼″, *p. 560*

557 Mountain Winterberry, ⅜–½″, *p. 562*

558 Carolina Holly, ¼″, *p. 557*

559 American Holly, ¼–⅜″, *p. 564*

560 Possumhaw, ¼″, *p. 561*

562 Anacua, 5/16", *p. 657*

563 Western Soapberry, 3/8–1/2", *p. 589*

564 American Mountain-ash, 1/4", *p. 510*

565 Southern Magnolia, 3–4″, *p. 440*

566 Smooth Sumac, ⅛″, *p. 548*

568 American Plum, ¾–1″, *p. 493*

569 Peach, 2–3″, *p. 505*

571 Common Persimmon, ¾–1½", *p. 635*

572 Sweet Crab Apple, 1–1¼", *p. 490*

574 Pawpaw, 3–5″, *p. 446*

575 Pear, 2½–4″, *p. 509*

Autumn Leaves

Many broadleaf trees are noted for their brilliant fall foliage. These displays are most dramatic in the Northeast but vary with the species and the year, depending on the weather. The leaves in this section are grouped according to the major fall colors: red, orange, and yellow. Red is produced by warm, sunny fall days followed by cool nights that transform leftover food in the leaves into red pigment. Foliage turns orange or yellow when the chlorophyll, which masks other colors, is destroyed; deep orange is a blend of hues. On a single tree, such as a Sugar Maple, leaves of several colors may appear at the same time. The foliage of other trees, such as Sweetgum, may show different hues at different times, depending upon soil and climatic conditions.

577 Shining Sumac, *pinnate,* 12″, *p. 547*

578 Staghorn Sumac, *pinnate,* 12–24″, *p. 551*

579 Flowering Dogwood, *opp.,* 2½–5″, *p. 615*

580 Tallowtree, 1½–3″, *p. 543*

581 Eastern Burningbush, *opp.,* 2–4½″, *p. 566*

583 Black Tupelo, 2–5″, *p. 620*

584 Pin Cherry, 2½–4½″, *p. 504*

586 Pin Oak, 3–5″, *p. 403*

587 Black Oak, 4–9″, *p. 410*

588 Turkey Oak, 4–8″, *p. 393*

589 Sassafras, 3–5″, *p. 450*

590 Blackjack Oak, 2½–5″, *p. 397*

591 Scarlet Oak, 3–7″, *p. 385*

592 Sugar Maple, *opp.*, 3½–5½″, *p.* 579

593 Red Maple, *opp.*, 2½–4″, *p.* 577

595 Downy Serviceberry, 1½–4", *p. 460*

596 American Hornbeam, 2–4½", *p. 372*

598 Gray Birch, 2–3″, *p. 370*

599 Bigtooth Aspen, 2½–4″, *p. 323*

600 Silver Maple, *opp.*, 4–6″, *p. 578*

601 American Beech, 2½–5″, *p. 380*

602 White Ash, *opp., pinnate,* 8–12″, *p. 647*

603 Shingle Oak, 3–6″, *p. 391*

604 Chestnut Oak, 4–8″, *p. 405*

605 Common Chokecherry, 1½–3¼″, *p. 508*

607 White Mulberry, 2½–7″, *p. 430*

608 Shagbark Hickory, *pinnate,* 8–14″, *p. 352*

610 Glossy Buckthorn, 1½–2¾″, *p. 595*

611 Eastern Cottonwood, 3–7″, *p. 322*

613 Sweet Birch, 2½–5″, *p. 365*

614 American Beech, 2½–5″, *p. 380*

616 Speckled Alder, 2–4″, *p. 362*

617 Yellow Birch, 3–5″, *p. 364*

618 Yellow-poplar, 3–6″, *p. 436*

619 Ginkgo, 1–2″, *p. 271*

620 Quaking Aspen, 1¼–3″, *p. 326*

621 Sassafras, 3–5″, *p. 450*

622 Black Maple, *opp.,* 4–5½″, *p. 573*

623 Striped Maple, *opp.,* 5–7″, *p. 574*

625 Horsechestnut, *opp., palmate,* 3–7″, *p.* 585

626 Shellbark Hickory, *pinnate,* 12–20″, *p.* 350

628 Eastern Hophornbeam, 2–5″, *p. 374*

629 Witch-hazel, 3–5″, *p. 452*

630 European Larch, ¾–1¼″, *p. 280*

Part II
Family and Species Descriptions

The numbers preceding the species descriptions in the following pages correspond to the plate numbers in the color section.

GINKGO FAMILY
(Ginkgoaceae)

The following species is the only one of its family worldwide.

235, 619 Ginkgo
"Maidenhair-tree"
Ginkgo biloba L.

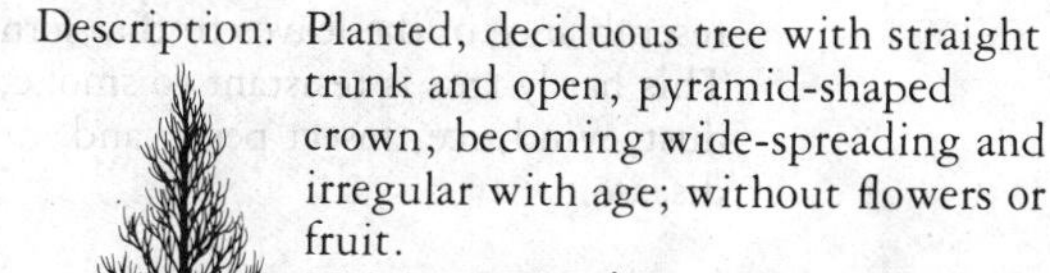

Description: Planted, deciduous tree with straight trunk and open, pyramid-shaped crown, becoming wide-spreading and irregular with age; without flowers or fruit.

Height: 50–70′ (15–21 m).
Diameter: 2′ (0.6 m).
Leaves: *3–5 in cluster* on spurs, or alternate; 1–2″ (2.5–5 cm) long, 1½–3″ (4–7.5 cm) wide. Oddly *fan-shaped,* slightly thickened, slightly wavy on broad edge, often 2-lobed, with fine forking parallel veins but no midvein. Dull light green, turning yellow and shedding in autumn. Long leafstalk.
Bark: gray; becoming rough and deeply furrowed.
Twigs: light green to light brown; hairless, long and stout, with *many spurs* or short side twigs bearing crowded leaf scars.
Seeds and Male Cones: on separate trees in early spring. Seeds 1″ (2.5 cm) long; *elliptical,* naked, *yellowish,* with thin juicy pulp of *bad odor* and large thick-walled edible kernel; *1–2 at end of long stalk;* maturing and shedding in autumn. Male or pollen cones ¾″ (19 mm) long.

Habitat: Lawns and along streets in moist soil, in humid temperate regions.
Range: Apparently native in SE. China. Planted in E. United States and on Pacific Coast.

Ginkgo is best known as a living fossil related to conifers and the sole survivor of its ancient and formerly widespread family. This sacred tree has long been cultivated and possibly preserved from extinction by Buddhist priests on temple grounds in China, Japan, and Korea. Female trees are objectionable because of the litter of seeds, which reek like rancid butter; they should not be touched because the smell lingers after contact. The seeds, or Ginkgo nuts, are eaten in the Orient. The name "Maidenhair-tree" alludes to the resemblance of the leaves to that fern. This hardy tree is resistant to smoke, dust, wind, ice, insect pests, and disease.

YEW FAMILY
(Taxaceae)

Trees or sometimes shrubs, slightly aromatic and resinous, without flowers or fruit; mostly in northern temperate regions. About 20 species worldwide, including 4 native tree and 1 shrub species in North America in the genera yew (*Taxus*) and torreya (*Torreya*).
Leaves: evergreen, alternate, spreading in 2 rows; needlelike, flattened, stiff, with short leafstalk extending down twig.
Twigs: slender, much-branched, with 2 lines below each leaf, rough with scars from fallen leaves, ending in scaly buds.
Seeds and Cones: mainly on separate plants, usually solitary from scaly bud at leaf base. Seeds naked, not in cones, nutlike, elliptical, hard with soft outer coat or cuplike base. Male cones rounded, with several pollen-sacs on a stalk.

---

29 Florida Yew
Taxus floridana Nutt. ex Chapm.

Description: A faintly aromatic but nonresinous, poisonous shrub or small bushy tree with many, stout, spreading branches.
Height: 13–20′ (4–6 m).
Diameter: 1′ (0.3 m).
Needles: *evergreen;* ¾–1″ (2–2.5 cm) long, 1/16″ (1.5 mm) or more wide. Spreading in *2 rows;* short-pointed at both ends; soft and *flexible, flattened* and slightly curved. Dark green above, light green with *2 gray bands* beneath. Very short leafstalk.
Bark: dark purplish-brown, very thin, smooth, shedding in irregular plates.
Twigs: yellow-green, turning red-brown; slender, with 2 lines below base of each leaf.
Seeds and Male Cones: on separate

trees. Seeds small and blunt-pointed; *egg-shaped* or elliptical; brown, nearly enclosed by a *scarlet cup* ⅜″ (10 mm) in diameter; soft, juicy, and sweet; *borne singly;* stalkless. Male or pollen cones ⅛″ (3 mm) in diameter; pale yellow; short-stalked; single at leaf bases.

Habitat: Moist ravines in hardwood forests.

Range: Local in NW. Florida in Gadsden and Liberty counties, mainly along eastern side of Apalachicola River; at less than 100′ (30 m).

This rare species is threatened with extinction because of its very limited distribution. Torreya State Park near Bristol, Florida, preserves it as well as Florida Torreya, and it is in cultivation in botanical gardens. Like other yews, the seeds and foliage are poisonous and can be fatal when eaten by people or livestock. The red juicy cup around the seed, however, is apparently harmless. Canada Yew (*Taxus canadensis* Marsh.), is a related poisonous low shrub with shorter straight needles and similar seeds in a red cup. It ranges from southeastern Canada and northeastern United States west to Minnesota and south locally to Virginia and Tennessee.

30 Florida Torreya

"Stinking-cedar"
Torreya taxifolia Arn.

Description: Small tree with unpleasant odor and open, conical crown of rows of horizontal, spreading branches.
Height: 30′ (9 m).
Diameter: 1′ (0.3 m), sometimes larger.
Needles: *evergreen;* 1–1½″ (2.5–4 cm) long, ⅛″ (3 mm) wide. Spreading in *2 rows;* with *long sharp point* at tip, short-pointed at base; *stiff,* slightly curved, *flattened* but slightly rounded on back. Shiny dark green above, light green

with *2 broad whitish bands* beneath. Almost stalkless.

Bark: brown, thin, irregularly fissured into broad scaly ridges.

Twigs: mostly paired, yellow-green, becoming gray; slender, stiff, with 2 lines below base of each leaf.

Seeds and Male Cones: on separate trees. Seeds 1–1¼″ (2.5–3 cm) long; *elliptical* and broadest beyond middle; *borne singly;* stalkless; *fleshy outer layer dark green* with purplish markings and whitish bloom, and shedding; inner layer light red-brown, thick-walled, maturing in 2 seasons. Male or pollen cones ¼″ (6 mm) long; pale yellow; single at leaf bases.

Habitat: River bluffs, slopes, and moist ravines; in hardwood forests.

Range: Extreme SW. Georgia and NW. Florida, mostly on east side of Apalachicola River; at less than 100′ (30 m).

Torreya State Park near Bristol, Florida, was established to preserve this species and Florida Yew. Florida Torreya is endangered by a fungal disease of the stems, which has destroyed many trees. Healthy plants are, however, growing in botanical gardens elsewhere. The name "Stinking-cedar" refers to the disagreeable resinous odor of crushed foliage and wood.

PINE FAMILY
(Pinaceae)

Large to very large trees, without flowers or fruit, including pines (*Pinus*), larches (*Larix*), spruces (*Picea*), hemlocks (*Tsuga*), firs (*Abies*), and Douglas-firs (*Pseudotsuga*). Resinous, mostly evergreen with straight axis and narrow crown, usually with soft lightweight wood. About 200 species worldwide in north temperate and tropical mountain regions. In North America, 61 native and 1 naturalized species; others southward.
Leaves: mostly alternate or whorled, sometimes of 2 forms, very narrow and needlelike.
Cones: pollen and seeds borne on same plant in separate cones. Male cones small and herbaceous; female cones large and woody, composed of many spirally arranged flattened cone-scales each above a bract. Usually 2 naked seeds at base of a cone-scale, mostly with wing at end.

The large genus of pines (*Pinus*) is characteristic of acid soils, often sprouting after fire. They occur nearly throughout temperate continental North America (although not native in Kansas) and also from Mexico to Nicaragua, West Indies, and Eurasia. They are among the most important native softwood timbers, producing lumber and pulpwood. Two groups (subgenera) include soft pines with needles commonly 5 (rarely 4–1) in a bundle, with the sheath shedding. Hard pines have needles mostly 2 or 3 in a bundle and the sheath persistent. Usually large trees, pines are resinous and evergreen. Leaves are of 2 kinds: mostly scalelike on long twigs; and needles on dwarf twigs or spurs in bundles of 2–5 (rarely 1), with sheath of bud-scales at base. Bark varies from furrowed into ridges to fissured into scaly plates. Twigs are stout, ending in compound bud with many bud-scales. Pollen and seeds are on same plant: male cones are numerous, clustered, small, and herbaceous; female cones are solitary to a few in a cluster, large and woody, composed of many flattened cone-scales each above a bract, maturing the second year, with 2 seeds at base of cone-scale, mostly with long wing.

22, 480 **Balsam Fir**
"Canada Balsam" "Eastern Fir"
Abies balsamea (L.) Mill.

Description: The only fir native to the Northeast, with *narrow, pointed, spirelike crown* of spreading branches and *aromatic foliage.* Height: 40–60′ (12–18 m). Diameter: 1–1½′ (0.3–0.5 m). Needles: *evergreen;* ½–1″ (1.2–2.5 cm) long. Spreading almost at right angles in 2 rows on hairy twigs, curved upward on upper twigs; flat, with rounded tip (sometimes notched or short-pointed). *Shiny dark green above, with 2 narrow whitish bands beneath.* Bark: brown, *thin, smooth, with many resin blisters,* becoming scaly. Cones: 2–3¼″ (5–8 cm) long; cylindrical; dark purple; *upright* on topmost twigs; cone-scales finely hairy, bracts mostly short and hidden; paired long-winged seeds.

Habitat: Coniferous forests; often in pure stands.

Range: Alberta east to Labrador and south to Pennsylvania, west to Minnesota and NE. Iowa; local in West Virginia and Virginia; to timberline in north and above 4000′ (1219 m) in south.

A major pulpwood species. Interior knotty pine paneling is a special product; Christmas trees, wreaths, and balsam pillows utilize the aromatic foliage. Canada balsam, an aromatic oleoresin obtained from swellings or resin blisters in the bark, is used for mounting microscopic specimens and for optical cement. Deer and moose browse the foliage in winter.

---

21 **Fraser Fir**
"Balsam" "She-balsam"
Abies fraseri (Pursh) Poir.

Description: The only native Southeastern fir, a handsome tree with *pointed crown* of silvery white *aromatic foliage.*
Height: 30–50′ (9–15 m).
Diameter: 1–2′ (0.3–0.6 m).
Needles: *evergreen;* ½–1″ (1.2–2.5 cm) long. Spreading almost at right angles in 2 rows on slender hairy twigs, crowded and curved upward on upper twigs; flat, with tip usually rounded. Shiny dark green above, *2 broad silvery-white bands beneath.*
Bark: gray or brown; *thin, smooth, with many resin blisters,* becoming scaly.
Cones: 1½–2½″ (4–6 cm) long; cylindrical; dark purple; *upright* on topmost twigs; cone-scales finely hairy, partly covered by *yellow-green pointed and toothed bracts;* paired long-winged seeds.

Habitat: Coniferous forests with Red Spruce in high mountains.

Range: Appalachian Mountains in SW. Virginia, W. North Carolina, and E. Tennessee; at 4000–6600′ (1219–2012 m).

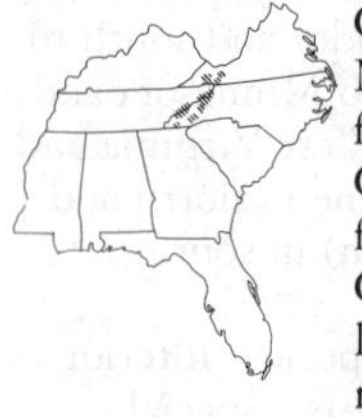

Common in Great Smoky Mountains National Park in virgin spruce-fir forests and at Mount Mitchell, North Carolina. With its silvery and green foliage, this species is grown for Christmas trees and ornament. Known locally as "She-balsam," because of the resin produced in the bark. In contrast, Red Spruce in the same forest but without resin blisters is often called "He-balsam." John Fraser (1750–

1811), the Scottish explorer, discovered this fir and introduced it and many other plants to Europe.

16 **Cedar-of-Lebanon**
Cedrus libani A. Rich.

Description: Large, planted, cone-bearing, evergreen tree with straight stout trunk and narrow pointed crown, becoming irregular and broad or flattened with *spreading horizontal branches.*

Height: 80′ (24 m).
Diameter: 3′ (0.9 m).
Needles: *evergreen;* 1–1¼″ (2.5–3 cm) long. *10–15 clustered* and crowded on spurs or alternate on leading twigs; 3-angled; mostly dark green.
Bark: dark gray; becoming thick and furrowed into scaly plates.
Twigs: abundant, spreading, long and stout, mostly hairless or slightly hairy, with *many spurs* or short side twigs.
Cones: 3–4½″ (7.5–11 cm) long, 1¾–2½″ (4.5–6 cm) in diameter. *Barrel-shaped* with flat top; reddish-brown; *upright;* almost *stalkless;* resinous; composed of many hard cone-scales; maturing in second year, axis remaining attached; paired broad-winged seeds.

Habitat: Moist soil in parks and gardens and around homes in temperate regions.

Range: Native of Asia Minor from S. Turkey to Lebanon and Syria at high altitudes. Planted mainly in SE. United States and on Pacific Coast; a hardy variety also in Northeast.

A handsome, picturesque ornamental. Developing large cones often are conspicuous on lower branches. Its association with the Bible and the Holy Land make Cedar-of-Lebanon of special interest. The fragrant durable wood is used for construction timbers, lumber, furniture, and paneling in its native lands. The northeastern range was extended by collecting seed from native trees at their highest altitude.

17, 630 **European Larch**
Larix decidua Mill.

Description: Introduced, cone-bearing, deciduous tree with a straight trunk and open, broadly pyramid-shaped crown, becoming irregular with age.

Height: 70′ (21 m).
Diameter: 2′ (0.6 m).
Needles: *deciduous;* ¾–1¼″ (2–3 cm) long; soft, flattened, with ridge beneath; *30–40 unequal in length, and crowded on spurs or alternate on leader twigs.* Light green, turning yellow in autumn before shedding.
Bark: gray, turning brown; thick, furrowed, and scaly.
Twigs: gray or yellowish; stout, hairless, with *many dark brown spurs* or short side twigs.
Cones: 1–1¼″(2.5–3 cm) long; egg-shaped; *upright, stalkless;* reddish and rosebudlike when young; maturing in autumn, remaining attached and turning dark gray; cone-scales numerous, rounded, thin, light brown, finely hairy, longer than 3-pointed bracts; paired broad-winged seeds.

Habitat: Moist soil in humid cool temperate regions.

Range: Native of N. and central Europe, at high altitudes. Grown in NE. United States and SE. Canada.

Grown as a handsome ornamental and in forestry plantations, it has escaped and naturalized locally. In its native range, it is an important timber tree and a source of Venetian turpentine, an aromatic resin formerly popular in medicine. The durable wood is used for utility poles, posts, railway cross-ties, shipbuilding, construction, and charcoal. The bark has served in tanning and medicine.

18, 482 **Tamarack**
"Hackmatack" "Eastern Larch"
Larix laricina (Du Roi) K. Koch

Description: Deciduous tree with straight, tapering trunk and thin, open, conical crown of horizontal branches; a shrub at timberline.
Height: 40–80′ (12–24 m).
Diameter: 1–2′ (0.3–0.6 m).
Needles: *deciduous;* ¾–1″ (2–2.5 cm) long, 1/32″ (1 mm) wide. Soft, very slender, 3-angled; crowded in cluster *on spur twigs,* also *scattered* and *alternate on leader twigs.* Light blue-green, turning yellow in autumn before shedding.
Bark: reddish-brown; scaly, thin.
Twigs: orange-brown; stout, hairless, with *many spurs* or short side twigs.
Cones: ½–¾″ (12–19 mm) long; *elliptical;* rose red turning brown; *upright, stalkless;* falling in second year; several overlapping rounded cone-scales; paired brown long-winged seeds.

Habitat: Wet peaty soils of bogs and swamps; also in drier upland loamy soils; often in pure stands.

Range: Across N. North America near northern limit of trees from Alaska east to Labrador, south to N. New Jersey, and west to Minnesota; local in N. West Virginia and W. Maryland; from near sea level to 1700–4000′ (518–1219 m) southward.

One of the northernmost trees, the hardy Tamarack is useful as an ornamental in very cold climates. Indians used the slender roots to sew together strips of birch bark for their canoes. Roots bent at right angles served the colonists as "knees" in small ships, joining the ribs to deck timbers. The durable lumber is used as framing for houses, railroad cross-ties, poles, and pulpwood. The larch sawfly defoliates stands in infrequent years, causing damage or death.

26, 477 **Norway Spruce**
Picea abies (L.) Karst.

Description: Large, introduced, cone-bearing tree with straight trunk and pyramid-shaped crown of spreading branches.
Height: 80′ (24 m).
Diameter: 2′ (0.6 m).
Needles: *evergreen;* ½–1″ (1.2–2.5 cm) long. Stiff, *4-angled, sharp-pointed;* spreading on all sides of twig from very short leafstalks. Shiny dark green with whitish lines.
Bark: reddish-brown; scaly.
Twigs: reddish-brown; slender, drooping, mostly hairless, rough with peglike bases.
Cones: 4–6″ (10–15 cm) long; *cylindrical;* light brown; *hanging down;* cone-scales numerous, thin, slightly pointed, *irregularly toothed,* opening and shedding year after maturing; paired long-winged seeds.

Habitat: Moist soil in humid cool temperate regions.

Range: Native of N. and central Europe, at high altitudes. Widely planted in SE. Canada, NE. United States, Rocky Mountains, and Pacific Coast region. Escaped in Northeast and perhaps naturalized locally.

Norway Spruce has been widely cultivated for ornament, shade, shelterbelts, Christmas trees, and forest plantations. The showy cones are the largest of the spruces. Numerous horticultural varieties include trees with narrow columnar shape, drooping or weeping branches, dwarf habit, and yellowish or variegated needles. The common spruce of northern Europe.

23, 479 **White Spruce**
"Canadian Spruce" "Skunk Spruce"
Picea glauca (Moench) Voss

Description: Tree with rows of horizontal branches forming a conical crown; smaller and shrubby at tree line.
Height: 40–100′ (12–30 m).
Diameter: 1–2′ (0.3–0.6 m).
Needles: *evergreen;* ½–¾″ (12–19 mm) long. Stiff, *4-angled, sharp-pointed;* spreading mainly on upper side of twig, from very short leafstalks. Blue-green, with whitish lines; exuding *skunklike odor* when crushed.
Bark: gray or brown; thin; smooth or scaly; cut surface of inner bark whitish.
Twigs: orange-brown; slender, hairless, rough with peglike bases.
Cones: 1¼–2½″ (3–6 cm) long; *cylindrical; shiny light brown;* hanging at end of twigs; falling at maturity; cone-scales thin and flexible, *margins nearly straight* and without teeth; paired brown long-winged seeds.

Habitat: Many soil types in coniferous forests; sometimes in pure stands.

Range: Across N. North America near northern limit of trees from Alaska and British Columbia east to Labrador, south to Maine, and west to Minnesota; local in NW. Montana, South Dakota, and Wyoming; from near sea level to timberline at 2000–5000′ (610–1524 m).

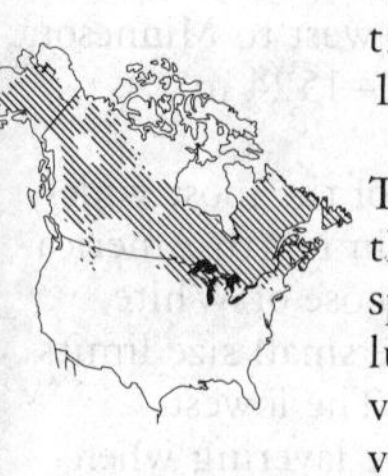

The foremost pulpwood and generally the most important commercial tree species of Canada. As well as providing lumber for construction, the wood is valued for piano sounding boards, violins, and other musical instruments. White Spruce and Black Spruce are the most widely distributed conifers in North America after Common Juniper, which rarely reaches tree size. Various kinds of wildlife, including deer, rabbits, and grouse, browse spruce foliage in winter.

24, 481 **Black Spruce**
"Bog Spruce" "Swamp Spruce"
Picea mariana (Mill.) B.S.P.

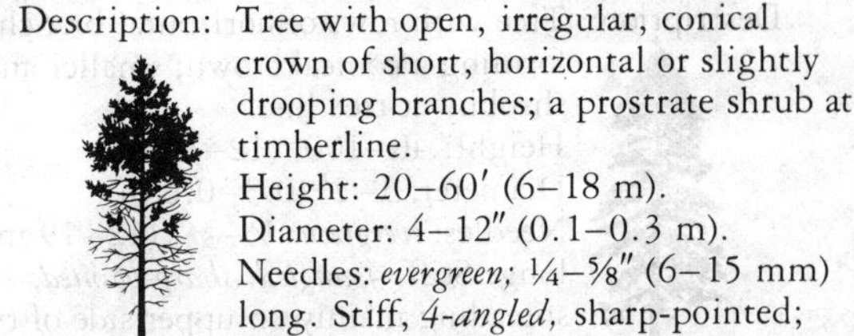

Description: Tree with open, irregular, conical crown of short, horizontal or slightly drooping branches; a prostrate shrub at timberline.
Height: 20–60′ (6–18 m).
Diameter: 4–12″ (0.1–0.3 m).
Needles: *evergreen;* ¼–⅝″ (6–15 mm) long. Stiff, *4-angled,* sharp-pointed; spreading on all sides of twig from very short leafstalks. Ashy blue-green with whitish lines.
Bark: gray or blackish, thin, scaly; brown beneath; cut surface of inner bark yellowish.
Twigs: brown; slender, hairy, rough, with peglike bases.
Cones: ⅝–1¼″ (1.5–3 cm) long; *egg-shaped* or rounded; dull gray; *curved downward* on short stalk and *remaining attached,* often *clustered* near top of crown; cone-scales stiff and brittle, rounded and finely toothed; paired brown long-winged seeds.

Habitat: Wet soils and bogs including peats, clays, and loams; in coniferous forests; often in pure stands.

Range: Across N. North America near northern limit of trees from Alaska and British Columbia east to Labrador, south to N. New Jersey, and west to Minnesota; at 2000–5000′ (610–1524 m).

Black Spruce is one of the most widely distributed conifers in North America. Uses are similar to those of White Spruce; however, the small size limits lumber production. The lowest branches take root by layering when deep snows bend them to the ground, forming a ring of small trees around a large one. Spruce gum and spruce beer were made from this species and Red Spruce.

25, 478 Red Spruce
"Eastern Spruce" "Yellow Spruce"
Picea rubens Sarg.

Description: The only spruce southward in eastern mountains, a handsome tree with broad or narrow, conical crown.
Height: 50–80′ (15–24 m).
Diameter: 1–2′ (0.3–0.6 m).
Needles: *evergreen;* ½–⅝″ (12–15 mm) long. Stiff, *4-angled, sharp-pointed;* spreading on all sides of twig from very short leafstalks. Shiny green, with whitish lines.
Bark: reddish-brown; thin, scaly.
Twigs: brown; slender, finely hairy, rough with peglike bases.
Cones: 1¼–1½″ (3–4 cm) long; *cylindrical; reddish-brown;* hanging down on short, straight stalk; falling at maturity; cone-scales *stiff,* rounded, often finely toothed; paired brown long-winged seeds.

Habitat: Rocky mountain soils; often in pure stands.

Range: Ontario east to Nova Scotia; from New England south in mountains to W. North Carolina and E. Tennessee; to 4500–6500′ (1372–1981 m) in south.

Extensive virgin spruce-fir forests are preserved in Great Smoky Mountains National Park. This species is a handsome ornamental; the wood has uses similar to White Spruce. Spruce gum, a forerunner of modern chewing gum made from chicle (gum from a tropical American tree), was obtained commercially from resin of both Red and Black spruce trunks. The young leafy twigs were boiled with flavoring and sugar to prepare spruce beer. Where the ranges overlap, Black Spruce is distinguishable from Red by its smaller dull gray cones curved downward on short stalks and remaining attached.

15, 476 **Jack Pine**
"Scrub Pine" "Gray Pine"
Pinus banksiana Lamb.

Description: Open-crowned tree with spreading branches and very short needles; sometimes a shrub.
Height: 30–70′ (9–21 m).
Diameter: 1′ (0.3 m).
Needles: *evergreen;* ¾–1½″ (2–4 cm) long. *2 in bundle; stout,* slightly flattened and twisted, *widely forking;* shiny green.
Bark: gray-brown or dark brown; thin, with narrow scaly ridges.
Cones: 1¼–2″ (3–5 cm) long; narrow, *long-pointed,* and *curved* upward; shiny light yellow; usually *remaining closed* on tree many years; cone-scales slightly raised and rounded, keeled, *mostly without prickle.*

Habitat: Sandy soils, dunes, and on rock outcrops; often in extensive pure stands.

Range: Mackenzie and Alberta, east to central Quebec and Nova Scotia, southwest to New Hampshire, and west to N. Indiana and Minnesota; to about 2000′ (610 m).

Jack Pine is a pioneer after fires and logging, although it is damaged or killed by fires. The cones usually remain closed for many years until opened by heat of fires or exposure after cutting. The northernmost New World pine, it extends beyond 65° northern latitude in Mackenzie and nearly to the limit of trees eastward. Kirtland's warbler is dependent upon Jack Pine; this rare bird breeds only in north-central Michigan, where it is confined to dense stands of young pines following forest fires.

13 **Sand Pine**
"Scrub Pine" "Spruce Pine"
Pinus clausa (Engelm.) Vasey ex Sarg.

Description: Florida's small to medium-sized pine, a tree with rounded or flattened crown.
Height: 30–70′ (9–21 m).
Diameter: 1–1½′ (0.3–0.5 m).
Needles: *evergreen;* 2–3½″ (5–9 cm) long. *2 in bundle;* slender, slightly twisted; dark green.
Bark: dark gray or reddish-brown, furrowed into narrow scaly ridges; on small trunks, gray and smooth.
Cones: 2–3½″ (5–9 cm) long; *narrowly egg-shaped; yellow-brown;* short-stalked; bent downwards, *often clustered* in 2–3 rows or whorls; mostly remaining closed on tree many years and becoming gray; cone-scales slightly keeled and raised, with *short stout prickle.*

Habitat: Well-drained sandy soils; often in pure stands.

Range: NE. to S. Florida (Ocala race) and in NW. Florida and extreme S. Alabama (Choctawhatchee race); to 200′ (61 m).

The two geographic races differ in time of cone opening. When dense stands of the Ocala race with closed cones are killed by a crown fire, the cones open and release seeds for another forest. The cones of the Choctawhatchee race open at maturity, producing open, uneven-aged stands burned by less destructive surface fires.

9, 474 **Shortleaf Pine**
"Shortstraw Pine" "Southern Yellow Pine"
Pinus echinata Mill.

Description: The most widely distributed of the southern yellow pines, a large tree with broad, open crown.
Height: 70–100′ (21–30 m).
Diameter: 1½–3′ (0.5–0.9 m).

Needles: *evergreen;* 2¾–4½″ (7–11 cm) long. *2 or sometimes 3 in bundle;* slender, flexible; dark blue-green.
Bark: reddish-brown, with large irregular flat scaly plates.
Cones: 1½–2½″ (4–6 cm) long; conical or *narrowly egg-shaped, dull brown;* short-stalked; opening at maturity but *remaining attached;* cone-scales thin, keeled, with *small prickle.*

Habitat: From dry rocky mountain ridges to sandy loams and silt loams of flood plains, and in old fields; often in pure stands or with other pines and oaks.

Range: Extreme SE. New York and New Jersey south to N. Florida, west to E. Texas, and north to S. Missouri; to 3300′ (1006 m).

Shortleaf Pine is native in 21 southeastern states. An important timber species, producing lumber for construction, millwork, and many other uses, as well as plywood and veneer for containers. This and other southern pines are the major native pulpwoods and leading woods in production of barrels. Seedlings and small trees will sprout after fire damage or injury.

---

6 Slash Pine

"Yellow Slash Pine" "Swamp Pine"
Pinus elliottii Engelm.

Description: Large tree with narrow, regular, pointed crown of horizontal branches and long needles.
Height: 60–100′ (18–30 m).
Diameter: 2–2½′ (0.6–0.8 m).
Needles: *evergreen;* 7–10″ (18–25 cm) long; *2 and 3 in bundle;* stout, stiff; slightly shiny green.
Bark: purplish-brown, with large, flattened, scaly plates; on small trunks, blackish-gray, rough and furrowed.
Cones: 2½–6″ (6–15 cm) long;

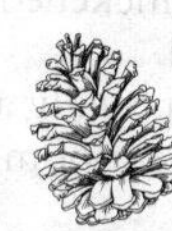

narrowly egg-shaped; shiny dark brown; short-stalked; opening and shedding at maturity, leaving a few cone-scales on twig; cone-scales flat, slightly keeled, with *short stout prickle.*

Habitat: Low areas such as pond margins, flatwoods, swamps or "slashes," including poorly drained sandy soils; also uplands and old fields. In pure stands as a subclimax after fires and in mixed forests.

Range: Coastal Plain from S. South Carolina to S. Florida, and west to SE. Louisiana; mostly near sea level, locally to 500′ (152 m).

An important species both for lumber and naval stores and one of the fastest-growing southern pines, Slash Pine is extensively grown in forest plantations both in its natural range and farther north. Its beauty makes it popular as a shade and ornamental tree. A variety, South Florida Slash Pine (var. *densa* Little & Dorman) is a medium-sized tree of south and central Florida with needles mostly 2 in a bundle and with a grasslike seedling stage.

---

11 Spruce Pine

"Cedar Pine" "Walter Pine"
Pinus glabra Walt.

Description: Large tree with long narrow crown and horizontal branches.
Height: 80–90′ (24–27 m).
Diameter: 2–2½′ (0.6–0.8 m).
Needles: *evergreen;* 1½–4″ (4–10 cm) long; *2 in bundle;* slender, often slightly flattened and twisted; dark green.
Bark: *gray, smooth,* thin; becoming dark gray and furrowed into flat scaly ridges.
Cones: 1¼–2½″ (3–6 cm) long; *conical* or narrowly egg-shaped; *reddish-brown;* nearly stalkless; pointing backward or *downward;* opening at maturity, shedding or sometimes persistent; cone-

scales thin or slightly thickened, with *tiny prickle* usually shed.

Habitat: Moist lowland soils, especially along rivers; a minor component of mixed swamp forests.

Range: Coastal Plain from E. South Carolina to N. Florida and west to SE. Louisiana; to about 500′ (152 m).

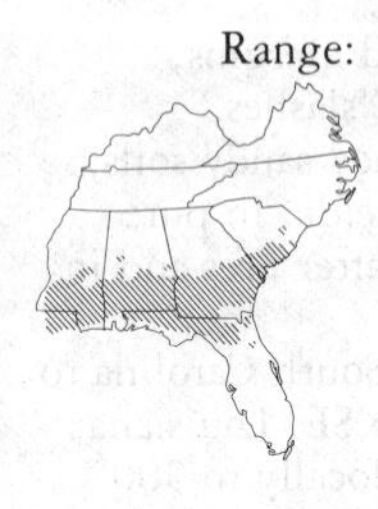

The wood of this uncommon species is used for pulpwood and infrequently for lumber. The smooth gray bark makes it suitable as an ornamental.

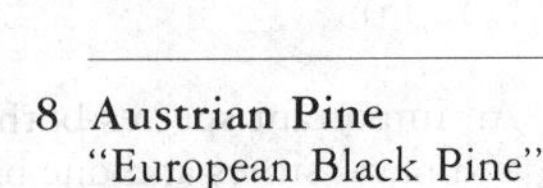

8 **Austrian Pine**
"European Black Pine"
Pinus nigra Arnold

Description: Introduced, cone-bearing, resinous tree with straight trunk and dense, rounded, spreading crown of dark green foliage.

Height: 60′ (18 m).
Diameter: 2′ (0.6 m).
Needles: *evergreen;* 3½–6″ (9–15 cm) long; crowded, *2 in bundle; stiff; shiny dark green.*
Bark: dark gray; thick, rough, furrowed into irregular scaly plates.
Twigs: light brown or gray; stout, hairless.
Cones: 2–3″ (5–7.5 cm) long, 1–1¼″ (2.5–3 cm) wide; *egg-shaped; shiny yellow-brown;* almost stalkless; opening and shedding after maturity; cone-scales numerous, raised and keeled, ending in *short prickle.*

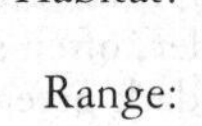

Habitat: Hardy except in coldest, hottest, and driest regions.

Range: Native of central and S. Europe and Asia Minor; also local in NW. Africa. Planted across the United States; escaped locally but apparently not naturalized.

Austrian Pine is one of the more common introduced ornamental trees, with several geographical and horticultural varieties. Used also in shelterbelts and screens, it is fast-growing and tolerant of city smoke, dust, and dry soil.

2 Longleaf Pine

"Longleaf Yellow Pine" "Southern Yellow Pine"
Pinus palustris Mill.

Description: Large tree with the *longest needles and largest cones of any eastern pine* and an open, irregular crown of a few spreading branches, 1 row added each year.
Height: 80–100′ (24–30 m).
Diameter: 2–2½′ (0.6–0.8 m).
Needles: *evergreen;* mostly 10–15″ (25–38 cm) long, on small plants to 18″ (46 cm). *Densely crowded, 3 in bundle;* slightly *stout,* flexible; spreading to *drooping;* dark green.
Bark: orange-brown, furrowed into scaly plates; on small trunks, gray and rough.
Twigs: dark brown; very stout, ending in *large white bud.*

Cones: 6–10″ (15–25 cm) long; narrowly conical or cylindrical; dull brown; almost stalkless; opening and shedding at maturity; cone-scales raised, keeled, with *small prickle.*

Habitat: Well-drained sandy soils of flatlands and sandhills; often in pure stands.

Range: Coastal Plain from SE. Virginia to E. Florida, and west to E. Texas. Usually below 600′ (183 m); to 2000′ (610 m) in foothills of Piedmont.

Longleaf Pine is a leading world producer of naval stores. The trees are tapped for turpentine and resin and then logged for construction lumber, poles and pilings, and pulpwood. Frequent fires caused by man or by lightning have perpetuated subclimax, pure stands of this species. The seedlings pass through a "grass" stage for a few years, in which the stem grows in thickness rather than height and the taproot develops rapidly. Later, the elongating, unbranched stem produces very long needles.

10 Table Mountain Pine
"Hickory Pine" "Mountain Pine"
Pinus pungens Lamb.

Description: Tree with rounded or irregular crown of stout, horizontal branches and abundant spiny cones in clusters along branches.
Height: 20–40′ (6–12 m).
Diameter: 1–2′ (0.3–0.6 m).
Needles: *evergreen;* 1¼–2½″ (3–6 cm) long; *2 in bundle* (sometimes 3); stout, stiff, usually *twisted;* dark green.
Bark: dark brown; thick, furrowed into scaly plates.
Cones: 2–3½″ (5–9 cm) long; *egg-shaped; shiny light brown;* usually in *clusters of 3–4;* stalkless; pointing backward or downward; opening partly at maturity but remaining attached many years; cone-scales thickened and keeled, with *stout curved spine.*

Habitat: Dry gravelly and rocky slopes and ridges of mountains with other pines; sometimes in pure stands.

Range: Appalachian region from Pennsylvania south to NE. Georgia and E. Tennessee; local in New Jersey, Delaware, and District of Columbia. To 4000′ (1219 m); rarely down almost to sea level.

Easily seen at Shenandoah and Great Smoky Mountains National Parks, this tree of mountain ridges is the only pine restricted to the Appalachian Mountains. Called "Hickory Pine" because of the very tough hickorylike branches.

7 **Red Pine**
"Norway Pine"
Pinus resinosa Ait.

Description: A common, large tree with small cones and broad, irregular or rounded crown of spreading branches, 1 row added a year.
Height: 70–80′ (21–24 m).
Diameter: 1–3′ (0.3–0.9 m), often larger.
Needles: *evergreen;* 4¼–6½″ (11–16.5 cm) long; *2 in bundle;* slender; dark green.
Bark: reddish-brown or gray; with broad, flat, scaly plates; becoming thick.

Cones: 1½–2¼″ (4–6 cm) long; *egg-shaped;* shiny *light brown;* almost stalkless; opening and shedding soon after maturity; cone-scales slightly thickened, keeled, *without prickle.*

Habitat: Well-drained soils; particularly sand plains; usually in mixed forests.

Range: SE. Manitoba east to Nova Scotia, south to Pennsylvania, and west to Minnesota. Local in Newfoundland, N. Illinois and E. West Virginia. At 700–1400′ (213–427 m) northward; to 2700′ (823 m) in Adirondacks; and at 3800–4300′ (1158–1311 m) in West Virginia.

The misleading name "Norway Pine" for this New World species may be traced to confusion with Norway Spruce by early English explorers. Another explanation is that the name comes from the tree's occurrence near Norway, Maine, founded in 1797. Because the name was in usage before this time, the former explanation is more likely. Red Pine is an ornamental and shade tree; the wood is used for general construction, planing-mill products, millwork, and pulpwood.

5, 472 Pitch Pine
Pinus rigida Mill.

Description: Medium-sized tree often bearing tufts of needles on trunk, with a broad, rounded or irregular crown of horizontal branches.
Height: 50–60′ (15–18 m).
Diameter: 1–2′ (0.3–0.6 m).
Needles: *evergreen;* 3–5″ (7.5–13 cm) long; *3 in bundle; stout,* stiff, often twisted; yellow-green.
Bark: dark gray; thick, rough, deeply furrowed into broad scaly ridges, exposing brown inner layers.
Cones: 1¼–2¾″ (3–7 cm) long; *egg-shaped; yellow-brown;* opening at maturity but remaining attached; cone-scales raised and keeled, with *slender sharp prickle.*

Habitat: Shallow sands and gravels on steep slopes and ridges, also in river valleys and swamps. Forms temporary pure stands, gradually replaced by hardwoods; also in mixed forests.

Range: S. Maine west to New York and southwest mostly in mountains to N. Georgia; local in extreme S. Quebec and extreme SE. Ontario. From sea level in Coastal Plain to about 2000′ (610 m) in north; 1400–4500′ (427–1372 m) in upper Piedmont and southern mountains.

Now used principally for lumber and pulpwood, Pitch Pine was once a source of resin. Colonists produced turpentine and tar used for axle grease from this species before naval stores were developed from the southern pines. Pine knots, when fastened to a pole, served as torches at night. The common name refers to the high resin content of the knotty wood. Pitch Pine is suitable for planting on dry rocky soil that other trees cannot tolerate, becoming open and irregular in shape in exposed situations. This hardy species is resistant to fire and injury, forming sprouts from roots and stumps. It is the pine at Cape Cod; and the New Jersey pine barrens are composed of dwarf sprouts of Pitch Pine following repeated fires.

---

4 Pond Pine
"Marsh Pine" "Pocosin Pine"
Pinus serotina Michx.

Description: Medium-sized tree with open, rounded or irregular crown of stout, often crooked branches.
Height: 40–70′ (12–21 m).
Diameter: 1–2′ (0.3–0.6 m).
Needles: *evergreen;* 5–8″ (13–20 cm) long; *3 in bundle;* slender, stiff; yellow-green.
Bark: blackish-gray or reddish-brown; furrowed into scaly plates.
Cones: 2–2½″ (5–6 cm) long; *nearly round* or egg-shaped; shiny *yellow;* almost stalkless; *remaining closed* on tree many years; cone-scales slightly raised and keeled, with weak prickle usually shed.

Habitat: Swamps, shallow bays, and ponds; often in nearly pure stands.

Range: S. New Jersey and Delaware south to central and NW. Florida, near sea level.

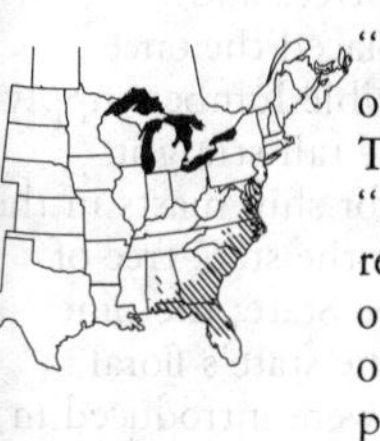

"Pocosin" is an Indian name for pond or bog, alluding to this species' habitat. The Latin name *serotina,* meaning "late," refers to the cones, which remain closed for years before opening, often following a fire. After fires or other damage, seedlings and trees will produce sprouts from roots.

1, 475 **Eastern White Pine**
"White Pine" "Northern White Pine"
Pinus strobus L.

Description: The largest northeastern conifer, a magnificent evergreen tree with straight trunk and crown of horizontal branches, 1 row added a year, becoming broad and irregular.
Height: 100′ (33 m), formerly 150′ (46 m) or more.
Diameter: 3–4′ (0.9–1.2 m) or more.
Needles: *evergreen;* 2½–5″ (6–13 cm) long, *5 in bundle; slender; blue-green.*
Bark: gray; smooth becoming rough; thick and deeply furrowed into narrow scaly ridges.
Cones: 4–8″ (10–20 cm) long; *narrowly cylindrical; yellow-brown;* long-stalked; cone-scales thin, rounded, flat.

Habitat: Well-drained sandy soils; sometimes in pure stands.

Range: SE. Manitoba east to Newfoundland, south to N. Georgia, and west to NE. Iowa; a variety in Mexico. From near sea level to 2000′ (610 m); in the southern Appalachians to 5000′ (1524 m).

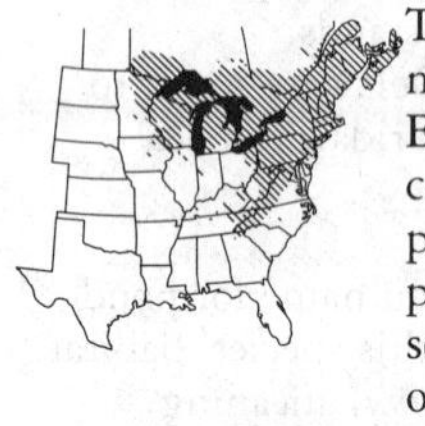

The largest conifer and formerly the most valuable tree of the Northeast, Eastern White Pine is used for construction, millwork, trim, and pulpwood. Younger trees and plantations have replaced the once seemingly inexhaustible lumber supply of virgin forests. The tall straight trunks were prized for ship masts in the colonial period. It is the state tree of Maine, the Pine Tree State; the pine cone and tassel are the state's floral emblem. The seeds were introduced in England (where it is called Weymouth Pine) from Maine in 1605 by Captain George Weymouth of the British Navy

12 **Scotch Pine**
"Scots Pine"
Pinus sylvestris L.

Description: Beautiful, large, introduced tree with crown of spreading branches that become rounded and irregular, and rich blue-green foliage.
Height: 70′ (21 m).
Diameter: 2′ (0.6 m) and much larger with age.
Needles: *evergreen;* 1½–2¾″ (4–7 cm) long. *2 in bundle;* stiff, slightly flattened; *twisted* and spreading; blue-green.
Bark: *reddish-brown,* thin; becoming gray and shedding in papery or *scaly plates.*

Cones: 1¼–2½″ (3–6 cm) long; *egg-shaped;* pale *yellow-brown;* short-stalked; opening at maturity; cone-scales thin, flattened, often with minute prickle.

Habitat: Various soils from loams to sand; tolerating city smoke.

Range: Native across Europe and N. Asia, south to Turkey. Naturalized locally in SE. Canada and NE. United States from New England west to Iowa.

The native pine of the Scottish Highlands, this is the most widely distributed pine in the world and one of the most important European timber trees. In the United States, native pines are better adapted for forestry plantations, but Scotch Pine is commonly grown for shelterbelts, ornament, and Christmas trees.

---

3 **Loblolly Pine**
"Oldfield Pine" "North Carolina Pine"
Pinus taeda L.

Description: The principal commercial southern pine, a large, resinous, and fragrant tree with rounded crown of spreading branches.

Height: 80–100′ (24–30 m).
Diameter: 2–3′ (0.6–0.9 m).
Needles: *evergreen;* 5–9″ (13–23 cm) long. *3 in bundle;* stout, stiff, often twisted; green.
Bark: blackish-gray; thick, deeply furrowed into scaly ridges exposing brown inner layers.
Cones: 3–5″ (7.5–13 cm) long; *conical; dull brown;* almost stalkless; opening at maturity but remaining attached; cone-scales raised, keeled, with short *stout spine.*

Habitat: From deep, poorly drained flood plains to well-drained slopes of rolling, hilly uplands. Forms pure stands, often on abandoned farmland.

Range: S. New Jersey south to central Florida, west to E. Texas, north to extreme SE. Oklahoma; to 1500–2000′ (457–610 m).

Loblolly Pine is native in 15 southeastern states. Among the fastest-growing southern pines, it is extensively cultivated in forest plantations for pulpwood and lumber. One of the meanings of the word *loblolly* is "mud puddle," where these pines often grow. It is also called "Bull Pine," from the giant size, and "Rosemary Pine," from the fragrant resinous foliage.

14, 473 **Virginia Pine**
"Scrub Pine" "Jersey Pine"
Pinus virginiana Mill.

Description: Short-needled tree with open, broad, irregular crown of long spreading branches; often a shrub.

Height: 30–60′ (9–18 m).
Diameter: 1–1½′ (0.3–0.5 m).
Needles: *evergreen;* 1½–3″ (4–7.5 cm) long; *2 in bundle;* stout, slightly flattened and twisted; dull green.
Bark: brownish-gray, thin, with narrow

scaly ridges, becoming shaggy; on small trunks, smoothish, peeling off in flakes.
Cones: 1½–2¾″ (4–7 cm) long; *narrowly egg-shaped,* shiny *reddish-brown;* almost stalkless; opening at maturity but *remaining attached;* cone-scales slightly raised and keeled, with *long slender prickle.*

Habitat: Clay, loam, and sandy loam on well-drained sites. Forms pure stands, especially on old fields or abandoned farmland, even in poor or severely eroded soil. Also in mixed forest types.

Range: SE. New York (Long Island) south to NE. Mississippi, and north to S. Indiana; at 100–2500′ (30–762 m).

Used principally for pulpwood and lumber, it is hardier than most pines and suitable for planting in poor dry sites. Common in old fields as a pioneer after grasses on hills of the Piedmont, growing rapidly and forming thickets. Later this pine is replaced by taller, more valuable hardwoods.

20, 483 **Eastern Hemlock**
"Canada Hemlock" "Hemlock Spruce"
Tsuga canadensis (L.) Carr.

Description: Evergreen tree with conical crown of long, slender, horizontal branches often drooping down to the ground, and a slender, curved, and *drooping leader.*

Height: 60–70′ (18–21 m).
Diameter: 2–3′ (0.6–0.9 m).
Needles: *evergreen;* ⅜–⅝″ (10–15 mm) long. *Flat, flexible, rounded at tip;* spreading in *2 rows* from very short leafstalks. Shiny dark green above, with *2 narrow whitish bands* beneath and green edges often minutely toothed.
Bark: cinnamon brown; thick, deeply furrowed into broad scaly ridges.
Twigs: yellow-brown; very slender, finely hairy, *rough with peglike bases.*

Cones: ⅝–¾″ (15–19 mm) long; *elliptical;* brown; short-stalked; hanging down at ends of twigs; composed of numerous *rounded cone-scales;* paired, light brown, long-winged seeds.

Habitat: Acid soils; often in pure stands. Characteristic of moist cool valleys and ravines; also rock outcrops, especially north-facing bluffs.

Range: S. Ontario east to Cape Breton Island, south in mountains to N. Alabama, and west to E. Minnesota. To 3000′ (914 m) in north; at 2000–5000′ (610–1524 m) in south.

The bark was once a commercial source of tannin in the production of leather. Pioneers made tea from leafy twigs and brooms from the branches. A graceful shade tree and ornamental, it can also be trimmed into hedges.

19 Carolina Hemlock
Tsuga caroliniana Engelm.

Description: Evergreen tree with compact, conical crown of short, stout, often drooping branches and very slender *drooping leader.*

Height: 40–60′ (12–18 m).

Diameter: 2′ (0.6 m).

Needles: *evergreen;* ⅜–¾″ (10–19 mm) long. *Flat, flexible, slightly notched* at tip, not toothed; *spreading in all directions* from very short leafstalks. Shiny dark green above, with *2 broad whitish bands* beneath.

Bark: red-brown; thick, deeply furrowed into broad, flat, connected scaly ridges.

Twigs: orange-brown; very slender, slightly hairy, *rough with peglike bases.*

Cones: ¾–1⅜″ (2–3.5 cm) long; *narrowly elliptical;* short-stalked; hanging down at ends of twigs; composed of numerous oblong cone-scales spreading widely, almost at right angles at maturity; paired, long-winged seeds.

Habitat: Dry slopes, rocky ridges, ledges, and cliffs; in mixed forests.

Range: SW. Virginia, NE. Tennessee, W. North Carolina, extreme NW. South Carolina, and extreme NE. Georgia; at 2500–4000′ (762–1219 m).

This popular ornamental has a handsome, symmetrical form and is hardy in cities. The plants can be trimmed into hedges or clipped to grow in various shapes. A local species unnoticed by early botanical collectors, Carolina Hemlock was first distinguished in 1850 and named in 1881.

REDWOOD FAMILY
(Taxodiaceae)

Medium-sized to very large, aromatic and resinous trees, without flowers or fruit. About 15 species in temperate North America, east Asia, and Tasmania; 4 native species in North America, 2 eastern and 2 western.
Leaves: needlelike or scalelike, often of both kinds, mostly evergreen (deciduous in baldcypress, *Taxodium*), mainly alternate in spirals.
Twigs: often very slender and shedding with leaves.
Cones: pollen and naked seeds borne on same plant. Male cones small and herbaceous, with 2–9 pollen-sacs. Female cones hard or woody, rounded or slightly elliptical, at ends of twigs, with many flat cone-scales lacking bracts; 2–9 naked seeds angled or narrowly winged at base of cone-scale.

---

27 Baldcypress
"Cypress" "Swamp-cypress"
Taxodium distichum (L.) Rich.

Description: Large, needle-leaf, aquatic, *deciduous tree* often with cone-shaped "*knees*" projecting from submerged roots, with trunks enlarged at base and spreading into ridges or buttresses, and with a crown of widely spreading branches, flattened at top.

Height: 100–120′ (30–37 m) or more.
Diameter: 3–5′ (0.9–1.5 m), rarely 10′ (3 m) or more.
Needles: *deciduous;* ⅜–¾″ (10–19 mm) long. Borne singly in *2 rows* on slender green twigs, crowded and featherlike; *flat,* soft, and *flexible.* Dull light green above, whitish beneath; turning brown and shedding with twig in fall.
Bark: brown or gray; with long fibrous or scaly ridges, peeling off in strips.
Cones: ¾–1″ (2–2.5 cm) in diameter;

round; gray; 1–2 at end of twig; several flattened, 4-angled, hard cone-scales shed at maturity in autumn; 2 brown, 3-angled seeds nearly ¼″ (6 mm) long, under cone-scale. Tiny pollen cones in narrow drooping cluster 4″ (10 cm) long.

Habitat: Very wet, swampy soils of riverbanks and floodplain lakes that are sometimes submerged; often in pure stands.

Range: S. Delaware to S. Florida, west to S. Texas and north to SE. Oklahoma and SW. Indiana. Below 500′ (152 m); locally in Texas to 1700′ (518 m).

Called the "wood eternal" because of the heartwood's resistance to decay, Baldcypress is used for heavy construction, including docks, warehouses, boats, bridges, as well as general millwork and interior trim. The trees are planted as ornamentals northward in colder climates and in drier soils. Easily seen in Big Cypress National Preserve near Naples, Florida. Pondcypress (var. *nutans* (Ait.) Sweet), a variety with shorter scalelike leaves, is found in shallow ponds and poorly drained areas from southeastern Virginia to southeastern Louisiana below 100′ (30 m).

28 Montezuma Baldcypress

"Mexican-cypress" "Sabino"
Taxodium mucronatum Ten.

Description: Large, needle-leaf, aquatic tree with tall, straight trunk and broad crown of spreading branches and drooping twigs, evergreen or nearly so. Trunk enlarged at base with ridges above; sometimes small "knees" project from submerged roots.

Height: 50–100′ (15–30 m).
Diameter: 3–6′ (0.9–1.8 m).
Needles: ¼–½″ (6–12 mm) long. In *2 rows* on slender green twigs, crowded

and *featherlike; flat,* soft, and *flexible.* Dull light green; turning yellow and brown and shedding with twig in spring when new leaves appear.

Bark: reddish-brown; smooth to shallowly furrowed, fibrous and shreddy.

Cones: ⅝–1″ (1.5–2.5 cm) in diameter; *round;* gray or brown; at end of twig; several flattened, 4-angled, hard cone-scales shed at maturity; 2 dark reddish-brown, 3-angled seeds ¼″ (6 mm) long, under cone-scale. Tiny pollen cones, in open drooping clusters 2–6″ (5–15 cm) long.

Habitat: Wet soil of stream banks and swamps.

Range: Extreme S. Texas in Cameron and Hidalgo counties, near sea level; also Mexico.

The national tree of Mexico, Montezuma Baldcypress is closely related to Baldcypress of southeastern United States but is usually evergreen and not hardy in cold climates. The Big Tree of Tule, near Oaxaca, Mexico, is a famous giant. Apparently formed by the fusion of 3 trees, it has a trunk circumference of 112′ (34 m) and a height of 141′ (43 m). The majestic veterans in Chapultepec Park, Mexico City, are taller, reaching 165′ (50 m), and are among the oldest cultivated trees in the New World, perhaps exceeding 600 years.

CYPRESS FAMILY
(Cupressaceae)

Resinous, evergreen trees and shrubs without flowers or fruits, mostly with straight axis and narrow crown, usually with soft, lightweight wood; including cypresses (*Cupressus*), junipers (*Juniperus*), and some "cedars." About 130 species worldwide; 26 native and 1 naturalized tree species, and 1 native shrub species in North America.
Leaves: opposite or whorled, usually of 2 forms, mostly small and scalelike or awl-shaped and producing flattened or angled twigs.
Cones: pollen and seeds borne mostly on same plant. Male cones small and herbaceous; female cones woody (berrylike in juniper, *Juniperus*), usually composed of few cone-scales opposite or whorled and flattened or attached at mid-point; 1–2 naked seeds at a cone-scale, often with 2 lateral wings.

---

33 Sawara False-cypress
"Sawara Cypress" "Retinospora"
Chamaecyparis pisifera (Sieb. & Zucc.) Endl.

Description: Large, introduced, cone-bearing, evergreen tree with straight trunk and open, narrow, pyramid-shaped crown.

Height: 70′ (21 m).
Diameter: 2′ (0.6 m).
Leaves: *evergreen;* nearly ⅛″ (3 mm) long. Mostly *scalelike, sharp-pointed;* overlapping, *opposite in 4 rows;* top pair flattened, side pair keeled; shiny green above, whitish-green beneath. Leaves on young plants ¼″ (6 mm) long; *needlelike;* spreading in 4 rows; blue-green above, whitish beneath.
Bark: reddish-brown; fibrous and shreddy, peeling in long thin strips.
Twigs: slender, spreading in *horizontal flattened fernlike sprays.*
Cones: ¼″ (6 mm) in diameter; *rounded;*

whitish-green, turning dark brown; short-stalked; usually composed of 10 cone-scales with tiny point in center; 1–2 broad-winged seeds under cone-scale.

Habitat: Moist soil in humid temperate regions.

Range: Native of Japan. Planted across the United States.

Commonly grown as an ornamental in many cultivated varieties, characterized by golden-yellow foliage, very small leaves, weeping threadlike twigs, dwarf spreading shrub habit, and very feathery or mosslike, spreading, needlelike leaves. In Japan this is a timber tree as well as an ornamental.

34, 486 **Atlantic White-cedar**
"Southern White-cedar" "Swamp-cedar"
Chamaecyparis thyoides (L.) B.S.P.

Description: Evergreen, aromatic tree with narrow, pointed, spirelike crown and slender, horizontal branches.
Height: 50–90′ (15–27 m).
Diameter: 1½–2′ (0.5–0.6 m).
Leaves: *evergreen; opposite;* 1⁄16–⅛″ (1.5–3 mm) long. *Scalelike;* dull blue-green, with gland-dot.
Bark: reddish-brown; thin, fibrous, with narrow connecting or forking ridges, becoming scaly and loose.
Twigs: *very slender,* slightly flattened or partly 4-angled, irregularly branched.
Cones: *tiny,* ¼″ (6 mm) in diameter; bluish-purple with a bloom, becoming dark red-brown; with 6 *cone-scales* ending in short point; maturing in 1 season; 1–2 gray-brown seeds under cone-scale.

Habitat: Wet, peaty, acid soils; forming pure stands in swamp forests.

Range: Central Maine south to N. Florida and west to Mississippi in narrow coastal belt; to 100′ (30 m).

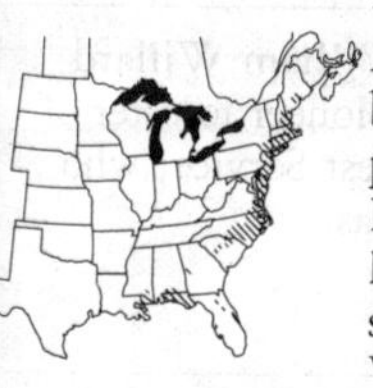

Ancient logs buried in swamps have been mined and found to be well preserved and suitable for lumber. Pioneers prized the durable wood for log cabins, including floors and shingles. During the Revolutionary War, the wood produced charcoal for gunpowder. One fine forest is preserved at Green Bank State Forest in southern New Jersey. As an ornamental, this species is the hardiest of its genus northward.

37 Ashe Juniper

"Rock-cedar" "Post-cedar"
Juniperus ashei Buchholz

Description: Evergreen tree with trunk often grooved and twisted or branched from base, and with rounded or irregular, open crown; sometimes forming thickets.

Height: 20–30′ (6–9 m).
Diameter: 1′ (0.3 m).
Leaves: *evergreen; scalelike,* 1/16″ (1.5 mm) long. Mostly *opposite in 4 rows,* forming crowded slender 4-angled twigs; minutely toothed on edges. *Dark blue-green,* usually without gland-dot.
Bark: gray-brown; fibrous and shreddy, fissured into long narrow scaly ridges.
Cones: 5/16″ (8 mm) in diameter; berrylike; *dark blue* with a bloom; *juicy,* sweetish, and resinous; usually 1-seeded.

Habitat: Limestone outcrops of plateaus and mountains; in oak-hickory forest.

Range: Ozark Mountains of Missouri, Arkansas, and NE. Oklahoma, S. Oklahoma in Arbuckle Mountains, and central Texas including Edwards Plateau; also NE. Mexico; at 800–2000′ (244–610 m).

The durable wood is a local source of fenceposts. Many kinds of wildlife eat the sweetish "berries."

Named in honor of William Willard Ashe (1872–1932), pioneer forester of the United States Forest Service, who collected it in Arkansas.

39, 543 **Common Juniper**
"Dwarf Juniper"
Juniperus communis L.

Description: Usually a spreading low shrub, sometimes forming broad or prostrate clumps; rarely a small tree with an open irregular crown.
Height: 1–4′ (0.3–1.2 m), rarely 15–25′ (4.6–7.6 m).
Diameter: 8″ (20 cm).
Leaves: *evergreen;* ⅜–½″ (10–12 mm) long. *Awl-shaped;* stiff, very *sharp-pointed,* jointed at base; *in 3's,* spreading at right angles. *Whitish* and grooved above, shiny *yellow-green* beneath.
Bark: reddish-brown to gray; thin, rough, scaly, and shreddy.
Twigs: light yellow; slender, 3-angled, hairless.
Cones: ¼–⅜″ (6–10 mm) in diameter; berrylike; *whitish blue* with a bloom; hard; *mealy;* sweetish and resinous; aromatic; maturing in 2–3 years and remaining attached; 1–3 brown, pointed seeds. Pollen cones mostly on same plant.

Habitat: Rocky slopes in coniferous forests of mountains and plains.

Range: Widespread from Alaska east to Labrador and S. Greenland, south to New York, and west to Minnesota and Wyoming; also south in mountains to NW. South Carolina and central Arizona; also Iceland and across N. Eurasia; to 8000–11,500′ (2438–3505 m) in south.

Although commonly a tree in Eurasia, Common Juniper is only rarely a small tree in New England and other northeastern States. In the West, it

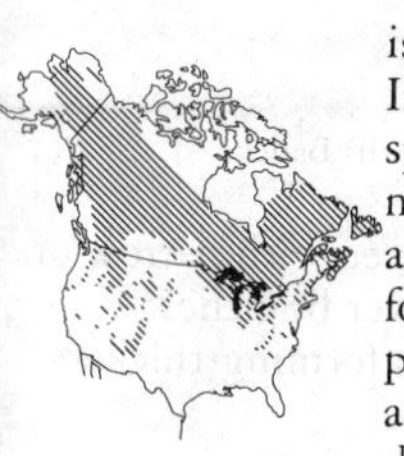

is a low shrub, often at timberline. Including geographic varieties, this species is the most widely distributed native conifer in both North America and the world. Juniper "berries" are food for wildlife, especially grouse, pheasants, and bobwhites. They are an ingredient in gin, producing the distinctive aroma and tang.

38 Pinchot Juniper

"Redberry Juniper"
Juniperus pinchotii Sudw.

Description: Evergreen shrub or small tree, usually with several branches from ground, sometimes with a single trunk, and a broad, irregular crown.

Height: 20′ (6 m).
Diameter: 1′ (0.3 m).
Leaves: *evergreen; mostly in 3's* on slender twigs; 1/16″ (1.5 mm) long. *Scalelike; yellow-green,* with gland-dot.
Bark: light brown or gray; thin, furrowed into scaly ridges.
Cones: 3/8″ (10 mm) in diameter; berrylike; *reddish;* hard and *dry;* mealy; 1–2 long, light brown, pointed, prominently angled seeds. Pollen cones on separate trees.

Habitat: Plains, including hills and canyons on gravelly and rocky soils on limestone and gypsum; in open juniper woodlands.

Range: SW. Oklahoma, Texas, and SE. New Mexico; also NE. Mexico; at 1000–5000′ (305–1524 m).

This juniper was named in honor of Gifford Pinchot (1865–1946), first chief of the United States Forest Service. Well displayed at Palo Duro State Park near Canyon, Texas, where it was discovered. The hardy plants will sprout from stumps after cutting or burning. The wood is used principally for fenceposts and fuel.

36 **Southern Redcedar**
"Sand-cedar"
Juniperus silicicola (Small) Bailey

Description: Evergreen aromatic tree with narrow or spreading crown, lower branches drooping; sometimes forming thickets.
Height: 50′ (15 m).
Diameter: 2′ (0.6 m).
Leaves: *evergreen; opposite in 4 rows* forming very *slender drooping 4-angled twigs;* 1/16″ (1.5 mm) long, to 3/8″ (10 mm) long on leaders. Scalelike; *dark green,* with gland-dot.
Bark: reddish-brown; thin, fibrous and shreddy.
Cones: 3/16″ (5 mm) in diameter; berrylike; *dark blue* with a bloom; soft, *juicy,* sweetish, and resinous; 1 or sometimes 2 pointed, ridged seeds. Pollen cones on separate trees.

Habitat: From dry uplands, especially on limestone, to wet soils of riverbanks and swamps and sandy soils near beaches; often in old fields and fence rows.

Range: Chiefly near coast, from North Carolina south to Florida and west to SE. Texas; near sea level.

This southeastern coastal relative of Eastern Redcedar is distinguished by its often drooping foliage and smaller "berries" and is planted as an ornamental. The wood is similarly used for fenceposts, cedar chests, cabinetwork, and carvings. The Latin species name means "growing in sand."

35 **Eastern Redcedar**
"Red Juniper"
Juniperus virginiana L.

Description: Evergreen, aromatic tree with trunk often angled and buttressed at base and narrow, compact, columnar crown;

sometimes becoming broad and irregular.
Height: 40–60′ (12–18 m).
Diameter: 1–2′ (0.3–0.6 m).
Leaves: *evergreen; opposite in 4 rows* forming slender 4-angled twigs; 1/16″ (1.5 mm) long, to 3/8″ (10 mm) long on leaders. *Scalelike,* not toothed; *dark green,* with gland-dot.
Bark: reddish-brown; thin, fibrous and shreddy.

Cones: 1/4–3/8″ (6–10 mm) in diameter; berrylike; *dark blue* with a bloom; soft, *juicy,* sweetish, and resinous; 1–2 seeds. Pollen cones on separate trees.

Habitat: From dry uplands, especially limestone, to flood plains and swamps; also abandoned fields and fence rows; often in scattered pure stands.

Range: S. Ontario and widespread in eastern half of United States from Maine south to N. Florida, west to Texas, and north to North Dakota.

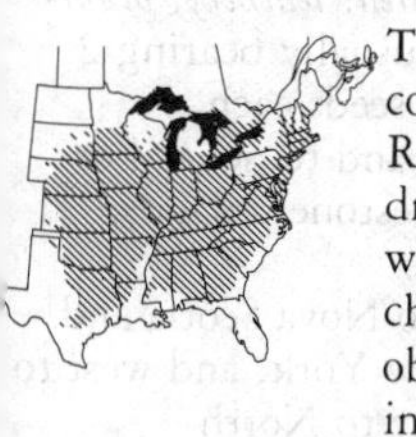

The most widely distributed eastern conifer, native in 37 states, Eastern Redcedar is resistant to extremes of drought, heat, and cold. The aromatic wood is used for fenceposts, cedar chests, cabinetwork, and carvings. First observed at Roanoke Island, Virginia, in 1564, it was prized by the colonists for building furniture, rail fences, and log cabins. Cedar oil for medicine and perfumes is obtained from the wood and leaves. The heartwood was once almost exclusively the source of wood for pencils; Incense-cedar (*Libocedrus decurrens* Torr.) is now used instead. Grown for Christmas trees, shelterbelts, and in many cultivated varieties for ornament. The juicy "berries" are consumed by many kinds of wildlife, including the cedar waxwing, named for this tree. Redcedar can be injurious to apple orchards because it is an alternate host for cedar-apple rust, a fungus disease.

32, 485 Northern White-cedar
"Eastern White-cedar" "Eastern Arborvitae"
Thuja occidentalis L.

Description: Resinous and aromatic evergreen tree with an angled, buttressed, often branched trunk and a narrow, conical crown of short, spreading branches.
Height: 40–70′ (12–21 m).
Diameter: 1–3′ (0.3–0.9 m).
Leaves: *evergreen;* opposite in 4 rows; 1/16–1/8″ (1.5–3 mm) long. *Scalelike;* short-pointed; side pair keeled, flat pair with gland-dot. Dull yellow-green above, paler blue-green beneath.
Bark: light red-brown; thin, fibrous and shreddy, fissured into narrow connecting ridges.
Twigs: branching in *horizontal plane;* much *flattened; jointed.*
Cones: 3/8″ (10 mm) long; *elliptical;* light brown; upright from short curved stalk; with *8–10 paired, leathery, blunt-pointed cone-scales,* 4 usually bearing 2 tiny narrow-winged seeds each.

Habitat: Adapted to swamps and to neutral or alkaline soils on limestone uplands; often in pure stands.

Range: SE. Manitoba east to Nova Scotia and Maine, south to New York, and west to Illinois; south locally to North Carolina; to 3000′ (914 m) in south.

Probably the first North American tree introduced into Europe, it was discovered by French explorers and grown in Paris about 1536. The year before, tea prepared from the foliage and bark, now known to be high in vitamin C, saved the crew of Jacques Cartier from scurvy. It was named *arborvitae,* Latin for "tree-of-life," in 1558. The trees grow slowly and reach an age of 400 years or more. The lightweight, easily split wood was preferred for canoe frames by Indians, who also used the shredded outer bark and the soft wood to start fires. Today, the wood is used principally for poles, cross-ties, posts, and lumber. Cedar oil for medicine is distilled from the twigs.

31 Oriental Arborvitae
"Chinese Arborvitae"
Thuja orientalis L.

Description: Ornamental, cone-bearing, resinous and aromatic evergreen shrub or small tree branching near the base, with few trunks and compact, narrow conical to rounded or irregular crown of many upright branches.
Height: 25′ (7.6 m).
Diameter: 6″ (15 cm).
Leaves: *evergreen;* opposite in 4 rows; 1/16–1/8″ (1.5–3 mm) long. *Scalelike;* short-pointed; side pair keeled, flat pair with gland-dot; *yellow-green.*
Bark: dark reddish-brown; thin, fibrous, finely fissured and shreddy.
Twigs: branching in *vertical plane; flattened; jointed* in fernlike sprays.
Cones: 5/8″ (15 mm) long; *egg-shaped;* whitish or bluish but becoming dark brown; *upright* at end of short twigs; opening widely at maturity and remaining attached; with usually 6 *paired cone-scales,* thick and ending in curved or *hooked point;* 2 tiny oblong wingless seeds at base of lower cone-scales.

Habitat: Moist soil in humid, warm temperate and subtropical regions.

Range: Native of China. Planted across the United States, especially in Southeast; sometimes escaped and naturalized in Florida.

Many horticultural varieties grown as shrubs around houses and in parks and gardens have different shapes; some have golden foliage. Often trimmed into hedges. Chinese use the fragrant evergreen branches for good luck at New Year celebrations.

PALM FAMILY
(Palmae)

Evergreen trees and shrubs, and sometimes vines. Stout or sometimes slender unbranched trunk, not divided into bark and wood and not increasing in diameter. About 2500 species worldwide in tropical and subtropical regions; 11 native and 1 naturalized tree species and 2 native shrub species in North America; many others southward.
Leaves: large, spreading, alternate and crowded at tip of trunk. 2 kinds: pinnately compound with many narrow leaflets with fine parallel veins along the axis; and fanlike or palmate-veined, thick and leathery.
Flowers: small, stalkless or short-stalked, generally whitish, commonly male and female on the same plant or bisexual, regular, with calyx of 3 sepals or lobes and corolla of 3 petals or lobes, mostly 6 stamens, and 1 pistil. In large branched clusters developing from a large bract among leaf bases or below.
Fruit: usually a 1-seeded berry or drupe.

---

363 Cabbage Palmetto
"Carolina Palmetto" "Cabbage-palm"
Sabal palmetto (Walt.) Lodd. ex Schult.

Description: Medium-sized, spineless, evergreen palm with stout, unbranched trunk and *very large, fan-shaped leaves* spreading around top.
Height: 30–50′ (9–15 m) or more.
Diameter: 1½′ (0.5 m).
Leaves: 4–7′ (1.2–2.1 m) long and nearly as broad. Folded into *many long, narrow segments;* long-pointed and drooping; coarse, stiff, and leathery; splitting apart nearly to the stout midrib; with *threadlike fibers* separating at edges; *shiny dark green*. Very stout,

stiff leafstalks 5–8′ (1.5–2.4 m) long; green, ridged above, with long, fibrous, shiny brown sheath at wedge-shaped base, which splits and hangs down with age.

Trunk: gray-brown; rough or ridged.

Flowers: 3/16″ (5 mm) long; with deeply *6-lobed whitish corolla;* fragrant; nearly stalkless; in curved or drooping, much-branched clusters arising from leaf bases; in early summer.

Fruit: 3/8″ (10 mm) in diameter; nearly round *berries;* shiny *black;* with thin, sweet, dry flesh; 1-seeded; maturing in autumn.

Habitat: Sandy shores, crowded in groves; inland in hammocks.

Range: Near coast from SE. North Carolina to S. and NW. Florida, including Florida Keys.

The trunks are used for wharf pilings, docks, and poles. Brushes and whisk brooms are made from young leafstalk fibers, and baskets and hats from the leaf blades. An ornamental and street tree, it is the northernmost New World palm and one of the hardiest. Formerly, plants were killed in order to eat the large leaf buds as a cabbagelike salad. The names are from the Spanish *palmito,* meaning "small palm."

LILY FAMILY
(Liliaceae)

Mostly perennial herbs, often with bulbs or tubers; sometimes shrubs or small trees with stout unbranched trunk or with a few branches ending in crowded, narrowed, pointed leaves. A very large family, 4000–5000 species worldwide; 12 native tree species mostly in the yucca genus (*Yucca*), about 30 native shrub, 10 woody vine, and numerous herb species in North America.

Leaves: generally alternate, narrow, not toothed, parallel-veined.

Flowers: mostly clustered and showy, often large and white, bisexual, regular, with 3 sepals and 3 petals nearly equal and mostly separate, usually 6 stamens separate or attached to corolla, and 1 pistil.

Fruit: a capsule or berry with many seeds.

361, 428, 576 **Aloe Yucca**
"Spanish-bayonet" "Spanish-dagger"
Yucca aloifolia L.

Description: Evergreen shrub or small tree often with stout clustered *trunks* that are *sometimes branched,* with sprouts at the slightly swollen base, and with *bayonetlike leaves* crowded and spreading at top.

Height: 20′ (6 m).

Diameter: 4″ (10 cm).

Leaves: 12–20″ (30–51 cm) long, 1–1¾″ (2.5–4.5 cm) wide. Grooved above; *thick and stiff;* without visible veins; bordered by *tiny teeth;* dull green, ending in long sharp brown *spine.*

Trunk: brown; becoming rough, fissured or scaly; upper part covered by dead leaves.

Flowers: 1½–2¼″ (4–6 cm) long; *bell-shaped* with 6 spreading *white* blunt-

pointed, fleshy *sepals;* long-stalked; showy; in upright, branched clusters 1½–2′ (0.5–0.6 m) long; in summer.
Fruit: 3–4″ (7.5–10 cm) long; *cylindrical* and slightly 6-angled berry, turning dark purple and black; containing bittersweet *juicy pulp,* becoming dry; hanging down; maturing in autumn; many flat, black seeds.

Habitat: Coastal sands, dunes, and mounds.

Range: SE. North Carolina to S. Florida including Florida Keys and west to S. Alabama; near sea level. Planted across S. United States, escaping and becoming naturalized.

Tolerant of salt and suitable for planting along sandy shores, Aloe Yucca is easily propagated from sprouts. Several cultivated varieties have striped or colored leaves. The fruit is eaten by birds and sometimes by humans, and the flowers can be served as a salad or cooked. Pioneers made rope and string from the fibrous leaves.

362, 429 **Moundlily Yucca**
"Spanish-bayonet" "Spanish-dagger"
Yucca gloriosa L.

Description: Evergreen shrub or in cultivation a small tree with stout, *unbranched trunks,* often clustered, and with *bayonetlike leaves* crowded and spreading at top.
Height: 17′ (5 m).
Diameter: 4–6″ (10–15 cm).
Leaves: 16–24″ (41–61 cm) long, 1¼–2¼″ (3–6 cm) wide. Broadest near middle; *flat* or grooved toward tip; *thick and stiff;* rough beneath; edges *without teeth* and with a few threads; ending in stout *sharp point;* dull gray-green.
Trunk: light gray; smooth; upper part covered by dead leaves.

Flowers: 1½–2″ (4–5 cm) long; *bell-shaped* with blunt-pointed fleshy *sepals,* white or tinged with purple; long-stalked; showy; in upright branched clusters 2–5′ (0.6–1.5 m) long; in autumn.

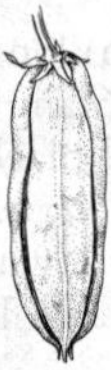

Fruit: 2–3″ (5–7.5 cm) long; *cylindrical* and slightly 6-angled berry; becoming blackish and *dry;* hanging down; infrequently produced; many flat, black seeds.

Habitat: Coastal sand dunes, beaches, and borders of brackish marshes. Rare and local.

Range: NE. North Carolina to SE. Georgia and extreme NE. Florida, near sea level.

Introduced into Europe around 1550 and one of the most common yuccas in cultivation. Varieties have leaves with yellow or white stripes. The common name apparently refers to its occurrence on sand mounds and to the lilylike flowers.

WILLOW FAMILY
(Salicaceae)

Deciduous, often aromatic trees and shrubs. About 350 species in the genera willow (*Salix*) and poplar (*Populus*); nearly worldwide, mostly in north temperate and arctic regions. 35 native and 5 naturalized tree species and about 60 native shrub species in North America.
Leaves: alternate, simple, mostly toothed, with paired stipules.
Flowers: tiny male and female on separate plants, regular, each above a scale, crowded in narrow catkins. Male flowers with cuplike disk or 1–2 glands and 1–40 stamens separate or united at base. Female flower with 1 pistil.
Fruit: a capsule opening in 2–4 parts, containing many tiny seeds with cottony hairs.

The large genus of willows (*Salix*), characteristic of wet soils, includes shrubs and mostly small trees, often with several stems or trunks from base and forming thickets. Leaves are narrow and commonly long-pointed and finely toothed, with distinct odor when crushed, turning yellow in autumn; leafstalks are very short with paired and often large stipules. Bark is gray or brown, smooth or becoming rough, scaly or furrowed, bitter, and aromatic. The slender or wiry twigs are tough, flexible, often shedding or easily detached at forks. The many tiny yellowish or greenish flowers usually appear in early spring before leaves; male and female are on separate plants, many crowded in mostly erect catkins. Each flower is above a hairy scale and has a glandlike disk, without calyx or corolla. Male flowers have 1–2 (sometimes to 12) stamens; female have a narrow pointed pistil. The many conical 1-celled long-pointed capsules along a slender stalk, are mostly light brown and mature in late spring or early summer, splitting into 2 parts. The numerous tiny seeds have tufts of white cottony hairs.

188, 238 **White Poplar**
"Silver Poplar"
Populus alba L.

Description: Large, much-branched, introduced tree with leaves that toss in the slightest breeze to reveal silvery-white undersides.
Height: 80′ (24 m).
Diameter: 2′ (0.6 m).
Leaves: 2½–4″ (6–10 cm) long, and nearly as wide. Ovate; *3- or 5-lobed and maplelike;* blunt-tipped with scattered small teeth. Dark green above; *densely white hairy and feltlike beneath;* turning reddish in autumn. Long leafstalks covered with white hairs.
Bark: *whitish-gray;* smooth, becoming rough and furrowed at base.
Twigs: densely covered with white hairs.
Flowers: catkins 1½–3″ (4–7.5 cm) long; densely covered with white hairs; male and female on separate trees; in early spring before leaves.
Fruit: 3⁄16″ (5 mm) long; *egg-shaped* capsules with many tiny, cottony seeds.

Habitat: Moist soils, especially along roadsides and borders of fields.

Range: Native of Europe and Asia. Planted and widely naturalized in S. Canada and across the United States.

Probably introduced in colonial times, White Poplar is hardy in cities and in dry areas. Handsome varieties, some with silvery leaves and one with a columnar form, are planted for shade and ornament. It grows rapidly and spreads readily by root sprouts, sometimes becoming an undesirable weed.

118 **Balsam Poplar**
"Tacamahac" "Balm"
Populus balsamifera L.

Description: Large tree with narrow, open crown of upright branches and fragrant, resinous buds with strong balsam odor.
Height: 60–80′ (18–24 m).
Diameter: 1–3′ (0.3–0.9 m).
Leaves: 3–5″ (7.5–13 cm) long, 1½–3″ (4–7.5 cm) wide. Ovate; pointed at tip; rounded or slightly notched at base; *finely wavy-toothed;* slightly thickened; hairless or nearly so. Shiny dark green above; *whitish, often with rusty veins beneath.* Leafstalks slender, *round,* hairy.
Bark: light brown, smooth; becoming gray, furrowed into flat, scaly ridges.

Twigs: brownish; stout, with large, gummy or sticky buds producing *fragrant yellowish resin.*
Flowers: catkins 2–3½″ (5–9 cm) long; brownish; male and female on separate trees; in early spring.
Fruit: 5/16″ (8 mm) long; *egg-shaped* capsules; *pointed;* light brown; *hairless;* maturing in spring and *splitting into 2 parts;* many tiny, cottony seeds.

Habitat: Moist soils of valleys, mainly stream banks, sandbars, and flood plains, also lower slopes; often in pure stands.

Range: Across N. North America along northern limit of trees from NW. Alaska south to SE. British Columbia and east to Newfoundland, south to Pennsylvania and west to Iowa; local south to Colorado and in eastern mountains to West Virginia; to 5500′ (1676 m) in Rocky Mountains.

The northernmost New World hardwood, Balsam Poplar extends in scattered groves to Alaska's Arctic Slope. Balm-of-Gilead Poplar, an ornamental with broad, open crown and larger, heart-shaped leaves, is a clone or hybrid. Balm-of-Gilead, derived from the resinous buds, has been used in home remedies.

185, 186, 512, 611

Eastern Cottonwood
"Carolina Poplar" "Southern Cottonwood"
Populus deltoides Bartr. ex. Marsh.

Description: Large tree with a massive trunk often forked into stout branches, and broad, open crown of spreading and slightly drooping branches.

Height: 100′ (30 m).
Diameter: 3–4′ (0.9–1.2 m), often larger.
Leaves: 3–7″ (7.5–18 cm) long, 3–5″ (7.5–13 cm) wide. *Triangular; long-pointed;* usually straight at base; *curved, coarse teeth;* slightly thickened; *shiny green,* turning yellow in autumn. Leafstalks long, slender, *flattened.*
Bark: yellowish-green and smooth; becoming light gray, thick, rough, and deeply furrowed.
Twigs: brownish; stout, with large resinous or sticky buds.
Flowers: catkins 2–3½″ (5–9 cm) long, brownish; male and female on separate trees; in early spring.
Fruit: ⅜″ (10 mm) long; *elliptical* capsules, light brown; maturing in spring and splitting into 3–4 parts; many on slender stalks in catkin to 8″ (20 cm) long; many tiny cottony seeds.

Habitat: Bordering streams and in wet soils in valleys; in pure stands or often with willows. Pioneers on new sandbars and bare flood plains.

Range: Widespread. S. Alberta east to extreme S. Quebec and New Hampshire, south to NW. Florida, west to W. Texas, and north to central Montana; to 1000′ (305 m) in east, to 5000′ (1524 m) in west.

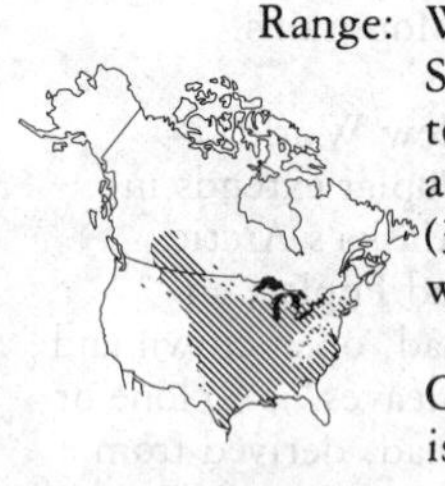

One of the largest eastern hardwoods, it is used for boxes and crates, furniture, plywood, woodenware, matches, and pulpwood. Planted as a shade tree and for shelterbelts. The common name refers to the abundant cottony seeds; another name, "Necklace Poplar," alludes to the resemblance of the long, narrow line of seed capsules to a string of beads. Although short-lived, it is one of the fastest-growing native trees; on favorable sites in the Mississippi Valley, trees average 5′ (1.5 m) in height growth annually with as much as 13′ (4 m) the first year. Plains Cottonwood (var. *occidentalis* Rydb.), a western variety, has slightly smaller leaves that are often broader than long and more coarsely toothed.

191, 599 **Bigtooth Aspen**
"Largetooth Aspen" "Poplar"
Populus grandidentata Michx.

Description: Medium-sized tree with narrow, rounded crown.
Height: 30–60′ (9–18 m).
Diameter: 1–1½′ (0.3–0.5 m).
Leaves: 2½–4″ (6–10 cm) long, 1¾–3½″ (4.5–9 cm) wide. *Broadly ovate;* short-pointed tip; rounded at base; *coarse, curved teeth; with white hairs when young. Dull green* above, paler beneath, turning pale yellow in autumn. Leafstalks long, slender, *flattened.*
Bark: greenish, smooth, thin; becoming dark brown and furrowed into flat, scaly ridges.
Twigs: brown, slender, hairy when young.
Flowers: catkins 1½–2½″ (4–6 cm) long; brownish; male and female on separate trees; in early spring.
Fruit: ¼″ (6 mm) long; narrowly *conical* capsules; light green; slightly curved; finely hairy; maturing in spring and splitting into 2 parts; many tiny cottony seeds.

Habitat: Sandy upland soils, also flood plains of

streams, often with Quaking Aspen.

Range: SE. Manitoba east to Cape Breton Island, south to Virginia, and west to NE. Missouri; local south to W. North Carolina; to 2000′ (610 m), or to 3000′ (914 m) in south.

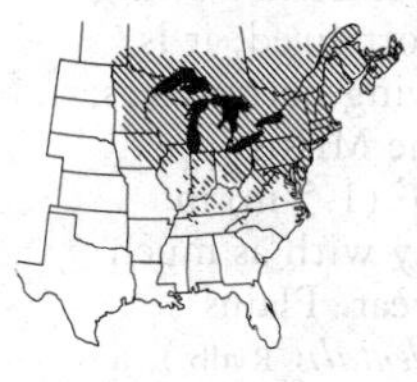

Easily distinguishable from Quaking Aspen by the large curved teeth of leaf edges, mentioned in both common and scientific names. Like that species, Bigtooth Aspen is a pioneer tree after fires and logging and on abandoned fields, short-lived and replaced by conifers. The foliage, twig buds, and bark are consumed by wildlife.

---

190 Swamp Cottonwood

"Swamp Poplar" "Black Cottonwood"
Populus heterophylla L.

Description: Tree with narrow, rounded crown and stout branches, found in wet soils.
Height: 80′ (24 m).
Diameter: 2′ (0.6 m).
Leaves: 4–7″ (10–18 cm) long, 3–6″ (7.5–15 cm) wide. *Broadly ovate;* blunt or rounded tip; heart-shaped or rounded at base; *fine, curved teeth. Densely covered with white hairs when unfolding;* becoming hairless and dark green above, remaining woolly and pale beneath. Leafstalks slender, *rounded,* hairless.
Bark: brown; furrowed into scaly ridges.
Twigs: stout; covered with white hairs when young.
Flowers: catkins 1–2½″ (2.5–6 cm) long; brownish; male and female on separate trees; in early spring.
Fruit: ½″ (12 mm) long; *egg-shaped* capsules; brown; long-stalked; maturing in spring and splitting into 2–3 parts; tiny cottony seeds.

Habitat: Wet sites, often submerged in flood plains and edges of swamps, with willows, Baldcypress, and Water Tupelo.

Range: Connecticut south to E. Georgia and from NW. Florida west to E. Louisiana, north to S. Michigan; to 800′ (244 m).

Because of its rapid growth, Swamp Cottonwood is sometimes planted as a shade tree.

187 Lombardy Poplar
Populus nigra L. 'Italica'

Description: Medium-sized, introduced tree with straight stout trunk often enlarged at base and *narrow columnar crown* of upright short brittle branches.
Height: 30–60′ (9–18 m).
Diameter: 1–2′ (0.3–0.6 m).
Leaves: 1½–3″ (4–7.5 cm) long and wide. *Triangular;* long-pointed at tip; straight at base; *wavy saw-toothed;* hairless or nearly so. Green above, light green beneath. Slender, flattened leafstalks 1–2″ (2.5–5 cm) long.
Bark: gray; thick, deeply furrowed.
Twigs: orange, turning gray; stout, hairless, with sticky scaly buds.
Flowers: catkins 2″ (5 cm) long; narrow; drooping; with many male flowers; in early spring before leaves.

Habitat: Moist soil in temperate regions.

Range: Widely planted nearly throughout the United States.

Lombardy Poplar is a clone or cultivated variety (designated as 'Italica') of Black Poplar, which is native of Europe and western Asia. Apparently this clone originated in northern Italy before 1750. The trees are male and bear no seeds; female flowers and fruits are not produced and propagation is by cuttings and root sprouts. Generally grown in rows for shelterbelts or windbreaks, screens, roadsides, and formal gardens. Especially common around rural homes in treeless, irrigated areas, and conspicuous in the landscape of the Great Basin. The trees grow rapidly but are short-lived. They are subject to European canker disease in the trunk and upper branches, which destroys the columnar form.

184, 620 **Quaking Aspen**
"Trembling Aspen" "Golden Aspen"
Populus tremuloides Michx.

Description: The most widely distributed tree in North America; with a narrow, rounded crown of thin foliage.

Height: 40–70′ (12–21 m).
Diameter: 1–1½′ (0.3–0.5 m).
Leaves: 1¼–3″ (3–7.5 cm) long. *Nearly round;* abruptly short-pointed; rounded at base; *finely saw-toothed;* thin. Shiny green above, dull green beneath; turning *golden-yellow* in autumn before shedding. Leafstalks slender, *flattened.*
Bark: *whitish, smooth,* thin; on very large trunks becoming dark gray, furrowed, and thick.
Twigs: shiny brown; slender, hairless.
Flowers: catkins 1–2½″ (2.5–6 cm) long; brownish; male and female on separate trees; in early spring before leaves.
Fruit: ¼″ (6 mm) long; narrowly *conical* light green capsules in drooping catkins to 4″ (10 cm) long; maturing in late spring and splitting in 2 parts. Many tiny cottony seeds; rarely produced in the West, where propagation is by root sprouts.

Habitat: Many soil types, espcially sandy and gravelly slopes; often in pure stands and in western mountains in an altitudinal zone below spruce-fir forest.

Range: Across N. North America from Alaska to Newfoundland, south to Virginia, and in Rocky Mountains south to S.

Arizona and N. Mexico; from near sea level northward to 6500–10,000′ (1981–3048 m) southward.

The names refer to the leaves, which in the slightest breeze tremble on their flattened leafstalks. The soft smooth bark is sometimes decorated with carved initials and marked by bear claws. A pioneer tree after fires and logging and on abandoned fields, it is short-lived and replaced by conifers. Sometimes planted as an ornamental. Principal uses of the wood include pulpwood, boxes, furniture parts, matches, excelsior, and particle-board. The twigs and foliage are browsed by deer, elk, and moose, also by sheep and goats. Beavers, rabbits, and other mammals eat the bark, foliage, and buds, and grouse and quail feed on the winter buds.

110 White Willow
"European White Willow"
Salix alba L.

Description: Naturalized tree with 1–4 trunks and open crown of spreading branches.
Height: 50–80′ (15–24 m).
Diameter: 2′ (0.6 m) or more.
Leaves: 2–4½″ (5–11 cm) long, ⅜–1¼″ (1–3 cm) wide. *Lance-shaped* to elliptical; *finely saw-toothed;* firm. *Shiny dark green above, whitish and silky beneath;* turning yellow in autumn.
Bark: gray; rough, furrowed into narrow ridges.
Twigs: *yellow to brown;* flexible and often slightly drooping; *silky when young.*
Flowers: catkins 1¼–2¼″ (3–6 cm) long; with yellow, hairy scales; at end of short, leafy twigs; in early spring.
Fruit: $\frac{3}{16}$″ (5 mm) long; hairless capsules; light brown; maturing in late spring or early summer.

Habitat: Wet soils of stream banks and valleys near cities.

Range: Native from Europe and N. Africa to central Asia. Naturalized in SE. Canada and E. United States.

Introduced in colonial times, this handsome willow is planted as a shade and ornamental tree and for shelterbelts, fenceposts, and fuel. Some cultivated varieties have golden-yellow or reddish twigs that have been used in basketmaking. One variety in England is prized for making cricket bats.

Peachleaf Willow
"Peach Willow" "Almond Willow"
Salix amygdaloides Anderss.

Description: Tree with 1 or sometimes several straight trunks, upright branches, and spreading crown.
Height: 60′ (18 m).
Diameter: 2′ (0.6 m).
Leaves: 2–4½″ (5–11 cm) long, ½–1¼″ (1–3 cm) wide. *Lance-shaped;* often slightly curved to one side; tapering to *long, narrow point; finely saw-toothed;* becoming hairless. Shiny green above, *whitish beneath.* With long, slender leafstalks.
Bark: dark brown; rough, furrowed into flat scaly ridges.
Twigs: shiny orange or brown; hairless.
Flowers: catkins 1¼–3″ (3–7.5 cm) long; with yellow hairy scales; on short leafy twigs; in spring with leaves.
Fruit: ¼″ (6 mm) long; reddish-yellow hairless capsules; long-stalked; maturing in late spring or early summer.

Habitat: Wet soil of valleys, often bordering stream banks with cottonwoods.

Range: SE. British Columbia east to extreme S. Quebec and New York, south to NW. Pennsylvania, and west to W. Texas;

also N. Mexico; at 500–7000′ (152–2134 m).

The common willow across the northern plains, where it is important in protecting riverbanks from erosion. Both common name and Latin species name refer to the leaf shape, which suggests that of Peach.

108 Weeping Willow
"Babylon Weeping Willow"
Salix babylonica L.

Description: A handsome, naturalized tree with short trunk and broad, open, irregular crown of *drooping branches.*
Height: 30–40′ (9–12 m).
Diameter: 2′ (0.6 m), sometimes much larger.
Leaves: 2½–5″ (6–13 cm) long, ¼–½″ (6–12 mm) wide. *Narrowly lance-shaped; with long-pointed tips; finely saw-toothed.* Dark green above, *whitish* or gray *beneath.* Hanging from short leafstalks.
Bark: gray; rough, thick; deeply furrowed in long, branching ridges.
Twigs: yellowish-green to brownish; *very slender, unbranched, drooping* vertically.

Flowers: catkins ⅜–1″ (1–2.5 cm) long; greenish; at end of short leafy twigs; in early spring; plants mostly female.
Fruit: 1⁄16″ (1.5 mm) long; light brown capsules; maturing in late spring or early summer.

Habitat: Parks, gardens, and cemeteries, especially near water.

Range: Native of China. Naturalized locally from extreme S. Quebec and S. Ontario south to Georgia and west to Missouri. Also planted in western states.

This willow is well known for its distinctive weeping foliage. It is among the first willows to bear leaves in spring and among the last to shed them in autumn. China, not Babylon, was its native home; when named, it was confused with Euphrates Poplar (*Populus euphratica* Olivier).

117 **Bebb Willow**
"Beak Willow" "Diamond Willow"
Salix bebbiana Sarg.

Description: Much-branched shrub or small tree with broad, rounded crown.
Height: 10–25′ (3–7.6 m).
Diameter: 6″ (15 cm).
Leaves: 1–3½″ (2.5–9 cm) long, ⅜–1″ (1–2.5 cm) wide. *Elliptical;* often broadest beyond middle; *short-pointed* at ends; slightly saw-toothed or wavy; firm; slightly hairy. *Dull green above, gray* or whitish and *net-veined* beneath.
Bark: gray; smooth, becoming rough and furrowed.
Twigs: *reddish-purple;* slender, widely forking; with pressed hairs when young.
Flowers: catkins ¾–1½″ (2–4 cm) long; with yellow or brown scales; on short, leafy stalks; before or with leaves.

Fruit: ⅜″ (10 mm) long; very slender capsules, hairy, light brown, *ending in long point;* long-stalked; maturing in early summer.

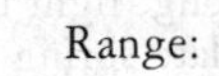

Habitat: Moist open uplands and borders of streams, lakes, and swamps.
Range: Central and SW. Alaska south to British Columbia and east to Newfoundland, south to Maryland, west to Iowa, and south in Rocky Mountains to S. New Mexico; to 11,000′ (3353 m) southward. Also in NE. Asia.

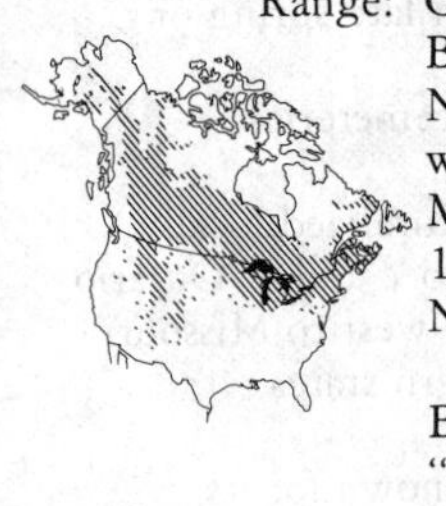

Bebb Willow is the most important "diamond willow," a term applied to several species which sometimes have diamond-shaped patterns on their trunks. These are caused by fungi, usually in shade or poor sites. The contrasting whitish and brownish stems are carved into canes, lamps, posts, furniture, and candleholders. Forms willow thickets as a weed on uplands after forest fires. Named for Michael Schuck Bebb (1833–95), U.S. specialist on willows.

---

109 **Coastal Plain Willow**
"Southern Willow" "Ward Willow"
Salix caroliniana Michx.

Description: Shrub or small tree with spreading or slightly drooping branches.
Height: 30′ (9 m).
Diameter: 1′ (0.3 m).
Leaves: 2–4″ (5–10 cm) long, ½–¾″ (12–19 mm) wide. *Lance-shaped; finely saw-toothed;* densely hairy when young. Green above; whitish and nearly hairless beneath. Leafstalks hairy.
Bark: gray to blackish; fairly smooth, furrowed into broad scaly ridges.
Twigs: brown; slender, limber; hairy when young.
Flowers: catkins 3–4″ (7.5–10 cm) long; greenish or yellowish; at ends of leafy twigs in spring.
Fruit: ¼″ (6 mm) long; long-pointed capsules; light reddish-brown; maturing in late spring or early summer.

Habitat: Wet soils of stream banks and swamps.

Range: S. Pennsylvania south to S. Florida, west to central Texas, and north to SE. Nebraska; to 2000′ (610 m).

This is the common small tree willow found at low altitudes in the southeastern United States.

116, 399 Pussy Willow
Salix discolor Muhl.

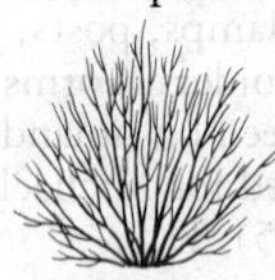

Description: Many-stemmed shrub or small tree with open rounded crown; silky, furry catkins appear in late winter and early spring.
Height: 20′ (6 m).
Diameter: 8″ (20 cm).
Leaves: 1½–4¼″ (4–11 cm) long, ⅜–1¼″ (1–3 cm) wide. *Lance-shaped* or narrowly elliptical; *irregularly wavy-toothed;* stiff; hairy when young; slender-stalked. *Shiny green above, whitish beneath.*
Bark: gray; fissured, scaly.
Twigs: brown; stout; hairy when young.

Flowers: catkins 1–2½″ (2.5–6 cm) long; cylindrical; thick with blackish scales; covered with *silky whitish hairs;* in *late winter and early spring* long before leaves.
Fruit: 5⁄16–½″ (8–12 mm) long; narrow capsules; light brown; finely hairy, in early spring before leaves.

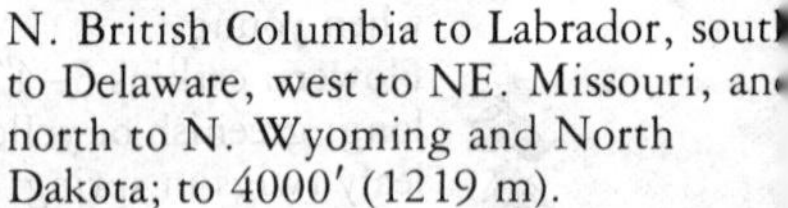

Habitat: Wet meadow soils and borders of streams and lakes; usually in coniferous forests.
Range: N. British Columbia to Labrador, south to Delaware, west to NE. Missouri, and north to N. Wyoming and North Dakota; to 4000′ (1219 m).

The large flower buds burst and expose their soft silky hair, or "pussy fur," early in the year. In winter, cut Pussy Willow twigs can be put in water and the flowers forced at warm temperatures. Some twigs will produce beautiful golden stamens, while others will bear slender greenish pistils. The Latin species name refers to the contrasting colors of the leaf surfaces, which aid in recognition.

107 Sandbar Willow

"Coyote Willow" "Narrowleaf Willow"
Salix exigua Nutt.

Description: Thicket-forming shrub with clustered stems or, rarely, a tree, with *very narrow leaves.*
Height: 3–10′ (1–3 m), sometimes to 20′ (6 m).
Diameter: 5″ (13 cm).
Leaves: 1½–4″ (4–10 cm) long, ¼″ (6 mm) wide. Linear; very long-pointed at ends; *few tiny, scattered teeth* or none; varying from *hairless to densely hairy* with pressed, silky hairs; almost stalkless. Yellow-green to gray-green on both surfaces.
Bark: gray; smooth or becoming fissured.
Twigs: reddish- or yellowish-brown; slender, upright; hairless or with gray hairs.
Flowers: catkins 1–2½″ (2.5–6 cm) long; with hairy yellow scales; at end of leafy twigs in spring.
Fruit: ¼″ (6 mm) long; light brown capsules; usually hairy; maturing in early summer.

Habitat: Wet soils, especially riverbanks, sandbars, and silt flats.

Range: Central Alaska east to Ontario and New York, southwest to Mississippi, and west to S. California; also local east to Quebec and Virginia, and in N. Mexico; to 8000′ (2438 m).

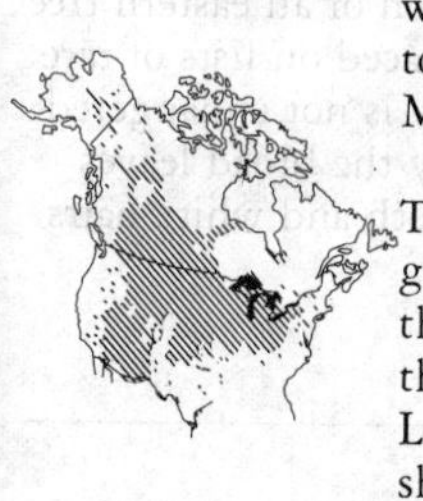

This hardy species has perhaps the greatest range of all tree willows: from the Yukon River in central Alaska to the Mississippi River in southern Louisiana. A common and characteristic shrub along streams throughout the interior, especially the Great Plains and Southwest, it is drought-resistant and suitable for planting on stream bottoms to prevent surface erosion. Livestock browse the foliage; Indians made baskets from the twigs and bark.

114 Florida Willow
Salix floridana Chapm.

Description: Shrub or, infrequently, a small tree with large broad leaves.
Height: 7–15′ (2–4.6 m).
Diameter: 4″ (10 cm).
Leaves: 2–5″ (5–13 cm) long, ¾–1½″ (2–4 cm) wide; larger on young twigs. *Elliptical* or ovate; very *short, scattered gland-teeth;* densely hairy when young; network of veins on both surfaces. *Dark green* and hairy only on yellowish midvein above; *whitish with soft hairs beneath,* especially on veins.
Bark: gray-brown; smooth, becoming rough at base.
Twigs: brown or purplish; brittle; sometimes slightly hairy.
Flowers: catkins 1½–2¼″ (4–6 cm) long; with yellowish hairy scales; at end of leafy twig; in early spring.
Fruit: about ¼″ (6 mm) long; long-pointed capsules; light brown; becoming nearly hairless; maturing in late spring.

Habitat: Wet soils along small streams and in wet limestone areas.

Range: S. Georgia to NW. and central Florida; to 200′ (61 m).

This rare, local species has the most restricted distribution of all eastern tree willows. Though placed on lists of rare plants, it apparently is not endangered. Easily recognized by the broad leaves with short gland-teeth and white hairs on lower surfaces.

111, 378 Crack Willow
"Brittle Willow" "Snap Willow"
Salix fragilis L.

Description: Large, naturalized tree with widely forking branches and very *brittle twigs.*
Height: 80′ (24 m).
Diameter: 2½′ (0.8 m).

Leaves: 4–6″ (10–15 cm) long, 1–1½″ (2.5–4 cm) wide. Lance-shaped; ending in *long point turned to one side; coarsely saw-toothed;* teeth gland-tipped; hairless. Shiny green above, *whitish beneath.*
Bark: gray; rough, thick, deeply furrowed into narrow ridges.
Twigs: shiny brownish; erect or spreading; easily broken at base, with gummy buds.
Flowers: catkins 1–2¼″ (2.5–6 cm) long; with yellow or greenish hairy scales; at ends of leafy twigs; in spring.
Fruit: 3/16″ (5 mm) long; light brown capsules; maturing in late spring and early summer.

Habitat: Escaping in moist soil along roadsides and streams and in clearings.

Range: Native of Europe and W. Asia. Naturalized from Newfoundland to Virginia, west to Kansas, and north to South Dakota.

Called Crack Willow because the twigs easily break off at the base, especially in spring; the Latin species name also refers to the fragile twigs. Introduced in colonial times to provide charcoal for gunpowder and as a shade tree. If partly covered by soil, detached twigs or cuttings will form roots and grow into new plants.

---

106 Black Willow
"Swamp Willow" "Goodding Willow"
Salix nigra Marsh.

Description: Large tree with 1 or more straight and usually leaning trunks, upright branches, and narrow or irregular crown.

Height: 60–100′ (18–30 m).
Diameter: 1½–2½′ (0.5–0.8 m).
Leaves: 3–5″ (7.5–13 cm) long, ⅜–¾″ (10–19 mm) wide. *Narrowly lance-shaped;* often slightly curved to one side; long-pointed; *finely saw-toothed;* hairless

or nearly so. *Shiny green above, paler beneath.*

Bark: dark brown or blackish; deeply furrowed into scaly, forking ridges.
Twigs: brownish; very slender, easily detached at base.
Flowers: catkins 1–3″ (2.5–7.5 cm) long; with yellow hairy scales; at end of leafy twigs in spring.
Fruit: 3/16″ (5 mm) long; reddish-brown capsules; hairless; maturing in late spring.

Habitat: Wet soils of banks of streams and lakes, especially flood plains; often in pure stands and with cottonwoods.

Range: S. New Brunswick and Maine south to NW. Florida, west to S. Texas, and north to SE. Minnesota; also from W. Texas west to N. California; local in N. Mexico; to 5000′ (1524 m).

The largest and most important New World willow with one of the most extensive ranges across the country. In the lower Mississippi Valley it attains commercial timber size, reaching 100–140′ (30–42 m) in height and 4′ (1.2 m) in diameter. The numerous uses of the wood include millwork, furniture, doors, cabinetwork, boxes, barrels, toys, and pulpwood. In pioneer times the wood of this and other willows was a source of charcoal for gunpowder. Large trees are valuable in binding soil banks, thus preventing soil erosion and flood damage. Mats and poles made from Black Willow trunks and branches provide further protection of riverbanks and levees. Also a shade tree and honey plant.

---

113 Balsam Willow
"Bog Willow"
Salix pyrifolia Anderss.

Description: Usually a shrub, sometimes a small tree, with clumps of slender stems

branched near the top, and a fragrance of balsam.
Height: 20′ (6 m).
Diameter: 4″ (10 cm).
Leaves: 2–3½″ (5–9 cm) long, 1–1½″ (2.5–4 cm) wide. *Ovate or elliptical; short-pointed;* base rounded and usually notched; finely saw-toothed; becoming hairless; aromatic. Dark green above; paler and whitish with yellow midvein and conspicuous *network of small veins* beneath.
Bark: gray; smooth, thin.
Twigs: *shiny reddish-brown;* slightly stout, *hairless,* with *shiny, bright red winter buds.*
Flowers: catkins 1–1½″ (2.5–4 cm) long; yellowish; on short leafy twigs; in late spring.
Fruit: ¼″ (6 mm) long; dark orange capsules; hairless; maturing in early summer.

Habitat: Cold wet bogs.

Range: Yukon south to E. British Columbia and across Canada to Labrador, south to Maine, and west to Minnesota; to 2000′ (610 m).

The common name refers to the aromatic, gland-toothed young leaves, while the Latin species name means "pear leaf." In winter, Balsam Willow is easily recognized by the shiny reddish buds and twigs.

---

105 Basket Willow
"Osier" "Silky Osier"
Salix viminalis L.

Description: Naturalized shrub or small tree with very long leaves and long slender green twigs.
Height: 25′ (7.6 m).
Diameter: 6″ (15 cm).
Leaves: 4–10″ (10–25 cm) long, ⅜″ (10 mm) wide. *Narrowly lance-shaped;* edges not toothed. Dull green above,

grayish-white with *silky hairs beneath.*
Bark: dark gray; smooth or slightly fissured.
Twigs: *green; long, slender, flexible;* finely *silky when young,* becoming hairless and shiny.
Flowers: catkins ¾–1½" (2–4 cm) long; greenish or yellowish; nearly stalkless; before leaves in early spring. 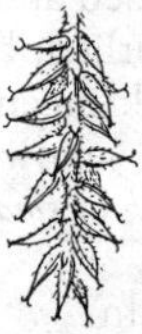
Fruit: ¼–⅝" (6–15 mm) long; light brown capsules; usually stalkless; finely hairy; maturing in late spring or early summer.

Habitat: Moist soils, spreading from cultivation near homes and towns.

Range: Native of Eurasia. Naturalized locally from Newfoundland to New England and other northeastern States.

One of the best of the basket willows, this species was introduced by the pioneers. The pliable twigs were cut to the ground each year and used for basketry and wickerwork. The Latin species name, meaning "osiers," "withes" or "flexible twigs," refers to this use.

BAYBERRY (WAXMYRTLE) FAMILY
(Myricaceae)

Nearly worldwide; about 40 species of small trees and shrubs, mostly in the bayberry genus (*Myrica*); 5 native tree species and 3 shrub species in North America.
Leaves: alternate, simple, often oblanceolate, toothed, and leathery, with orange or yellow resinous dots, very aromatic when crushed, mostly without stipules.
Flowers: tiny, greenish or yellowish, male and female usually on the same plant or on separate plants in short lateral clusters, regular, without calyx or corolla, each above a scale. Male flower usually with 4–8 (2–20) stamens, sometimes united. Female with 1 pistil.
Fruit: small rounded whitish drupe covered with wax, 1-seeded.

---

209 **Southern Bayberry**
"Candle-berry" "Southern Waxmyrtle"
Myrica cerifera L.

Description: Evergreen aromatic resinous shrub or small tree with narrow rounded crown.
Height: 30′ (9 m).
Diameter: 6″ (15 cm).
Leaves: 1½–3½″ (4–9 cm) long, ¼–¾″ (6–19 mm) wide; those toward end of twigs often smaller. *Reverse lance-shaped;* coarsely saw-toothed beyond middle; slightly thickened and stiff; aromatic when crushed; short-stalked. Shiny *yellow-green* with tiny dark brown gland-dots above, paler with tiny *orange gland-dots* and often hairy beneath.
Bark: light gray; smooth, thin.
Flowers: *tiny;* yellow-green; in narrowly cylindrical clusters ¼–¾″ (6–19 mm) long; at base of leaf. Male and female on separate trees; in early spring.

Fruit: 1/8″ (3 mm) in diameter; 1-seeded drupes; warty; light green, covered with *bluish-white wax;* several crowded in a cluster; maturing in autumn; remaining attached in winter.

Habitat: Moist, sandy soil, in fresh or slightly brackish banks, swamps, hammocks, flatwoods, pinelands, and upland hardwood forests.

Range: S. New Jersey south to S. Florida, west to S. Texas, and north to extreme SE. Oklahoma; to about 500′ (152 m).

One of the very few Puerto Rican trees native also in the United States north of Florida, this popular evergreen ornamental is used for screens, hedges, landscaping, and as a source of honey. Colonists separated the fruit's waxy covering in boiling water to make fragrant-burning candles, a custom still followed in some countries.

---

61 **Odorless Bayberry**
"Odorless Waxmyrtle" "Waxmyrtle"
Myrica inodora Bartr.

Description: A many-branched evergreen shrub or small tree and, unlike the rest of its family, not aromatic.
Height: 20′ (6 m).
Diameter: 8″ (20 cm).
Leaves: 2–4″ (5–10 cm) long, 3/4–1 1/2″ (2–4 cm) wide. *Elliptical* or reverse ovate; sides convex or cupped; mostly *toothless* and edges turned under; thick and leathery, with tiny gland-dots; *not aromatic* when crushed; short-stalked. Shiny *dark green* above, green beneath.
Bark: light gray, smooth, thin.
Flowers: Male and female *tiny,* on separate plants, in narrowly cylindrical crowded clusters 3/4–1″ (2–2.5 cm) long; almost stalkless at base of leaf. Male crowded and yellowish. Female reddish-green, usually paired.

Fruit: 1/4″ (6 mm) in diameter; 1-seeded

drupes; brown; rough with raised dots; *slightly waxy;* few along stalk at leaf bases.

Habitat: Bays, swamps, and pinelands.

Range: SW. Georgia and NW. Florida west to SE. Louisiana; to 100′ (30 m).

The common and Latin names emphasize that this uncommon species differs from other bayberries or waxmyrtles in not being aromatic. Discovered in 1778 by naturalist William Bartram, who noted that the French colonists made candles from the waxy fruits.

CORKWOOD FAMILY
(Leitneriaceae)

The following species is the only one of its family worldwide.

82 Corkwood
Leitneria floridana Chapm.

Description: Thicket-forming shrub or small tree with trunk swollen at base, and few branches.
Height: 20′ (6 m).
Diameter: 4″ (10 cm).
Leaves: 3½–6″ (9–15 cm) long, 1¼–2½″ (3–6 cm) wide. *Elliptical to lance-shaped; edges straight* or slightly turned under; *thick and leathery,* with prominent side veins and fine network; crowded near ends on upright twigs; hairy stalks. Shiny green and usually becoming hairless above, paler and with soft hairs beneath; remaining green into late autumn.
Bark: brown; thin, and smooth, becoming fissured into narrow ridges.
Twigs: reddish-brown; stout, densely hairy when young, with pale dots, raised, crescent-shaped leaf-scars, and large pith.
Flowers: *tiny; crowded,* stalkless in *upright cylindrical clusters* 1–1¼″ (2.5–3 cm) long, ⅜″ (10 mm) wide; covered with *pointed hairy brown scales;* in early spring before leaves. Male and female on separate plants.
Fruit: ¾″ (19 mm) long; *elliptical* drupes; flattened; short-pointed; brown *dry;* 1-seeded; 1–4 on short stalk.

Habitat: Wet, often flooded soils, including swamps, riverbanks, and bayous; sometimes on saline shores.

Range: Local in S. Georgia, N. Florida, SE. Texas, E. Arkansas, and SE. Missouri; to 300′ (91 m).

Rare and very scattered, this species is abundant locally and forms thickets from root sprouts. The soft, pale yellow wood is lighter than cork or any other native wood, weighing only 13 pounds per cubic foot. It has been used for fish net floats and bottle stoppers. Totally different from all other plants, Corkwood is placed alone in a distinct plant family. No other native North American tree species has its own separate family. Its small size and scarcity make it unsuitable for lumber. Cultivated plants are hardy in dry soils and north to Chicago.

WALNUT FAMILY (Juglandaceae)

Deciduous, aromatic trees including hickories and Pecan (*Carya*) and Butternut and walnuts (*Juglans*). About 50 species worldwide in north temperate and tropical regions; 17 in North America, others southward.
Leaves: mostly alternate, odd pinnately compound, without stipules; leaflets with toothed border and resin dots beneath.
Flowers: male and female on same tree; tiny, greenish. Male, usually many in long, narrow catkins, composed of 3 bracts, 4 or fewer sepals, no corolla, and 3–40 or more stamens. Female, few or only 1–2 composed of 3 bracts and 1 pistil in short erect clusters.
Fruit: a nut with hard shell often splitting open or sometimes winged, or a drupe with large oily edible seed.

---

324 **Water Hickory**
"Swamp Hickory" "Bitter Pecan"
Carya aquatica (Michx. f.) Nutt.

Description: Large tree with tall straight trunk, slender upright branches, narrow crown, and bitter inedible nuts.

Height: 70–100′ (21–30 m).
Diameter: 1½–2½′ (0.5–0.8 m).
Leaves: pinnately compound; 9–15″ (23–38 cm) long, with dark red, hairy axis. *Usually 9–13 leaflets* 2–5″ (5–13 cm) long; *lance-shaped;* long-pointed at tip; *slightly curved;* finely saw-toothed; mainly stalkless. Dark green and hairless above, often hairy beneath.
Bark: light brown; thin, fissured into long platelike red-tinged scales.
Twigs: brown; slender, becoming hairless.
Flowers: tiny; greenish; in early spring before leaves. Male, with 6–7 stamens, many in slender drooping

catkins, *3 hanging from 1 stalk.* Female, 2–10 flowers at tip of same twig.
Fruit: 1–1½″ (2.5–4 cm) long; broadly elliptical; much *flattened;* 4-winged; becoming dark brown; with *thin husk* splitting to middle; *4 or fewer* in cluster. Nut flattened, 4-angled, *thin-shelled,* with *bitter* seed.

Habitat: Low wet flatlands, especially clay and flats, often submerged, in flood plains and swamps; bottomland hardwood forests.

Range: SE. Virginia south to central Florida, west to E. Texas, and north to S. Illinois; to 400′ (122 m).

Both the common and scientific names describe this hickory occupying wettest soils. The bitter nuts are consumed by ducks and other wildlife. Water Hickory is the tallest of all hickories; the national champion measures 150′ (45.7 m).

---

332 Bitternut Hickory

"Bitternut" "Pignut"
Carya cordiformis (Wangenh.) K. Koch

Description: Tree with tall trunk, broad and rounded crown, and bitter inedible nuts.

Height: 60–80′ (18–24 m).
Diameter: 1–2′ (0.3–0.6 m).
Leaves: pinnately compound; 6–10″ (15–25 cm) long, with slender hairy axis. 7–9 *leaflets,* 2–6″ (5–15 cm) long; *stalkless; lance-shaped;* finely saw-toothed. Yellow-green above, light green and slightly hairy beneath; turning yellow in autumn.
Bark: gray or light brown; shallowly furrowed into narrow forking scaly ridges.
Twigs: slender, ending in *bright yellow* slightly flattened buds.
Flowers: tiny; greenish; in early spring before leaves. Male flowers with 4–5

stamens; many in slender drooping *catkins, 3 hanging from 1 stalk.* 1–2 female flowers at tip of same twig.
Fruit: ¾–1¼″ (2–3 cm) long; *nearly round* or slightly flattened; short-pointed; *husk thin,* with tiny yellow scales, and *splitting along 4 wings.* Nut nearly smooth, *thin-shelled,* with *bitter* seed.

Habitat: Moist soil of valleys and in north also on dry upland soil; in mixed hardwood forests.

Range: S. Quebec and SW. New Hampshire, south to NW. Florida, west to E. Texas, and north to Minnesota; to 2000′ (610 m).

One of the most widely distributed and most common hickories through eastern United States and also one of the easiest to identify because of the small bright yellow buds. Rabbits have been observed to eat the bitter seeds which may be unpalatable to most wildlife. Early settlers used oil extracted from the nuts for oil lamps; they also believed it was valuable as a cure for rheumatism.

---

338 Scrub Hickory
"Florida Hickory"
Carya floridana Sarg.

Description: Small tree with several trunks or a thicket-forming shrub with spreading crown, confined to central Florida.
Height: 10–20′ (3–6 m).
Diameter: 8″ (20 cm).
Leaves: pinnately compound; 4–8″ (10–20 cm) long. *3–5 leaflets* (sometimes 7), 1½–4″ (4–10 cm) long; stalkless; *lance-shaped* to elliptical; *saw-toothed* with small *scattered teeth;* covered with rust-colored hairs when young. Green above, yellowish-green with tiny brown gland-dots beneath.

Bark: gray; smooth, becoming furrowed into ridges.
Twigs: brown; slender; covered with rust-colored hairs when young.
Flowers: tiny; greenish; in early spring before leaves. Male, with 4–5 stamens, many in slender drooping *catkins, 3 hanging from 1 stalk.* 1–2 female flowers at tip of same twig.
Fruit: ¾–1¼″ (2–3 cm) long; *slightly pear-shaped* or reverse egg-shaped, narrowing to stalklike base; becoming brown; *thin husk* splits into 2–3 parts. Rounded hickory nut less than ⅝″ (15 mm) in diameter, with slightly thickened shell and edible seed.

Habitat: Dry sand ridges of old dunes and hammocks in scrub vegetation with evergreen oaks.

Range: Central Florida (Marion to Palm Beach counties); to 100′ (30 m).

This small scrubby local hickory is the most southeastern in range. It is closely related to Pignut Hickory, which extends to the same region. Often bears nuts abundantly from the time it is only 7′ (2 m) high.

343, 377, 534, 627 **Pignut Hickory**
"Pignut" "Smoothbark Hickory"
Carya glabra (Mill.) Sweet

Description: Tree with irregular, spreading crown and thick-shelled nuts.

Height: 60–80′ (18–24 m).
Diameter: 1–2′ (0.3–0.6 m).
Leaves: pinnately compound; 6–10″ (15–25 cm) long, with slender hairless axis. *Usually 5 leaflets,* 3–6″ (7.5–15 cm) long, largest toward tip; *lance-shaped;* nearly stalkless; finely saw-toothed; *hairless* or hairy on veins beneath. Light green, turning yellow in autumn.
Bark: light gray; smooth or becoming furrowed with forking ridges.

Twigs: brown; slender, hairless.
Flowers: tiny; greenish; in early spring before leaves. Male, with 4 stamens, many in slender drooping *catkins, 3 hanging from 1 stalk.* Female, 2–10 flowers at tip of same twig.
Fruit: 1–2″ (2.5–5 cm) long; *slightly pear-shaped* or rounded; *husk thin,* becoming dark brown and *opening late* and splitting usually to middle. Hickory nut usually not angled, *thick-shelled,* with small sweet or bitter seed.

Habitat: Dry and moist uplands in hardwood forests with oaks and other hickories.

Range: S. Ontario east to S. New England, south to central Florida, west to extreme E. Texas, and north to Illinois; to 4800′ (1463 m) in southern Appalachians.

One of the most common hickories in the southern Appalachians and an important timber source there, its wood is made into tool handles and skis. It was formerly used for wagon wheels and textile loom picker sticks because it could sustain tremendous vibration. Named in colonial times from the consumption of the small nuts by hogs. Early settlers, who also called it "Broom Hickory," made brooms from narrow splits of the wood. Red Hickory (var. *odorata* (Marsh.) Little), a variety with nearly the same range, has the fruit husk splitting to base, usually 7 leaflets, and often shaggy bark.

327, 535 Pecan

"Sweet Pecan"
Carya illinoensis (Wangenh.) K. Koch

Description: Large wild and planted tree with tall trunk, broad rounded crown of massive spreading branches, and familiar pecan nuts.
Height: 100′ (30 m).
Diameter: 3′ (0.9 m).

Leaves: pinnately compound; 12–20″ (30–51 cm) long; 11–17 *slightly sickle-shaped leaflets,* 2–7″ (5–18 cm) long; long-pointed at tip; finely saw-toothed; short-stalked; hairless or slightly hairy. Yellow-green above, paler beneath; turning yellow in autumn.

Bark: light brown or gray; deeply and irregularly furrowed into narrow forked scaly ridges.

Flowers: tiny; greenish; in early spring before leaves. Male, with 5–6 stamens, many in slender drooping *catkins, 3 hanging from 1 stalk.* Female, 2–10 flowers at tip of the same twig.

Fruit: 1¼–2″ (3–5 cm) long; *oblong;* short-pointed at tip, rounded at base; with thin husk becoming dark brown, splitting to base along 4 ridges; 3–10 in cluster. Pecan nut light brown with darker markings, thin-shelled, with edible seed.

Habitat: Moist well-drained loamy soils of river flood plains and valleys; in mixed hardwood forests.

Range: E. Iowa east to Indiana, south to Louisiana, west to S. Texas; to 1600′ (488 m); also mountains of Mexico.

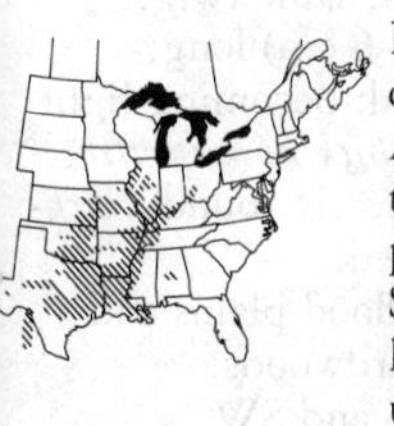

Pecan is one of the most valuable cultivated plants originating in North America. Improved varieties with large, thin-shelled nuts are grown in plantations or orchards in the Southeast; pecans are also harvested locally from wild trees. The wood is used for furniture, flooring, veneer, and charcoal for smoking meats. The word *pecan* is of Algonquian Indian origin. The Latin species name is from an old term, "Illinois nuts," and refers to the region where traders found wild trees and nuts. Indians may have extended the range by planting. This tree of the Mississippi valley was unknown to British colonists on the Atlantic coast. Thomas Jefferson planted seeds at Monticello and gave some to George

Washington; now these Pecans are the oldest trees in Mount Vernon.

345, 536, 626 **Shellbark Hickory**
"Big Shagbark Hickory" "Kingnut"
Carya laciniosa (Michx. f.) Loud.

Description: Large tree with straight trunk, narrow rounded crown, large leaves, and the largest hickory nuts.

Height: 70–100′ (21–30 m).
Diameter: 2½′ (0.8 m).
Leaves: pinnately compound; 12–20″ (30–51 cm) long. *Usually 7 broadly lance-shaped leaflets,* 2–8″ (5–20 cm) long; finely saw-toothed; nearly stalkless. Shiny dark green above, pale and covered with *soft hairs beneath.*
Bark: light gray; becoming *rough and shaggy,* separating into long narrow strips loosely attached.
Twigs: *pale orange;* stout, *hairy,* ending in large brown hairy buds.
Flowers: tiny; greenish; in early spring before leaves. Male, with 3–10 stamens, many in slender drooping *catkins, 3 hanging from 1 stalk.* 2–5 female flowers at tip of same twig.
Fruit: 1¾–2½″ (4.5–6 cm) long; nearly round; flattened; becoming light to dark brown; with *thick husk splitting to base.* Hickory nut nearly round, *thick-shelled,* with edible seed.

Habitat: Moist or wet soils of flood plains and valleys, with other hardwoods.

Range: SE. Iowa east to Ohio and SW. Pennsylvania, south to Tennessee, and west to NE. Oklahoma; also local to extreme S. Ontario, New York, N. Georgia, and Mississippi; to 1000′ (305 m).

This uncommon species is distinguished from other hickories by the large leaves, large nuts, and orange twigs; and from Shagbark Hickory by the larger number of leaflets and the thick-shelled nuts. The Latin species name, meaning "with flaps or folds," refers to the shaggy bark.

347 Nutmeg Hickory
"Swamp Hickory" "Bitter Water Hickory"
Carya myristiciformis (Michx. f.) Nutt.

Description: Tree with tall straight trunk, stout branches, narrow open crown of handsome bronze foliage, and edible nuts.
Height: 80′ (24 m).
Diameter: 2′ (0.6 m).
Leaves: pinnately compound; 7–14″ (18–36 cm) long, with slender scurfy hairy axis. *5–9 leaflets,* 2–5″ (5–13 cm) long; *broadly lance-shaped;* finely saw-toothed; short-stalked. Dark green and hairy above, shiny whitish beneath; becoming bright *golden-bronze* in autumn.
Bark: gray or brown; fissured, with long thin scales.
Twigs: brown, with tiny yellow or brown scales; slender.
Flowers: tiny; greenish; in early spring before leaves. Male, with 6–7 stamens, many in slender drooping *catkins, 3 hanging from 1 stalk.* Female, 2–10 flowers at tip of same twig.
Fruit: 1¼–1½″ (3–4 cm) long; *elliptical;* short-pointed; becoming *yellow-brown;* covered with *scurfy hairs;* with thin husk splitting along 4 ridges nearly to base. Hickory nut *thick-shelled,* with edible seed.

Habitat: Moist soil of valleys and lower uplands in hardwood forests.

Range: Scattered from South Carolina west to E. Texas and SE. Oklahoma; also NE. Mexico; to 500′ (152 m).

The common and scientific names of this patchily distributed hickory refer to the nutmeglike shape of the nut.

Nutmeg Hickory is easily recognized by the brownish hue produced by numerous tiny scales on various parts. The wood is marketed as Pecan and has similar uses.

344, 537, 608 **Shagbark Hickory**
"Scalybark Hickory" "Shellbark Hickory"
Carya ovata (Mill.) K. Koch

Description: Large tree with tall trunk, narrow irregular crown, and distinctive *rough shaggy bark.*

Height: 70–100′ (21–30 m).
Diameter: 2½′ (0.8 m).
Leaves: pinnately compound; 8–14″ (20–36 cm) long. 5 (*rarely* 7) *elliptical or ovate leaflets,* 3–7″ (7.5–18 cm) long; stalkless; edges *finely saw-toothed and hairy;* yellow-green above, paler (and hairy when young) beneath; turning golden-brown in autumn.
Bark: light gray; separating into long narrow curved strips loosely attached at middle.
Twigs: brown; stout; ending in large brown hairy buds.
Flowers: tiny; greenish; in early spring before leaves. Male, with 4 stamens, many in slender drooping *catkins, 3 hanging from 1 stalk.* 2–5 female flowers at tip of same twig.
Fruit: 1¼–2½″ (3–6 cm) long; *nearly round;* flattened at tip; with *husk thick,* becoming dark brown or blackish and *splitting to base.* Hickory nut elliptical or rounded, slightly flattened and angled, light brown, with edible seed.

Habitat: Moist soils of valleys and upland slopes in mixed hardwood forests.

Range: Extreme S. Quebec and SW. Maine, south to Georgia, west to SE. Texas, and north to SE. Minnesota; also NE. Mexico; to 2000′ (610 m) in north and 3000′ (914 m) in southern Appalachians.

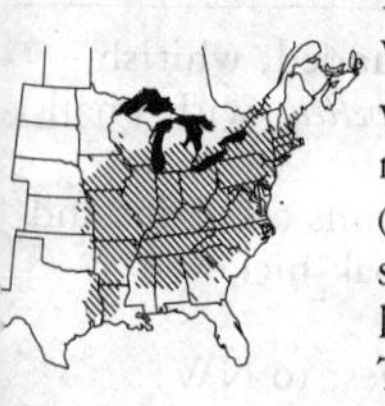

Wild trees and improved cultivated varieties produce commercial hickory nuts. Carolina Hickory (var. *australis* (Ashe) Little), a variety found in southeastern mountains, has small lance-shaped leaflets and small nuts. The name "hickory" is from *pawcohiccora,* the American Indian word for the oily food removed from pounded kernels steeped in boiling water. This sweet hickory milk was used in cooking corn cakes and hominy. Pioneers made a yellow dye from the inner bark. The nickname "Old Hickory" was given by his backwoods militia to General Andrew Jackson (afterwards our seventh President) because he was "tough as hickory."

---

337 **Sand Hickory**
"Pale Hickory" "Pignut Hickory"
Carya pallida (Ashe) Engl. & Graebn.

Description: Nut tree with *tiny, silvery or yellowish scales* on twigs, buds, and undersurface of leaflets.
Height: 30–80′ (9–24 m).
Diameter: 1–2′ (0.3–0.6 m).
Leaves: pinnately compound; 7–15″ (18–38 cm) long, *hairy axis with silvery scales; 7–9 leaflets,* 3½–6″ (9–15 cm) long; *lance-shaped* or elliptical; finely saw-toothed. Shiny green above, pale and *finely hairy* with tiny scales beneath.
Bark: dark gray or blackish; becoming rough and deeply furrowed.
Twigs: brown, with tiny silvery scales; slender.
Flowers: tiny; greenish; in early spring before leaves. Male, with 4–5 stamens; many in slender drooping *catkins, 3 hanging from 1 stalk.* Female, 1 flower at tip of same twig.

Fruit: ¾–1½″ (2–4 cm) long; *elliptical or round; thin husk* hairy and covered *with yellow scales,* splitting late on 2–4

lines. Hickory nut rounded, whitish, ridged, *slightly thick-shelled,* with small edible seed.

Habitat: Dry sandy and rocky soils on plains and mountain valleys; in oak-hickory forests.

Range: S. New Jersey southwest to NW. Florida and SE. Louisiana, north to SW. Indiana; to 2500′ (762 m).

Recognized by the tiny scales on various parts. The Latin species name, meaning "pale," refers to the lower surface of the leaflets.

333 Black Hickory
"Buckley Hickory" "Pignut Hickory"
Carya texana Buckl.

Description: Nut tree with irregular spreading or rounded crown, the common hickory from the Ozark region southwest to Texas.
Height: 20–30′ (6–9 m).
Diameter: 1′ (0.3 m), sometimes larger.
Leaves: pinnately compound; 6–12″ (15–30 cm) long, axis covered with rust-colored hairs when young. *Usually 7 leaflets* (or 5), 2–6″ (5–15 cm) long; *lance-shaped;* stalkless; finely saw-toothed; covered with rust-colored hairs when young. Becoming shiny dark green and hairless above, paler and hairless or slightly hairy on veins beneath.
Bark: black or becoming gray; rough, thick, and deeply furrowed.
Twigs: brown; slender; covered with rust-colored hairs when young.
Flowers: tiny; greenish; in early spring before leaves. Male, with 4–5 stamens; many in slender drooping *catkins, 3 hanging from 1 stalk.* Female, 1–2 flowers at tip of same twig.

Fruit: 1¼–1½″ (3–4 cm) long; *rounded* with *thin husk* often slightly winged, splitting to base. Hickory nut short-pointed, slightly angled, *thick-shelled,* with small edible or bitter seed.

Habitat: Dry rocky and sandy uplands, usually with oaks.

Range: SW. Indiana and Missouri south to S. Texas and Louisiana; at 100–2800′ (30–854 m).

This species is the hickory that ranges farthest west, to the Edwards Plateau of central Texas. Its small size makes the wood of minor commercial importance.

348 Mockernut Hickory
"White Hickory" "Mockernut"
Carya tomentosa (Poir.) Nutt.

Description: Nut tree with rounded crown and *leaves* that are *very aromatic* when crushed.
Height: 50–80′ (15–24 m).
Diameter: 2′ (0.6 m).
Leaves: pinnately compound; 8–20″ (20–51 cm) long, with hairy axis; *7 or 9 leaflets,* 2–8″ (5–20 cm) long; *elliptical or lance-shaped;* finely saw-toothed; nearly stalkless. Shiny dark yellow-green above, *pale and densely hairy and glandular beneath;* turning yellow in autumn.
Bark: gray; irregularly furrowed into narrow forking ridges.
Twigs: brown; stout, hairy, ending in large hairy bud.
Flowers: tiny; greenish; in early spring before leaves. Male, with 4–5 stamens; many in slender drooping *catkins, 3 hanging from 1 stalk* Female, 2–5 flowers at tip of same twig.
Fruit: 1½–2″ (4–5 cm) long; *elliptical or pear-shaped;* becoming brown; with *thick husk splitting to middle* or nearly to base. Hickory nut rounded or elliptical, slightly 4-angled, *thick-shelled,* with edible seed.

Habitat: Moist uplands and less frequently on

flood plains; usually with oaks, also pines.

Range: Extreme S. Ontario east to Massachusetts, south to N. Florida, west to E. Texas, and north to SE. Iowa; to 3000′ (914 m) in southern Appalachians.

The wood of this common hickory and related species is prized for furniture, flooring, tool handles, baseball bats, skis, and veneer. Hickory wood has a very high fuel value, both as firewood and as charcoal, and is the preferred wood for smoking hams. People must arrive early to gather hickory nuts before they are consumed by squirrels and other wildlife. The Latin species name, meaning "densely covered with soft hairs," describes the undersurfaces of leaflets, a characteristic that makes this tree easily identifiable.

330, 533 **Butternut**
"White Walnut" "Oilnut"
Juglans cinerea L.

Description: Tree with short straight trunk, stout branches, broad open crown, and butternut fruit with sticky husk.

Height: 40–70′ (12–21 m).
Diameter: 1–2′ (0.3–0.6 m).
Leaves: pinnately compound; 15–24″ (38–61 cm) long, with hairy axis, sticky when young. *11–17 leaflets,* 2–4½″ (5–11 cm) long; *broadly lance-shaped;* pointed at tip, unequal and rounded at base; *finely saw-toothed; stalkless.* Yellow-green and slightly hairy above, paler and covered with *soft hairs beneath;* turning yellow or brown in autumn.
Bark: light gray; smooth becoming rough and furrowed.
Twigs: brown; stout; with sticky hairs and a *hairy fringe above leaf-scars;* and with *chambered pith.*

Flowers: small; greenish; in early spring. Male, with 8–12 stamens; many in catkins. Female, with 2-lobed style; 6–8 at tip of same twig.
Fruit: 1½–2½″ (4–6 cm) long; *3–5 in drooping clusters;* narrowly *egg-shaped; long-pointed;* with 2 ridges, *rust-colored sticky hairs,* and thick husk. Shell of nut thick, *light brown,* rough with 8 ridges, containing a very oily edible seed, the *butternut.*

Habitat: Moist soils of valleys and slopes; also dry rocky soils; in hardwood forests.

Range: S. Quebec east to SW. New Brunswick, south to extreme N. Georgia, west to Missouri and Arkansas, and north to E. Minnesota; to 4800′ (1463 m).

The edible butternuts soon become rancid, and so must be harvested quickly after maturing. Indians made them into oil for many uses, including ceremonial anointing of the head. They are also eaten by wildlife. The husks of the nuts, which contain a brown stain that colors the fingers, yield a yellow or orange dye. The lumber serves as a cabinet wood.

316, 376 **Little Walnut**
"Texas Walnut" "Nogal"
Juglans microcarpa Berland.

Description: Large shrub or small tree, usually branching near ground, with broad rounded crown, *small nuts,* and characteristic walnut odor.
Height: 10–20′ (3–6 m).
Diameter: ½–1½′ (0.15–0.5 m).
Leaves: pinnately compound; 8–13″ (20–33 cm) long. *Usually 7–13 leaflets,* 2–3″ (5–7.5 cm) long; *narrowly lance-shaped;* long-pointed; usually *slightly curved; finely saw-toothed or almost without teeth;* becoming hairless or nearly so; very short-stalked. Yellow-green,

turning yellow in autumn.
Bark: gray; smooth to deeply furrowed.
Twigs: gray, slender, with brown *chambered pith.*
Flowers: small; greenish; in early spring. Male, with about 20 stamens; many in catkins. Female, with 2-lobed style; few at tip of same twig;
Fruit: ½–¾″ (12–19 mm) in diameter. Thin, hairy *husk becoming brown;* nut with hard, grooved, thick shell; and *very small* edible seed.

Habitat: Moist soils along streams in plains and foothills, and grasslands, and deserts.

Range: SW. Kansas west to New Mexico and south to S. Texas; also NE. Mexico; at 1500–4000′ (457–1219 m).

Squirrels and other rodents consume these nuts, which are mostly shell. The common and scientific names describe the tiny marblelike fruit, the smallest of the walnuts.

317, 532 **Black Walnut**
"Eastern Black Walnut" "American Walnut"
Juglans nigra L.

Description: Large walnut tree with open, rounded crown of dark green, *aromatic foliage.*
Height: 70–90′ (21–27 m).
Diameter: 2–4′ (0.6–1.2 m).
Leaves: pinnately compound; 12–24″ (30–61 cm) long. *9–21 leaflets* 2½–5″ (6–13 cm) long; *broadly lance-shaped;* finely saw-toothed; long-pointed; stalkless; nearly hairless above, covered with soft hairs beneath. Green or dark green, turning yellow in autumn.
Bark: dark brown; deeply furrowed into scaly ridges.
Twigs: brown, stout, with brown *chambered pith.*
Flowers: small; greenish; in early spring. Male, with 20–30 stamens, many in catkins. Female, with 2-lobed

style, 2–5 at tip of same twig.
Fruit: single or paired, 1½–2½″ (4–6 cm) in diameter; thick green or brown husk; irregularly ridged, thick-shelled inner layer covering sweet edible seed.

Habitat: Moist well-drained soils, especially along streams, scattered in mixed forests.

Range: Eastern half of United States except northern border; New York south to NW. Florida, west to central Texas, north to SE. South Dakota; local in S. New England and S. Ontario; to 4000′ (1219 m).

One of the scarcest and most coveted native hardwoods, Black Walnut is used especially for furniture, gunstocks, and veneer. Individual trees fetch attractive prices and a few prized trees have even been stolen. Since colonial days and before, Black Walnut has provided edible nuts and a blackish dye made from the husks. Tomatoes and apples do not survive near mature trees. The delicious nuts must be gathered early, before squirrels and other wildlife can consume them.

BIRCH FAMILY
(Betulaceae)

Trees, often large and some shrubs including alders (*Alnus*), hornbeams (*Carpinus*), and hophornbeams (*Ostrya*), as well as birches (*Betula*). About 135 species worldwide. About 20 native and 1 naturalized tree species and 8 shrub species in North America.
Leaves: deciduous, alternate, often spreading in 2 rows, simple, mostly ovate or elliptical, doubly saw-toothed with several nearly straight side veins; paired stipules shedding early.
Bark: mostly smooth (peeling in papery layers in birch, *Betula*).
Flowers: male and female on same plant, usually in early spring before or with leaves; tiny, greenish, with 4 to no sepals and no petals. Male in long narrow catkins, with 1–20 stamens; female in short conelike or headlike clusters, with 1 pistil.
Fruit: usually many in conelike cluster, small nuts or nutlets; often short-winged; 1-seeded.

---

233 **European Alder**
"Black Alder" "European Black Alder"
Alnus glutinosa (L.) Gaertn.

Description: Large, introduced tree with rounded or oblong crown and *gummy twigs and foliage*.

Height: 50–70′ (15–21 m).
Diameter: 1–2′ (0.3–0.6 m).
Leaves: *in 3 rows;* 1½–4″ (4–10 cm) long, 1–2½″ (2.5–6 cm) wide. *Elliptical* to nearly *round; doubly saw-toothed* (also lobed in cultivated varieties); with *5–7 parallel veins* on each side; gummy when young. *Shiny dark green* above, light green with tufts of rusty hairs beneath, remaining green and *shedding late*.
Bark: brown; smooth becoming

furrowed into broad plates.
Twigs: mostly hairless; very gummy when young; with 3-angled pith.
Flowers: tiny; in early spring before leaves. Male in catkins 1–1½″ (2.5–4 cm) long; upright and later drooping. Female in cones ¼″ (6 mm) long.

Cones: ⅝–⅞″ (15–22 mm) long; in clusters of 3–5; elliptical or egg-shaped, black, hard, gummy, long-stalked; remaining attached. Nutlets rounded, nearly ⅛″ (3 mm) long, flattened.

Habitat: Wet soils in humid, cool temperate regions.

Range: Native of Europe, N. Africa, and Asia. Naturalized locally in SE. Canada and NE. United States.

The Latin species name, meaning "gummy" or "gluey," describes the young twigs and leaves. This alder was introduced in colonial times and planted for shade and ornament.

228, 484 Seaside Alder
Alnus maritima Muhl. ex Nutt.

Description: Small, *autumn flowering* tree with straight trunk and narrow, rounded crown, or a much-branched shrub.
Height: 30′ (9 m).
Diameter: 4″ (10 cm).
Leaves: *in 3 rows;* 2½–4″ (6–10 cm) long, 1–2″ (2.5–5 cm) wide. *Elliptical;* finely saw-toothed; with *5–8 curved veins* on each side. *Shiny dark green* above, dull light green, sometimes slightly hairy and with tiny gland-dots beneath.
Bark: light brown or gray; smooth; thin.
Twigs: slender, slightly zigzag, hairy when young, with 3-angled pith.
Flowers: tiny; *opening in autumn.* Male yellowish, in narrowly cylindrical catkins 1½–2½″ (4–6 cm) long,

drooping from near ends of twigs. Female in cones ¼″ (6 mm) long.
Cones: ⅝–¾″ (15–19 mm) long; single or in clusters of 2–3; elliptical, dark brown, hard, short-stalked; maturing second autumn and remaining attached; with tiny flat egg-shaped nutlets having membranous border.

Habitat: Wet soil bordering streams and ponds.

Range: S. Delaware, E. shore of Maryland near sea level; S. Oklahoma at 700′ (213 m).

Both common and Latin species names refer to the local but not rare distribution near (but not on) the coast. The isolated occurrence in southern Oklahoma is unexplained.

229, 372, 616 **Speckled Alder**
"Tag Alder" "Gray Alder"
Alnus rugosa (Du Roi) Spreng.

Description: A low and clump-forming shrub; sometimes a small tree.
Height: 20′ (6 m).
Diameter: 4″ (10 cm).
Leaves: *in 3 rows;* 2–4″ (5–10 cm) long, 1¼–3″ (3–7.5 cm) wide. *Elliptical* or ovate, broadest near or below middle; *doubly and irregularly saw-toothed* and wavy-lobed; with *9–12* nearly straight *parallel veins* on each side; short, hairy stalks. *Dull dark green* with network of *sunken veins* above; whitish-green and often with soft hairs, and with prominent veins and veinlets arranged in rows like a ladder beneath.
Bark: gray, smooth.
Twigs: gray-brown, slender, slightly hairy when young; with 3-angled pith.

Flowers: tiny; in early spring before leaves. Male in drooping catkins 1½–3″ (4–7.5 cm) long. Female in cones ¼″ (6 mm) long.
Cones: ½–⅝″ (12–15 mm) long;

elliptical, blackish, hard, short-stalked; maturing in autumn; with tiny rounded flat nutlets.

Habitat: Wet soil along streams and lakes, and in swamps.

Range: Widespread across Canada from Yukon and British Columbia to Newfoundland, south to West Virginia, west to NE. Iowa, and north to NE. North Dakota; almost to northern limit of trees; in south to 2600′ (792 m).

The Latin species name, meaning "rugose" or "wrinkled," refers to the network of sunken veins prominent on the lower leaf surfaces. Planted as an ornamental at water edges. Alder thickets provide cover for wildlife, browse for deer and moose, and seeds for birds.

---

227 Hazel Alder

"Common Alder" "Tag Alder"
Alnus serrulata (Ait.) Willd.

Description: Large, spreading shrub with several trunks, sometimes a small tree, commonly found at edge of water.
Height: 20′ (6 m).
Diameter: 4″ (10 cm).
Leaves: *in 3 rows;* gummy and aromatic when immature; 2–4½″ (5–11 cm) long, 1¼–2¾″ (3–7 cm) wide. *Obovate* to elliptical, broadest usually beyond middle; finely saw-toothed with regular sharp teeth, sometimes also slightly wavy; with *9–12* nearly straight *parallel veins* on each side. *Dull green* above, light green and often hairy on veins beneath; turning red-brown in autumn.
Bark: dark gray or brown; smooth.
Twigs: covered with rust-colored hairs when young; with 3-angled pith.
Flowers: tiny; in early spring. Male in drooping catkins. Female in cones ¼″ (6 mm) long.

Cones: 3/8–5/8″ (10–15 mm) long; in clusters of 4–10; elliptical, dark brown, hard, short-stalked; maturing in late summer or autumn and remaining attached; with tiny egg-shaped flat brown nutlets.

Habitat: Wet soil bordering streams and lakes, and in swamps.

Range: SW. Nova Scotia and S. New Brunswick south to N. Florida; west to E. Texas and north to SE. Kansas; to 3000′ (914 m).

The only alder native in southeastern United States, where it is common and widespread, forming thickets.

180, 487, 617 Yellow Birch
"Gray Birch" "Silver Birch"
Betula alleghaniensis Britton

Description: Large, aromatic tree with broad, rounded crown of drooping branches and *slight odor of wintergreen in crushed twigs and foliage.*

Height: 70–100′ (21–30 m).
Diameter: 2½′ (0.8 m).
Leaves: 3–5″ (7.5–13 cm) long, 1½–2″ (4–5 cm) wide. *Elliptical,* short-pointed or rounded at base; sharply and doubly saw-toothed; mostly with *9–11 veins on each side;* hairy when young. Dark dull green above, light yellow-green beneath; turning bright yellow in autumn.
Bark: *shiny yellowish or silvery-gray;* separating into *papery* curly strips; becoming reddish-brown and fissured into scaly plates.
Twigs: greenish-brown, slender, hairy.
Flowers: tiny; in early spring. Male yellowish, with 2 stamens, many in long drooping catkins near tip of twigs. Female greenish, in short upright catkins back of tip of same twig.
Cones: ¾–1¼″ (2–3 cm) long; oblong; hairy; brownish; upright; nearly

stalkless; with many hairy scales and 2-winged nutlets; maturing in autumn.

Habitat: Cool moist uplands including mountain ravines; with hardwoods and conifers.

Range: Extreme SE. Manitoba east to S. Newfoundland, south to extreme NE. Georgia, and west to NE. Iowa; to 2500′ (762 m) in north and 3000–6000′ (914–1829 m) or higher in south.

One of the most valuable birches and one of the largest hardwoods in northeastern North America. Yellow Birch when fairly mature is easily recognized by its distinctive bark. Young specimens, which may be mistaken for Sweet Birch, are most readily identified by their hairy twigs and buds and more persistently hairy leaves with mostly unbranched side veins.

176, 379, 488, 613 **Sweet Birch**
"Black Birch" "Cherry Birch"
Betula lenta L.

Description: Aromatic tree with rounded crown of spreading branches and *odor of wintergreen in crushed twigs and foliage.*

Height: 50–80′ (15–24 m).
Diameter: 1–2½′ (0.3–0.8 m).
Leaves: 2½–5″ (6–13 cm) long, 1½–3″ (4–7.5 cm) wide. *Elliptical, long-pointed, often notched at base;* sharply and doubly saw-toothed; mostly with *9–11 veins on each side;* becoming nearly hairless. Dull dark green above, light yellow-green beneath; turning bright yellow in autumn.
Bark: *shiny, dark brown or blackish, smooth but not papery;* on large trunks fissured into scaly plates like Black Cherry.
Twigs: dark brown, slender, hairless.
Flowers: tiny; in early spring. Male yellowish, with 2 stamens, many in

long drooping catkins near tip of twigs. Female greenish, in short upright catkins back of tip of same twig.
Cones: ¾–1½″ (2–4 cm) long; oblong, brownish, upright, nearly stalkless; with hairless scales and many 2-winged nutlets; maturing in autumn.

Habitat: Cool, moist uplands; with hardwoods and conifers.

Range: S. Maine southwest to N. Alabama and north to Ohio; local in extreme S. Quebec and SE. Ontario; nearly to sea level in north; at 2000–6000′ (610–1829 m) in southern Appalachians.

Birch oil, or oil of wintergreen, used to flavor medicines and candy, was once obtained from the bark and wood of young trees. That wasteful process has been replaced by the manufacture of the same oil from wood alcohol and salicylic acid. The trees can be tapped like Sugar Maples in early spring and the fermented sap made into birch beer.

178 River Birch
"Red Birch" "Black Birch"
Betula nigra L.

Description: Often slightly leaning and forked tree with irregular, spreading crown.

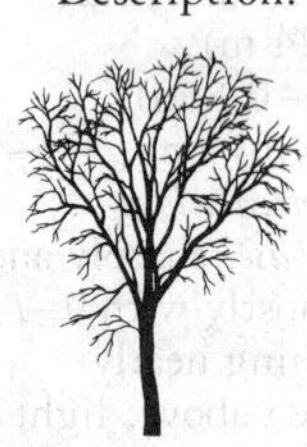

Height: 40–80′ (12–24 m).
Diameter: 1–2′ (0.3–0.6 m).
Leaves: 1½–3″ (4–7.5 cm) long, 1–2¼″ (2.5–6 cm) wide. *Ovate or nearly 4-sided;* coarsely doubly saw-toothed or slightly lobed; usually with 7–9 *veins on each side.* Shiny dark green above, *whitish and usually hairy beneath;* turning dull yellow in autumn.
Bark: *shiny pinkish-brown or silvery-gray;* separating into *papery* scales; becoming thick, fissured, and shaggy.
Twigs: reddish-brown, slender, hairy.
Flowers: tiny; in early spring. Male yellowish, with 2 stamens, many in long drooping catkins near tip of twigs.

Female greenish, in short upright catkins back of tip of same twig.

Cones: 1–1½″ (2.5–4 cm) long; cylindrical, brownish, upright, short-stalked; with many hairy scales and hairy 2-winged nutlets; maturing in late spring or early summer.

Habitat: Wet soil of stream banks, lakes, swamps, and flood plains; with other hardwoods.

Range: SW. Connecticut south to N. Florida, west to E. Texas, and north to SE. Minnesota; local in Massachusetts and S. New Hampshire; to 1000′ (305 m); to 2500′ (762 m) in southern Appalachians.

This is the southernmost New World birch and the only birch that occurs at low altitudes in the southeastern United States. Its ability to thrive on moist sites makes it useful for erosion control.

---

177 **Water Birch**
"Red Birch" "Black Birch"
Betula occidentalis Hook.

Description: Shrub or small tree with rounded crown of spreading and drooping branches, usually forming clumps and often in thickets.

Height: 25′ (7.6 m).
Diameter: 6–12″ (15–30 cm).
Leaves: ¾–2″ (2–5 cm) long, ¾–1″ (2–2.5 cm) wide. *Ovate;* sharply and often doubly saw-toothed; usually with *4–5 veins on each side.* Dark green above, pale yellow-green with tiny gland-dots beneath; turning dull yellow in autumn.
Bark: *shiny, dark reddish-brown; smooth,* with horizontal lines, not peeling.
Twigs: greenish, slender, with gland-dots.
Flowers: tiny; in early spring. Male yellowish, with 2 stamens, many in

long drooping catkins near tip of twigs. Female greenish, in short upright catkins back of tip of same twig.
Cones: 1–1¼″ (2.5–3 cm) long; cylindrical, brownish, upright or spreading on slender stalk; with many 2-winged nutlets; maturing in late summer.

Habitat: Moist soil along streams in mountain canyons, usually in coniferous forests and with cottonwoods and willows.

Range: NE. British Columbia, east to S. Manitoba, and south to N. New Mexico and California; at 2000–8000′ (610–2438 m).

This uncommon but widespread species is the only native birch in the Southwest and the southern Rocky Mountains. Sheep and goats browse the foliage.

179, 615 **Paper Birch**
"Canoe Birch" "White Birch"
Betula papyrifera Marsh.

Description: One of the most beautiful native trees, with narrow, open crown of slightly drooping to nearly horizontal branches; sometimes a shrub.
Height: 50–70′ (15–21 m).
Diameter: 1–2′ (0.3–0.6 m).
Leaves: 2–4″ (5–10 cm) long, 1½–2″ (4–5 cm) wide. *Ovate, long-pointed;* coarsely and doubly saw-toothed; usually with *5–9 veins on each side.* Dull dark green above, light yellow-green and nearly hairless beneath; turning light yellow in autumn.
Bark: *chalky to creamy white; smooth, thin,* with long horizontal lines; separating into *papery strips* to reveal orange inner bark; becoming brown, furrowed, and scaly at base; bronze to purplish in varieties.
Twigs: reddish-brown, slender, mostly hairless.

Flowers: tiny; in early spring. Male yellowish, with 2 stamens, many in long drooping catkins near tip of twigs. Female greenish, in short upright catkins back of tip of same twig.
Cones: 1½–2″ (4–5 cm); narrowly cylindrical, brownish, hanging on slender stalk; with many 2-winged nutlets; maturing in autumn.

Habitat: Moist upland soils and cutover lands; often in nearly pure stands.

Range: Transcontinental across North America near northern limit of trees from NW. Alaska east to Labrador, south to New York, and west to Oregon; local south to N. Colorado and W. North Carolina; to 4000′ (1219 m), higher in southern mountains.

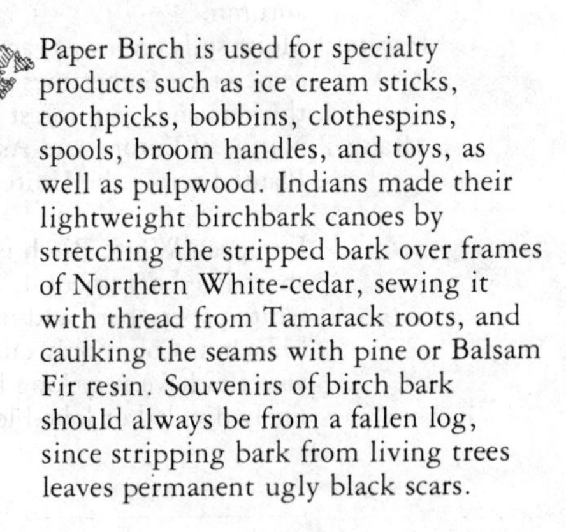

Paper Birch is used for specialty products such as ice cream sticks, toothpicks, bobbins, clothespins, spools, broom handles, and toys, as well as pulpwood. Indians made their lightweight birchbark canoes by stretching the stripped bark over frames of Northern White-cedar, sewing it with thread from Tamarack roots, and caulking the seams with pine or Balsam Fir resin. Souvenirs of birch bark should always be from a fallen log, since stripping bark from living trees leaves permanent ugly black scars.

---

181 European White Birch

"European Birch" "European Weeping Birch"
Betula pendula Roth

Description: Ornamental, planted tree with open, pyramid-shaped or spreading crown of long *drooping branches.*
Height: 50′ (15 m).
Diameter: 1′ (0.3 m).
Leaves: 1¼–2¾″ (3–7 cm) long, 1–1½″ (2.5–4 cm) wide. *Ovate* or nearly *triangular,* very long-pointed at

tip, blunt or almost straight at base; *doubly saw-toothed* with 6–9 veins on each side; sticky when young; long-stalked. Dull green above, paler beneath; turning yellow in autumn.

Bark: *white; smooth,* flaky, peeling in *papery* strips.

Twigs: slender, *drooping,* with tiny resin gland-dots.

Flowers: tiny; in early spring. Male yellowish, with 2 stamens, many in long drooping catkins near tip of twigs. Female greenish, in short upright catkins back of tip of same twig.

Cones: ¾–1¼″ (2–3 cm) long; cylindrical; hanging on slender stalks; composed of many small *2-winged nutlets* and 3-lobed bracts; maturing in autumn.

Habitat: Moist soils on lawns, and in parks and cemeteries. Sometimes an escape in thickets and open forest areas.

Range: Native of Europe and Asia Minor. Planted across the United States.

European White Birch is a graceful short-lived ornamental, grown for its white papery bark and drooping branches. Commonly cultivated varieties have very long branches and finely divided or lobed leaves.

182, 598 **Gray Birch**
"White Birch" "Wire Birch"
Betula populifolia Marsh.

Description: Small, bushy tree with open, conical crown of short slender branches reaching nearly to the ground; more often a clump of several slightly leaning trunks from an old stump.

Height: 30′ (9 m).

Diameter: 1′ (0.3 m).

Leaves: 2–3″ (5–7.5 cm) long, 1½–2½″ (4–6 cm) wide. *Triangular, tapering from near base to long-pointed tip;* sharply and doubly saw-toothed;

usually with *4–8 veins on each side;* leafstalks slender, with black gland-dots. Shiny dark green above, paler with tufts of hairs along midvein beneath; turning pale yellow in autumn.

Bark: *chalky or grayish-white; smooth,* thin, not papery; becoming darker and fissured at base.

Twigs: reddish-brown, slender, with warty gland-dots.

Flowers: tiny; in early spring. Male yellowish, with 2 stamens, many in long drooping catkins near tip of twigs. Female greenish, in short upright catkins back of tip of same twig.

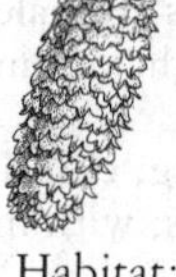

Cones: ¾–1¼″ (2–3 cm) long; cylindrical, brownish, spreading, short-stalked; with many hairy scales and hairy 2-winged nutlets; maturing in autumn.

Habitat: Dry barren uplands, also on moist soils, in mixed woodlands.

Range: S. Ontario east to Cape Breton Island, south to Pennsylvania and New Jersey; local to W. North Carolina and NW. Indiana; to 2000′ (610 m).

A pioneer tree on clearings, abandoned farms, and burned areas, Gray Birch grows rapidly but is short-lived. A nurse tree, it shades and protects seedlings of the larger, long-lived forest trees. The wood is used for spools and other turned articles and for firewood. Its trunks are so flexible that when weighted with snow, the upper branches may bend to the ground without breaking. The long-stalked leaves dance in the slightest breeze.

---

202 **Virginia Roundleaf Birch**
"Ashe Birch" "Virginia Birch"
Betula uber (Ashe) Fern.

Description: Small tree with narrow trunk and many slender spreading branches and twigs.

Height: 15–33′ (4.6–10 m).
Diameter: 4″ (10 cm).
Leaves: 1½–2½″ (4–6 cm) long, 1¼–2″ (3–5 cm) wide. *Nearly round,* ovate, or broadly elliptical; *blunt or rounded at tip, slightly notched at base; irregularly saw-toothed;* with *3–6 veins on each side;* short slender hairy stalks. Dark green and hairless except along veins, paler beneath.
Bark: gray; thin, smooth, with horizontal lines; aromatic, with strong *odor of wintergreen.*
Twigs: brown, slender, hairless.
Flowers: tiny. Male yellowish, with 2 stamens, many in long drooping catkins near tip of twigs. Female greenish, in short upright catkins back of tip of same twig.
Cones: ½″ (12 mm) long; elliptical, upright, nearly stalkless; with many dark brown nutlets.

Habitat: Very rare; along a stream in understory of mixed hardwood forest.

Range: Only in SW. Virginia in Smyth County; at about 2750′ (838 m).

This birch was discovered in 1914, named in 1918 and, when it did not appear in later searches, was listed as extinct. In 1975, it was rediscovered and officially classed as endangered. About 15 trees, and some seedlings as well, were found on private land; 2 others were located in the adjacent National Forest.

147, 596 **American Hornbeam**
"Blue-beech" "Water-beech"
Carpinus caroliniana Walt.

Description: Small, shrubby tree with one or more short *trunks angled* or fluted, long, slender, spreading branches, and broad, rounded crown.
Height: 30′ (9 m).
Diameter: 1′ (0.3 m).

Leaves: 2–4½" (5–11 cm) long, 1–2½" (2.5–6 cm) wide. *Elliptical,* long-pointed at tip; sharply *doubly saw-toothed;* with many nearly straight parallel side veins. *Dull dark blue-green* above, paler with hairs on veins and vein angles below; turning orange to red in autumn.
Bark: *blue-gray;* thin, smooth.
Twigs: brown, slender, slightly zigzag.
Flowers: tiny; in early spring before leaves. Male greenish, in drooping catkins 1¼–1½" (3–4 cm) long. Female reddish-green, paired in narrow catkins ½–¾" (12–19 mm) long.
Fruit: ¼" (6 mm) long; paired, *egg-shaped,* hairy greenish *nutlets,* with leaflike *3-pointed,* toothed, *greenish scale;* in clusters 2–4" (5–10 cm) long, hanging on slender stalks; maturing in late summer.

Habitat: Moist rich soils, mainly along streams and in ravines; in understory of hardwood forests.

Range: SE. Ontario east to SW. Quebec and central Maine, south to central Florida, west to E. Texas, and north to Minnesota; to 3000′ (914 m). Also in Mexico.

The word "hornbeam," originally given to the European Hornbeam (*Carpinus betulus* L.), is from the words "horn" (for toughness) and "beam" (for tree) and refers to the very hard tough wood. The small size of this species limits uses to tool handles and wooden articles. The name beech has been misapplied to this member of the birch family, because of the similar bark. Deer browse the twigs and foliage, and grouse, pheasants, and quail eat the nutlets.

222, 371, 491, 628 **Eastern Hophornbeam**
"American Hophornbeam" "Ironwood"
Ostrya virginiana (Mill.) K. Koch

Description: A tree with a trunk that looks like sinewy muscles and a rounded crown of slender, spreading branches.
Height: 20–50′ (6–15 m).
Diameter: 1′ (0.3 m).
Leaves: 2–5″ (5–13 cm) long, 1–2″ (2.5–5 cm) wide. Ovate or elliptical; sharply *doubly saw-toothed;* with many nearly straight *parallel side veins;* short, hairy leafstalks. Dull yellow-green and nearly hairless above, paler and hairy chiefly on veins beneath; turning yellow in autumn.
Bark: light brown; thin, finely fissured into *long narrow scaly ridges.*
Flowers: tiny; in early spring before leaves. Male greenish, in 1–3 drooping, narrowly cylindrical clusters 1½–2½″ (4–6 cm) long. Female reddish-green, in narrowly cylindrical clusters ½–¾″ (12–19 mm) long.
Fruit: 1½–2″ (4–5 cm) long, ¾–1″ (2–2.5 cm) wide; conelike hanging clusters maturing in late summer; composed of many flattened, small, egg-shaped brown nutlets, each within a swollen egg-shaped flattened light brown cover that is *papery* and *sacklike.*

Habitat: Moist soil in understory of upland hardwood forests.

Range: SE. Manitoba east to Cape Breton Island, south to N. Florida, and west to E. Texas; to 4500′ (1372 m).

The common name refers to the resemblance of the fruit clusters to hops, an ingredient of beer. The nutlets and buds are eaten by wildlife, such as bobwhites, pheasants, grouse, deer, and rabbits. Called "Ironwood," for its extremely hard tough wood, which is used for tool handles, small wooden articles, and fenceposts. Planted as an ornamental but slow-growing.

BEECH FAMILY
(Fagaceae)

Often large trees and some shrubs including chestnuts and chinkapins (*Castanea*) and oaks (*Quercus*) as well as beeches (*Fagus*). About 700–900 species nearly worldwide except tropical South America and tropical Africa. About 65 native and 1 naturalized tree species and 10 of shrubs in North America.

Leaves: usually deciduous or in warm climates evergreen; alternate, simple, mostly toothed or lobed, with narrow paired stipules shedding early.

Flowers: male and female on same plant, usually in early spring before or at same time as the leaves; tiny, without petals. Male flowers mostly in long narrow catkins composed of 4- to 7-lobed calyx and 4–40 stamens. 1–3 female flowers, often in spikes, consisting of cup of scales and 1 pistil.

Fruit: a 1-seeded nut or 2–3 nuts within a cup; often edible acorns, chestnuts, beechnuts, etc.

The oak genus (*Quercus*), including the most important native hardwoods, is mainly deciduous or in warmer climates evergreen. Leaves are alternate in 5 rows, often variable in shape, with lobed, toothed, or straight edges, often prominently veined, and short-stalked. Bark is light gray and scaly, or blackish and furrowed. Twigs are slender, often slightly 5-angled and hairy, with star-shaped pith. Small, greenish flowers appear in early spring, male and female on same twig. Male flowers are composed of a bell-shaped calyx and usually 6 stamens; many are clustered in slender drooping catkins; 1 or a few female flowers at leaf bases have a cup of many overlapping scales and a pistil with 3 styles protruding. The hard-shelled acorns, containing 1 large bitter seed, are borne within a cup of many overlapping scales. Two groups (subgenera) of oaks are distinguishable. The leaves and lobes of red (or black) oaks are bristle-tipped; bark is usually blackish and furrowed; the bitter-seeded acorns mature in autumn of the second year, with two sizes usually present. The leaves of white oaks are not bristle-tipped; bark is usually light gray and scaly; the less bitter and sometimes edible acorns mature the first year.

151, 527 **Florida Chinkapin**
"Trailing Chinkapin"
Castanea alnifolia Nutt.

Description: Low spreading shrub or small tree with irregular, spreading crown.
Height: 40′ (12 m).
Diameter: 1′ (0.3 m).
Leaves: 3–4″ (7.5–10 cm) long, 1–2″ (2.5–5 cm) wide. *Obovate* or elliptical; with many *straight parallel side veins* each ending in short tooth; leafstalks short, hairless. *Dark green* above, *pale green* and sometimes hairy beneath.
Bark: gray-brown; smooth, becoming rough and furrowed into long ridges.
Twigs: reddish-brown, slender.
Flowers: in early summer. Many *tiny* whitish male flowers in catkins 4–5″ (10–13 cm) long at base of leaf. Several tiny female flowers at base of shorter catkins.
Fruit: ¾–1″ (2–2.5 cm) in diameter; *burs* with *few scattered* branched hairy *spines;* maturing in autumn and splitting open. Single egg-shaped nut, shiny brown, edible.

Habitat: Uplands, including dry sandy soils in oak forests.
Range: North Carolina south to N. Florida, west to Louisiana; near sea level.

This species is typically a low shrub that forms thickets from underground rootstalks. The tree form occurs mainly in northern Florida. Chinkapins or nuts of this species are eaten chiefly by wildlife.

150, 526, 609 **American Chestnut**
"Chestnut"
Castanea dentata (Marsh.) Borkh.

Description: Formerly a large tree with a massive trunk and a broad, rounded, dense crown; now small sprouts from base of long-dead trees. 
Height: 20′ (6 m); formerly 60–100′ (18–30 m).
Diameter: 4″ (10 cm); formerly 2–4′ (0.6–1.2 m).
Leaves: 5–9″ (13–23 cm) long, 1½–3″ (4–7.5 cm) wide. *Narrowly oblong, long-pointed;* with *many straight parallel side veins,* each ending in *curved tooth;* short-stalked. *Shiny yellow-green* above, paler green below with a few hairs along midvein; turning yellow in autumn.
Bark: dark gray-brown; furrowed into flat ridges; on sprouts smooth.
Twigs: green, slender, hairless.
Flowers: in early summer. Many whitish male flowers 3⁄16″ (5 mm) long, in upright catkins 6–8″ (15–20 cm) long at base of leaf. Few female flowers ⅜″ (10 mm) long, bordered by narrow greenish scales, at base of shorter catkins.
Fruit: 2–2½″ (5–6 cm) in diameter; short-stalked *burs* covered with many stout *branched spines* about ½″ (12 mm) long; maturing in autumn and splitting open along 3–4 lines; 2–3 chestnuts ½–¾″ (12–19 mm) long, broadly *egg-shaped,* becoming *shiny dark brown,* flattened and pointed; edible.

Habitat: Moist upland soils in mixed forests.

Range: Extreme S. Ontario east to Maine, south to SW. Georgia, west to Mississippi, north to Indiana; to 4000′ (1219 m).

American Chestnut is gone from the forests, a victim of the chestnut blight caused by an introduced fungus. This disease began in New York City in 1904, spread rapidly, and within 40 years had virtually wiped out this once abundant species. Fortunately, there is no threat of extinction; sprouts continue from roots until killed back by the blight, and cultivated trees grow in western states and other areas where the parasite is absent. Blight-resistant chestnuts such as hybrids between American and Chinese species are being developed for ornament, shade, and wildlife. The wood of this species was once the main domestic source of tannin, the edible chestnuts were a commercial crop, and the leaves were used in home medicines.

---

149 **Ozark Chinkapin**
"Ozark Chestnut"
Castanea ozarkensis Ashe

Description: Small to medium-sized tree of Ozark region bearing very spiny burs.
Height: 20–50′ (6–15 m).
Diameter: 4–18″ (0.1–0.5 m).
Leaves: 5–8″ (13–20 cm) long, 1½–3″ (4–7.5 cm) wide. *Narrowly oblong* or lance-shaped; with many *straight parallel side veins* each ending in *long* straight or curved *tooth;* leafstalks short, hairless or nearly so. *Yellow-green* above, paler and with fine *whitish hairs* or nearly hairless beneath.
Bark: gray-brown; smooth becoming furrowed into scaly plates.
Twigs: gray, slender, hairy when young.
Flowers: in early summer. Many tiny whitish male flowers, in catkins 2–8″ (5–20 cm) long at base of leaf. Few female flowers ³⁄₁₆″ (5 mm) long at base of smaller catkins.

Fruit: 1–1¼″ (2.5–3 cm) in diameter;

burs with many long *branched hairy spines;* maturing in autumn and splitting open. Single rounded to egg-shaped nut, dark brown, edible.

Habitat: Acid soils in dry rocky ridges, slopes, and ravines.

Range: S. Missouri, Arkansas, and E. Oklahoma; at 500–2800′ (152–854 m).

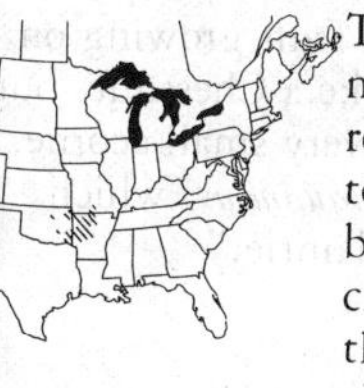

This is the only tree species with a natural range limited to the Ozark region, for which it is named. Related to American Chestnut, bearing similar but smaller edible chestnuts or chinkapins. The trees are attacked by the same fungus parasite that causes chestnut blight but are less susceptible and not threatened with extinction.

148, 403 **Allegheny Chinkapin**
Castanea pumila Mill.

Description: Shrub-forming thicket or tree with rounded crown, bearing very spiny burs.

Height: 40′ (12 m).
Diameter: 1′ (0.3 m).
Leaves: 3–6″ (7.5–15 cm) long, 1¼–2″ (3–5 cm) wide. *Oblong* or elliptical, short-pointed; with many *straight parallel side veins,* each ending in *short tooth;* leafstalks short, hairy. *Yellow-green* above, with *velvety white hairs* beneath.
Bark: reddish-brown; furrowed into scaly plates.
Twigs: gray, hairy.
Flowers: in early summer. Many tiny whitish male flowers in upright catkins 4–6″ (10–15 cm) long at base of leaf. Few female flowers ⅛″ (3 mm) long at base of smaller catkins.
Fruit: ¾–1¼″ (2–3 cm) in diameter; *burs* with many *branched hairy spines;* maturing in autumn and splitting open. Single *egg-shaped* nut, *shiny dark*

brown with whitish hairs, edible.

Habitat: Dry sandy and rocky uplands; in oak and hickory forests.

Range: New Jersey and S. Pennsylvania south to central Florida, west to E. Texas, and north to SE. Oklahoma, local in S. Ohio; to 4500′ (1372 m).

Captain John Smith published the first record of this nut in 1612: "They [the Indians] have a small fruit growing on little trees, husked like a Chestnut, but the fruit most like a very small acorne. This they call *Checkinquamins,* which they esteem a great daintie."

152, 528, 601, 614

American Beech
"Beech"
Fagus grandifolia Ehrh.

Description: Large tree with rounded crown of many long, spreading and horizontal branches, producing edible beechnuts.

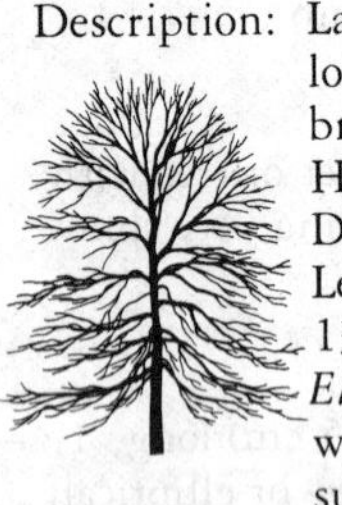

Height: 60–80′ (18–24 m).
Diameter: 1–2½′ (0.3–0.8 m).
Leaves: *spreading in 2 rows;* 2½–5″ (6–13 cm) long, 1–3″ (2.5–7.5 cm) wide. *Elliptical* or ovate, long-pointed at tip; with many *straight parallel* slightly sunken *side veins* and *coarsely saw-toothed* edges; short-stalked. *Dull dark blue-green* above, light green beneath, becoming hairless or nearly so; turning yellow and brown in fall.
Bark: light gray; smooth, thin.
Twigs: slender, ending in long narrow scaly buds, with short side twigs or spurs.
Flowers: with new leaves in spring. Male flowers small, yellowish with many stamens, crowded in ball ¾–1″ (2–2.5 cm) in diameter, hanging on slender hairy stalk to 2″ (5 cm). Female flowers about ¼″ (6 mm) long, bordered by narrow hairy reddish scales, 2 at end of short stalk.
Fruit: ½–¾″ (12–19 mm) long; short-

stalked light brown *prickly burs;* maturing in autumn and splitting into 4 parts. Usually *2 nuts,* about ⅝″ (15 mm) long, 3-angled, shiny brown, known as *beechnuts.*

Habitat: Moist rich soils of uplands and well-drained lowlands; often in pure stands.

Range: S. Ontario, east to Cape Breton Island, south to N. Florida, west to E. Texas and north to N. Michigan; a variety in mountains of NE. Mexico; to 3000′ (914 m) in north and to 6000′ (1829 m) in southern Appalachians.

American Beech was recognized by the colonists, who already knew the famous, closely related European Beech. American Beech is a handsome shade tree and bears similar edible beechnuts, which are consumed in quantities by wildlife, especially squirrels, raccoons, bears, other mammals, and game birds. Unlike most trees, beeches retain smooth bark in age. The trunks are favorites for carving and preserve initials and dates indefinitely.

174 European Beech
Fagus sylvatica L.

Description: Cultivated tree with stout trunk and dense, rounded crown of spreading branches extending almost to ground; producing edible beechnuts.

Height: 70′ (21 m).
Diameter: 2½′ (0.8 m).
Leaves: spreading in *2 rows;* 2–4″ (5–10 cm) long, 1½–3″ (4–7.5 cm) wide. *Broadly elliptical* or ovate; edges with small teeth; with 5–9 straight parallel veins on each side; hairy when young; short-stalked. *Shiny dark green* above, light green beneath; turning reddish-brown or bronze in autumn.
Bark: dark gray; smooth.
Twigs: gray or brown, finely hairy when young, slender, ending in long

narrow brown scaly buds; with short side twigs or spurs.
Fruit: 1″ (2.5 cm) long; a light brown *prickly bur;* maturing in autumn, splitting into 4 parts; usually containing 2 triangular shiny brown edible seeds, known as *beechnuts.*

Habitat: Tolerates most soils, best in calcareous or deep sandy loam; hardy in cool, moist temperate regions.

Range: Native in central and S. Europe to high altitudes. Planted in NE. United States and Pacific states.

European Beech is one of the most popular large shade trees. Numerous horticultural varieties include purple, copper, fernleaf, cutleaf, oakleaf, and roundleaf foliage and columnar and weeping habits. It can be pruned and clipped into arbors and hedges. An important hardwood in its native range, forming extensive forests. Beechnuts serve as food for people, livestock, and wildlife. The words *beech* and *book* come from the same root, because ancient Saxons and Germans wrote on pieces of beech board.

282, 380, 524, 597 **White Oak**
"Stave Oak"
Quercus alba L.

Description: The classic eastern oak, with widespreading branches and a rounded crown, the trunk irregularly divided into spreading, often horizontal, stout branches.
Height: 80–100′ (24–30 m) or more.
Diameter: 3–4′ (0.9–1.2 m) or more.
Leaves: 4–9″ (10–23 cm) long, 2–4″ (5–10 cm) wide. *Elliptical;* 5- to 9-*lobed;* widest beyond middle and tapering to base; *hairless.* Bright green above, whitish or gray-green beneath; turning red or brown in fall, often remaining attached in winter.

Bark: light gray; shallowly fissured into long broad scaly plates or ridges, often loose.
Acorns: ⅜–1¼″ (1–3 cm) long; egg-shaped; about ¼ enclosed by *shallow cup;* becoming light gray; with warty, finely hairy scales; maturing first year.
Habitat: Moist well-drained uplands and lowlands, often in pure stands.
Range: S. Ontario and extreme S. Quebec east to Maine, south to N. Florida, west to E. Texas, and north to E. central Minnesota; to 5500′ (1676 m), or above in southern Appalachians.

The most important lumber tree of the white oak group, its high-grade wood is useful for all purposes. Called "Stave Oak" because the wood is outstanding in making tight barrels for whiskey and other liquids. In colonial times the wood was important in shipbuilding.

268 Arkansas Oak
"Water Oak"
Quercus arkansana Sarg.

Description: Medium-sized tree with tall trunk and narrow crown.
Height: 40–60′ (12–18 m).
Diameter: 1′ (0.3 m).
Leaves: 2–5″ (5–13 cm) long, 1–2½″ (2.5–6 cm) wide. *Broadly obovate;* broadest toward *3-toothed* or *slightly 3-lobed* and *rounded tip,* and gradually narrowed to short-pointed base. Light *yellow-green* above, paler with tufts of hairs in vein angles beneath.
Bark: black, thick, rough; deeply furrowed into long narrow scaly ridges.
Acorns: ½″ (12 mm) long; *nearly round,* less than ¼ enclosed by *shallow cup;* green becoming brown; 1–2 on short stalk or stalkless; maturing first year.
Habitat: Well-drained sandy soils in hardwood forests.
Range: SW. Georgia and NW. Florida west to

Louisiana and SW. Arkansas; to 400′ (107 m).

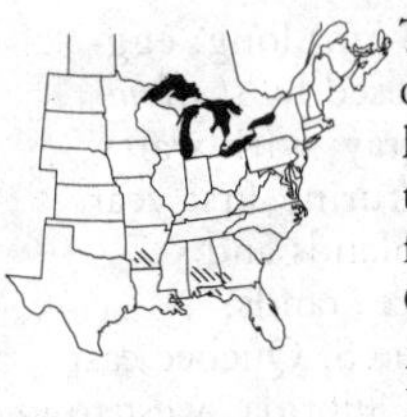

This minor and uncommon species occurs scattered with other oaks in a local range of 5 southeastern states. It is thought to be an ancient species of formerly wider distribution in the Coastal Plain. First discovered in Arkansas, hence the common name and Latin species name given in 1911.

280, 515 **Swamp White Oak**
Quercus bicolor Willd.

Description: Large tree with a narrow, rounded, open crown of often-drooping branches.

Height: 60–70′ (18–21 m).
Diameter: 2–3′ (0.6–0.9 m).
Leaves: 4–7″ (10–18 cm) long, 2–4½″ (5–11 cm) wide. *Obovate,* rounded or blunt at tip, broadest beyond middle, gradually narrowed to pointed base; *edges wavy* with *5–10 shallow rounded lobes* on each side. Green and slightly shiny above, *soft whitish hairs beneath;* turning brown to red in fall.
Bark: light gray; with large thin scales, becoming furrowed into plates.
Acorns: ¾–1¼″ (2–3 cm) long; *egg-shaped;* ⅓ or more enclosed by deep cup of *many distinct scales,* becoming *light brown;* usually 2 on long slender stalk, maturing first year.

Habitat: Wet soils of lowlands, including stream borders, flood plains, and swamps subject to flooding; in mixed forests.

Range: Extreme S. Ontario east to extreme S. Quebec and Maine, south to Virginia, west to Missouri, and north to SE. Minnesota; local to SW. Maine, North Carolina, and NE. Kansas; to 1000′ (305 m), locally to 2000′ (610 m).

The Latin species name, meaning "two-colored," refers to the leaves, which are green above and whitish beneath.

269 **Chapman Oak**
"Chapman White Oak" "Scrub Oak"
Quercus chapmanii Sarg.

Description: Shrub or small tree with a rounded crown of stout, spreading branches; often a low shrub.
Height: 30′ (9 m).
Diameter: 8″ (20 cm).
Leaves: 1½–3½″ (4–9 cm) long and ¾–1½″ (2–4 cm) wide. *Obovate* or oblong, broadest beyond middle and narrowed toward blunt or rounded base; edges straight or *slightly wavy* toward rounded or *3-lobed tip;* slightly thickened. Shiny dark green above, *dull light green* and often hairy on midvein beneath; turning yellow or reddish in fall and winter, and shedding gradually by early spring.
Bark: light gray; shedding in scaly plates.
Acorns: ⅝–¾″ (15–19 mm) long; *egg-shaped;* partly enclosed by *deep half-round* cup; becoming brown; stalkless or nearly so; maturing first year.

Habitat: Sandy hills, ridges, and coastal dunes; with Sand Pine and evergreen oaks.

Range: Extreme S. South Carolina and SE. Georgia to S. and NW. Florida and S. Alabama; near sea level.

Named for Alvan Wenworth Chapman (1809–99), physician and botanist of Apalachicola, Florida, who first distinguished this oak in his *Flora of the Southern United States.*

287, 518, 591 **Scarlet Oak**
"Red Oak" "Black Oak"
Quercus coccinea Muenchh.

Description: Large tree with a rounded, open crown of glossy foliage, best known for its brilliant autumn color.
Height: 60–80′ (18–24 m).
Diameter: 1–2½′ (0.3–0.8 m).

Leaves: 3–7″ (7.5–18 cm) long, 2–5″ (5–13 cm) wide. *Elliptical; deeply divided* nearly to midvein into 7 (rarely 9) *lobes,* broadest toward tip, each lobe ending in several bristle-tipped teeth; the *wide round sinuses* between lobes often forming more than a half-circle; long, slender stalks. *Shiny green* above; pale yellow-green, slightly shiny, and with tufts of hairs in vein angles along midvein beneath; turning *scarlet* in fall.

Bark: dark gray, smooth; becoming blackish, thick, rough, and furrowed into scaly ridges or plates; inner bark reddish.

Acorns: ½–1″ (1.2–2.5 cm) long; *egg-shaped;* becoming brown with *2–4 faint rings;* ⅓–½ enclosed by thick *deep top-shaped cup* of tightly pressed scales, tapering to stalklike base; maturing second year.

Habitat: Various soils, especially poor and sandy, on upland ridges and slopes; with other oaks and in mixed forests.

Range: SW. Maine south to Georgia, west to NE. Mississippi, north to Missouri and Indiana; local in Michigan; to 3000′ (914 m), locally to 5000′ (1524 m).

A popular and handsome shade and street tree. The lumber is marketed as Red Oak, which differs in its shallowly lobed, dull green leaves, and acorns with a shallow cup. Black Oak is also similar, but has yellow-green leaves with brown hairs beneath and acorns with a deep cup of loose hairy scales.

---

273 **Durand Oak**
"Bluff Oak" "White Oak"
Quercus durandii Buckl.

Description: Tree with rounded crown, or a shrub.
Height: 70′ (21 m).
Diameter: 2′ (0.6 m).
Leaves: 1½–4″ (4–10 cm) long,

⅝–1½″ (1.5–4 cm) wide, sometimes larger. *Obovate* to elliptical, and variable in shape; sometimes *slightly 3-lobed* toward rounded tip, or wavy-lobed; thin or slightly thickened. Shiny dark green above, *gray-green* and covered with *tiny star-shaped hairs* beneath.
Bark: light gray, thin, scaly.
Acorns: ½–⅝″ (12–15 mm) long; *egg-shaped;* less than ¼ enclosed by *saucer-shaped scaly cup;* becoming brown; maturing first year.

Habitat: Various soils, including limestone hills; in hardwood forests.

Range: North Carolina to N. Florida, west to S. and central Texas, and north to S. Oklahoma; also NE. Mexico; to 2000′ (620 m).

Named for Elias Magloire Durand (1794–1873), Philadelphia pharmacist and botanist.

---

286 Northern Pin Oak
"Black Oak" "Jack Oak"
Quercus ellipsoidalis E. J. Hill

Description: Tree with short trunk, many small branches, and a narrow crown.
Height: 50–70′ (15–21 m).
Diameter: 1–2½′ (0.3–0.8 m).
Leaves: 3–5″ (7.5–13 cm) long, 2½–4″ (6–10 cm) wide. *Elliptical;* deeply divided into 5–7 *lobes* with few bristle-tipped teeth and wide rounded sinuses. *Shiny green* above; paler, often with tufts of hairs along midvein beneath; turning yellow, brown, or purple in fall, and frequently remaining attached in winter.
Bark: gray or dark brown; smooth, becoming shallowly fissured or furrowed into narrow plates; inner bark light yellow.
Acorns: ½–¾″ (12–19 mm) long; *elliptical* or nearly round; ⅓–½ enclosed by *deep cup* tapering to

stalklike base; becoming brown; maturing second year.

Habitat: Well-drained dry or moist sandy and clay soils; in pure stands, with other oaks, or in mixed forests.

Range: SW. Ontario, southeast to extreme NW. Ohio, southwest to extreme N. Missouri, and north to extreme SE. North Dakota; at 600–1300′ (183–396 m).

This Lake States species resembles Pin Oak and Scarlet Oak, which have more southern ranges.

283 **Southern Red Oak**
"Spanish Oak" "Swamp Red Oak"
Quercus falcata Michx.

Description: Tree with rounded, open crown of large spreading branches, and twigs with rust-colored hairs.
Height: 50–80′ (15–24 m).
Diameter: 1–2½′ (0.3–0.8 m).
Leaves: 4–8″ (10–20 cm) long, 2–6″ (5–15 cm) wide. *Elliptical; deeply divided* into long narrow end lobe and 1–3 shorter mostly curved lobes on each side, with 1–3 bristle-tipped teeth; sometimes slightly *triangular* with *bell-shaped base* and 3 broad lobes. *Shiny green* above, with rust-colored or *gray soft hairs* beneath; turning brown in fall.
Bark: dark gray; becoming furrowed into broad ridges and plates.
Acorns: ½–⅝″ (12–15 mm) long; *elliptical* or rounded; becoming brown; ⅓ or more enclosed by cup tapering to broad stalklike base; maturing second year.

Habitat: Dry, sandy loam and clay loam soils of uplands; in mixed forests.

Range: Long Island and New Jersey south to N. Florida, west to E. Texas, and nort to S. Missouri; to 2500′ (762 m).

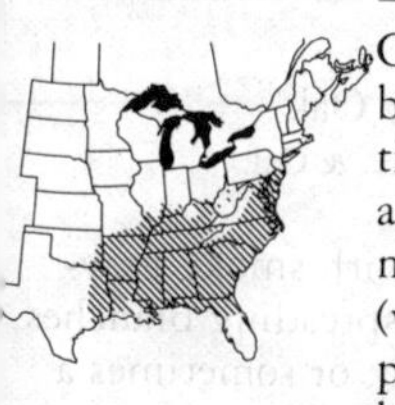

Often called Spanish Oak, possibly because it commonly occurs in areas of the early Spanish colonies. It is unlike any oaks native to Spain. The lumber is marketed as Red Oak. Cherrybark Oak (var. *pagodifolia* Ell.) is a variety with pagoda-shaped leaves having 5–11 broad shallow lobes, with whitish hairs beneath, and smooth cherrylike bark with short ridges. It is found on well-drained lowland soils from southeastern Virginia to northwestern Florida and eastern Texas.

---

290 Georgia Oak
Quercus georgiana M. A. Curtis

Description: Shrub or sometimes a small tree, rare and very local in Piedmont of Georgia and adjacent states.
Height: 6–10′ (1.8–3 m), sometimes to 30′ (9 m).
Diameter: 1′ (0.3 m).
Leaves: mostly 2–4″ (5–10 cm) long, 1–2″ (2.5–5 cm) wide. *Elliptical;* irregularly divided into *3–5 short-pointed* bristle-tipped *lobes;* larger lobes have 1–2 small teeth; base short-pointed; thick. *Shiny green* above, paler with tufts of hairs in vein angles beneath.

Bark: gray or light brown; thin; smooth becoming scaly.
Acorns: ⅜–½″ (10–12 mm) long; *nearly round,* ¼–⅓ enclosed by *shallow cup;* becoming brown; stalkless or short-stalked; maturing second year.

Habitat: Granite and sandstone outcrops of mountain slopes.

Range: Very local in South Carolina, N. Georgia, and N. Alabama; at about 1700′ (518 m).

Discovered at Stone Mountain, Georgia, this species was found afterwards in 2 adjacent states.

272 **Lacey Oak**
"Rock Oak" "Smoky Oak"
Quercus glaucoides Mart. & Gal.

Description: Medium-sized tree with smoky gray-green foliage, large spreading branches, and a rounded crown; or sometimes a shrub.
Height: 35′ (11 m).
Diameter: 20″ (0.5 m).
Leaves: 2–4½″ (5–11 cm) long, ¾–2¼″ (2–6 cm) wide. *Oblong* or elliptical; rounded at tip, rounded or blunt at base; with *few wavy teeth; thick* and leathery. *Dull blue-green* or gray-green above, paler or whitish and slightly hairy beneath.
Bark: gray; smooth, becoming furrowed into narrow ridges.
Acorns: ½–¾″ (12–19 mm) long; *egg-shaped,* ¼ enclosed by *shallow cup;* becoming brown; 1–2 on short stalk, or stalkless; maturing first year.

Habitat: Rocky limestone bluffs and in canyons; with other oaks, junipers, and mesquite.

Range: Central Texas on Edwards Plateau and NE. Mexico; at 1500–2000′ (457–610 m).

The distinctive color of the leaves aids recognition at a distance. The common name honors Howard Lacey, who collected a specimen on his ranch near Kerrville, Texas.

289, 606 **Bear Oak**
"Scrub Oak"
Quercus ilicifolia Wangenh.

Description: Much-branched shrub or sometimes small tree with rounded crown.
Height: 20′ (6 m).
Diameter: 5″ (13 cm).
Leaves: 2–4″ (5–10 cm) long, 1½–3″ (4–7.5 cm) wide. *Obovate;* with 3–7, *usually* 5, shallow short-pointed *lobes*

ending in 1–3 bristle-tipped teeth; short-pointed at base; slightly thickened and firm. *Dull dark green* above, densely covered with *light gray hairs* beneath; turning dull red or yellow in autumn and often remaining attached in winter.
Bark: dark gray, thin, smooth, becoming fissured and scaly.
Acorns: ⅜–⅝″ (10–15 mm) long; *egg-shaped* or *rounded;* brown-striped; *many clustered* and mostly paired on twigs; ½–¼ enclosed by *deep cup* with stalklike base and fringelike border of many overlapping brown scales; maturing second year.

Habitat: Dry sandy barrens and rocky ridges in mountains, forming thickets or with pines and other oaks.

Range: S. Maine southwest to W. North Carolina; to 3000′ (914).

A temporary scrub type after heavy cutting and repeated fires, Bear Oak is replaced by taller pines and oaks. The Latin species name, meaning "holly leaf," refers to the foliage. It is called "Bear Oak," reportedly because only bears like the very bitter acorns.

50, 519, 603

Shingle Oak
"Laurel Oak"
Quercus imbricaria Michx.

Description: A handsome tree with a symmetrical, conical to rounded crown.

Height: 50–60′ (15–18 m).
Diameter: 1–2′ (0.3–0.6 m).
Leaves: 3–6″ (7.5–15 cm) long, ¾–2″ (2–5 cm) wide. *Oblong* or lance-shaped, short-pointed or rounded at ends; bristle-tipped; *edges straight* or slightly wavy and turned under. *Shiny dark green* above; midvein yellow, light gray-green and with soft hairs beneath; turning yellow or reddish-brown in fall, often shedding late.

Bark: brown to gray; smooth, becoming rough and furrowed into scaly ridges.
Acorns: ½–⅝″ (12–15 mm) long; *nearly round,* ⅓–½ enclosed by *deep cup* of blunt hairy scales; becoming brown; 1–2 on stout stalks; maturing second year.

Habitat: Moist soils along streams and in uplands; scattered with Post and Black oaks.

Range: Pennsylvania south to North Carolina, west to Arkansas, and north to S. Iowa and S. Michigan; local in Louisiana and Alabama; to 2000′ (610 m).

The Latin species name, meaning "overlapping," and the common name both refer to use of the wood for shingles by the pioneers, a practice continued today. An ornamental and shade tree, it is also suitable for hedges, screens, and windbreaks.

46 **Bluejack Oak**
"Sandjack" "Upland Willow Oak"
Quercus incana Bartr.

Description: Thicket-forming shrub or small tree with an irregular crown of stout, crooked branches and distinctive, blue-green foliage.
Height: 20′ (6 m).
Diameter: 6″ (15 cm).
Leaves: 2–4″ (5–10 cm) long, ½–1″ (1.2–2.5 cm) wide. *Oblong; bristle-tipped;* edges straight (rarely, slightly lobed on young twigs); *slightly thickened* and leathery. *Shiny blue-green* above with prominent network of veins, *dull gray-green* and *finely hairy beneath;* turning reddish before shedding in late fall or early winter.
Bark: dark gray or blackish; rough, thick, furrowed in nearly square plates.
Acorns: ½–⅝″ (12–15 mm) long;

rounded; ¼–½ enclosed by *shallow or deep cup;* becoming brown; nearly stalkless; maturing second year.

Habitat: Dry sandy uplands, with other oaks and pines.

Range: SE. Virginia to central Florida, west to E. and central Texas, and north to SE. Oklahoma; to 500′ (152 m).

Easily recognized by the distinctive deciduous leaves. The common name refers to the shiny blue-green foliage, while the Latin species name, meaning "hoary," describes the gray-green undersurface.

284, 588 Turkey Oak
"Catesby Oak" "Scrub Oak"
Quercus laevis Walt.

Description: Tree with irregular, open crown of crooked branches, often a shrub.

Height: 20–40′ (6–12 m).
Diameter: 1′ (0.3 m).
Leaves: 4–8″ (10–20 cm) long, 3–6″ (7.5–15 cm) wide. *Nearly triangular,* spreading from pointed base into *3–5* (rarely 7) long *narrow lobes* each with 1–3 *long-pointed bristle-tipped teeth;* mostly *turned on edge; thick and stiff. Shiny yellow-green* above, light green beneath, with prominent veins and tufts of rust-colored hairs in vein angles; turning red before shedding in late fall or early winter.
Bark: gray to blackish; becoming thick, very rough, deeply furrowed into irregular ridges; inner bark reddish.
Acorns: ¾–1″ (2–2.5 cm) long; *egg-shaped,* about ⅓ enclosed by *top-shaped cup* with hairy scales *extending down inner surface;* short-stalked; becoming brown; maturing second year.

Habitat: Dry sandy ridges and dunes, especially near coast; often in pure stands.

Range: SE. Virginia to central Florida and west to SE. Louisiana; to 500′ (152 m).

The common name refers to the shape of the 3-lobed leaves suggesting a turkey's foot. The Latin species name, meaning "smooth," describes the nearly hairless leaves. Spreads by underground runners, especially after frequent fires.

---

51 Laurel Oak

"Darlington Oak" "Diamond-leaf Oak"
Quercus laurifolia Michx.

Description: Large, nearly evergreen tree with dense, broad, rounded crown.
Height: 60–80′ (18–24 m).
Diameter: 1–2½′ (0.3–0.8 m).
Leaves: 2–5½″ (5–14 cm) long, ⅜–1½″ (1–4 cm) wide. *Narrowly oblong;* diamond- or lance-shaped, often broadest near middle; *bristle-tipped; edges straight* (rarely, with few lobes or teeth); thin or *slightly thickened;* usually hairless. *Shiny green* or dark green above, light green and slightly shiny beneath; shedding in early spring and nearly evergreen.
Bark: brown to gray, smooth; becoming blackish, rough, and furrowed.
Acorns: ½″ (12 mm) long; nearly round, ¼ or less enclosed by *shallow cup* of blunt hairy scales; short-stalked or nearly stalkless; becoming brown; maturing second year.

Habitat: Moist to wet well-drained sandy soil along rivers and swamps; sometimes in pure stands.

Range: SE. Virginia to S. Florida, west to SE. Texas, and north locally to S. Arkansas; to 500′ (152 m).

Common and Latin species names refer to the resemblance of the foliage to Grecian Laurel (*Laurus nobilis* L.), of the Mediterranean region. A handsome shade tree, widely planted in the Southeast.

278 **Overcup Oak**
"Swamp Post Oak" "Water White Oak"
Quercus lyrata Walt.

Description: Tree with rounded crown of small, often drooping branches, with acorns almost covered by the cup, and narrow deeply lobed leaves.
Height: 60–80′ (18–24 m).
Diameter: 2–3′ (0.6–0.9 m).
Leaves: 5–8″ (13–20 cm) long, 1½–4″ (4–10 cm) wide. *Narrowly oblong;* deeply divided into *7–11 rounded or short-pointed lobes,* the longest near short-pointed tip; pointed base. Dark green and slightly shiny above, gray-green and with soft hairs or nearly hairless beneath; turning yellow, brown, or red in fall.
Bark: light gray; furrowed into scaly or slightly shaggy ridges or plates.
Acorns: ½–1″ (1.2–2.5 cm) long; *nearly round, almost enclosed by large rounded cup* of warty gray scales, the upper scales long-pointed; usually stalkless; maturing first year.

Habitat: Wet clay and silty clay soils, mostly on poorly drained flood plains and swamp borders; sometimes in pure stands.

Range: Delaware to NW. Florida, west to E. Texas, and north to S. Illinois; to 500′ (152 m), sometimes slightly higher.

The Latin species name, meaning "lyre-shaped," refers to the leaves.

281, 525 **Bur Oak**
"Blue Oak" "Mossycup Oak"
Quercus macrocarpa Michx.

Description: Tree with very large acorns, stout trunk, and broad, rounded, open crown of stout, often crooked, spreading branches; sometimes a shrub.
Height: 50–80′ (15–24 m).
Diameter: 2–4′ (0.6–1.2 m).

Leaves: 4–10″ (10–25 cm) long, 2–5″ (5–13 cm) wide. *Obovate,* broadest beyond middle, lower half deeply divided into 2–3 lobes on each side; upper half usually with 5–7 *shallow rounded lobes* on each side to *broad rounded tip.* Dark green and slightly shiny above, gray-green and with fine hairs beneath; turning yellow or brown in fall.

Bark: light gray; thick, rough, deeply furrowed into scaly ridges.

Acorns: *large;* ¾–2″ (2–5 cm) long and wide; *broadly elliptical,* ½–¾ enclosed by *large deep cup* with hairy gray scales, (the upper scales very long-pointed) forming *fringelike border;* maturing first year.

Habitat: From dry uplands on limestone and gravelly ridges, sandy plains, and loamy slopes to moist flood plains of streams; often in nearly pure stands.

Range: Extreme SE. Saskatchewan east to S. New Brunswick, south to Tennessee, west to SE. Texas and north to North Dakota; local in Lousiana and Alabama. Usually at 300–2000′ (91–610 m); to 3000′ (914 m) or above in northwest.

The acorns of this species, distinguished by very deep fringed cups, are the largest of all native oaks. The common name describes the cup of the acorn, which slightly resembles the spiny bur of Chestnut. Bur Oak is the northernmost New World oak. In the West, it is a pioneer tree, bordering and invading the prairie grassland. Planted for shade, ornament, and shelterbelts.

291, 590 Blackjack Oak
"Blackjack" "Jack Oak"
Quercus marilandica Muenchh.

Description: Tree with open irregular crown of crooked, spreading branches.
Height: 20–50′ (6–15 m).
Diameter: 6–12″ (15–30 cm).
Leaves: 2½–5″ (6–13 cm) long, 2–4″ (5–10 cm) wide. Slightly *triangular* or broadly obovate, *broadest near tip* with 3 shallow broad bristle-tipped lobes; gradually narrowed to rounded base; slightly thickened. *Shiny yellow-green* above, light yellow-green with brownish hairs (especially along veins) beneath; turning brown or yellow in fall.
Bark: blackish; rough, thick, deeply furrowed into broad, nearly square plates.
Acorns: ⅝–¾″ (15–19 mm) long; *elliptical,* ending in stout point; ⅓–⅔ enclosed by *deep* thick *top-shaped cup* of rusty-brown, hairy, loosely overlapping scales; short-stalked; maturing second year.

Habitat: Dry sandy and clay soils in upland ridges and slopes with other oaks and with pines.

Range: Long Island and New Jersey south to NW. Florida, west to central and SE. Texas, and north to SE. Iowa; local in S. Michigan; to 3000′ (914 m).

This species and Post Oak form the Cross Timbers in Texas and Oklahoma, the forest border of small trees and transition zone to prairie grassland. The wood is used for railroad cross-ties, firewood, and charcoal. This tree was first described in 1704 from a specimen in the colony of Maryland, referred to in the Latin species name. Virginia, where earlier plant collections were made, is honored in the species names of Live Oak (*Quercus virginiana* Mill.) and several other important trees.

277 Swamp Chestnut Oak
"Basket Oak" "Cow Oak"
Quercus michauxii Nutt.

Description: Large tree with compact, rounded crown and chestnutlike foliage.
Height: 60–80′ (18–24 m).
Diameter: 2–3′ (0.6–0.9 m).
Leaves: 4–9″ (10–23 cm) long, 2–5½″ (5–14 cm) wide. *Obovate,* broadest beyond middle; *edges wavy* with 10–14 *rounded teeth* on each side; abruptly pointed at tip; gradually narrowed to base. Shiny dark green above, gray-green and with soft hairs beneath; turning brown or dark red in fall.
Bark: light gray; fissured into scaly plates.
Acorns: 1–1¼″ (2.5–3 cm) long; *egg-shaped,* ⅓ or more enclosed by *deep thick cup* with broad base, composed of *many overlapping* hairy brown scales; stalkless or short-stalked; maturing first year.

Habitat: Moist sites including well-drained, sandy loam and silty clay flood plains along streams; sometimes in pure stands.

Range: New Jersey south to N. Florida, west to E. Texas, and north to S. Illinois; to 1000′ (305 m).

Called "Basket Oak" because baskets were woven from fibers and splints obtained by splitting the wood. These strong containers were used to carry cotton from the fields. The sweetish acorns can be eaten raw, without boiling. Cows consume the acorns, hence the name "Cow Oak."

274 Mohr Oak
"Shin Oak" "Scrub Oak"
Quercus mohriana Buckl. ex Rydb.

Description: Thicket-forming shrub or sometimes small tree with rounded crown, usually evergreen.

Height: 20′ (6 m).
Diameter: 8″ (20 cm).
Leaves: *evergreen* or deciduous; 1–3″ (2.5–7.5 cm) long, ½–1″ (1.2–2.5 cm) wide. *Oblong* or elliptical; *edges straight or wavy* lobed with few teeth; *thick and leathery;* leafstalks very short. Shiny dark green above, with *dense gray hairs* and prominent veins beneath.
Bark: gray-brown, thin, furrowed.

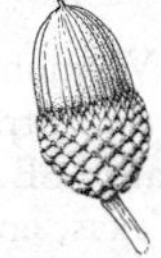

Acorns: ⅜–⅝″ (10–15 mm) long; *broadly elliptical;* green becoming brown; about ½ enclosed by *deep cup;* 1–2 on short stalk; maturing first year.

Habitat: Plains and hills, especially in limestone soils; forming thickets in oak brush or "shinnery."

Range: W. Oklahoma, central W. and Trans-Pecos Texas, and NE. New Mexico; also NE. Mexico; at 2000–4000′ (610–1219 m).

The names Shin Oak and "shinnery" refer to the dense thickets, scarcely knee-high, of dwarf evergreen oaks of this and related species on uplands of western Texas and borders of adjacent states.

275, 381, 522 **Chinkapin Oak**
"Chestnut Oak" "Rock Oak"
Quercus muehlenbergii Engelm.

Description: Tree with narrow, rounded crown; characteristic of limestone uplands.
Height: 50–80′ (15–24 m).
Diameter: 2–3′ (0.6–0.9 m).
Leaves: 4–6″ (10–15 cm) long, 1½–3″ (4–7.5 cm) wide. *Narrowly elliptical* to obovate; slightly thickened; pointed at tip; narrowed to base; with *many straight, parallel side veins,* each ending in curved tooth on *wavy edges.* Shiny green above, *whitish-green* and covered with tiny hairs beneath; turning brown or red in fall.

Bark: light gray; thin, fissured and scaly.
Acorns: ½–1″ (1.2–2.5 cm) long; *egg-shaped,* ⅓ or more enclosed by *deep thin cup* of *many overlapping* hairy long-pointed gray-brown *scales;* usually stalkless; maturing first year.

Habitat: Mostly on limestone outcrops in alkaline soils, including dry bluffs and rocky river banks; often with other oaks.

Range: S. Ontario east to W. Vermont, south to NW. Florida, west to central Texas, and north to Iowa; local in SE. New Mexico, Trans-Pecos Texas, and NE. Mexico; at 400–3000′ (122–914 m).

The common name refers to the resemblance of the foliage to chinkapins (*Castanea*), while the Latin species name honors Henry Ernst Muehlenberg (1753–1815), a Pennsylvania botanist.

48, 520 **Myrtle Oak**
"Scrub Oak"
Quercus myrtifolia Willd.

Description: Evergreen, much-branched, thicket-forming shrub or small tree with short, crooked branches and rounded crown.
Height: 30′ (9 m).
Diameter: 1′ (0.3 m).
Leaves: *evergreen;* ¾–2″ (2–5 cm) long, ½–1″ (1.2–2.5 cm) wide. Usually *elliptical* to obovate but varying in shape; rounded or sometimes pointed at tip, gradually narrowed to blunt or rounded base; edges *turned under* and sometimes wavy or toothed; *thick* and leathery; leafstalks very short, hairy. *Shiny dark green* and *hairless* with *prominent network of veins* above, dull light green beneath with tufts of hairs in vein angles.
Bark: light gray; smooth, becoming furrowed.

Acorns: ⅜–½″ (10–12 mm) long; *nearly round,* ¼–⅓ enclosed by *shallow cup;* becoming brown; stalkless or short-stalked; usually maturing second year.

Habitat: Dry sandy ridges and sand dunes, especially near coast and on islands; usually with other oaks and pines.

Range: S. South Carolina to S. Florida and west to S. Mississippi; near sea level.

Common and Latin species names refer to the resemblance of the leaves to those of Myrtle (*Myrtus communis* L.), an evergreen shrub from the Mediterranean region, introduced in Florida and California.

---

271 **Water Oak**
"Spotted Oak" "Possum Oak"
Quercus nigra L.

Description: Tree with conical or rounded crown of slender branches, and fine textured foliage of small leaves.

Height: 50–100′ (15–30 m).
Diameter: 1–2½′ (0.3–0.8 m).
Leaves: 1½–5″ (4–13 cm) long and ¾–2″ (2–5 cm) wide. Obovate or *wedge-shaped;* broadest near rounded and slightly *3-lobed tip;* bristle-tipped; gradually narrowed to long-pointed base; sometimes with small lobes on each side. *Dull blue-green* above, paler with tufts of hairs along vein angles beneath; turning yellow in late fall and shedding in winter.
Bark: dark gray, smooth; becoming blackish and furrowed into narrow scaly ridges.
Acorns: ⅜–⅝″ (10–15 mm) long and broad; *nearly round,* with *shallow, saucer-shaped cup;* becoming brown; maturing second year.

Habitat: Moist or wet soils of lowlands, including flood plains or bottomlands of streams and borders or swamps; also moist uplands; often with Sweetgum.

Range: S. New Jersey south to central Florida, west to E. Texas, and north to SE. Missouri; to about 1000′ (305 m).

A handsome, rapidly growing shade tree for moist soils in the Southeast; however, Water Oak is short-lived.

---

294 Nuttall Oak
"Red Oak" "Pin Oak"
Quercus nuttallii Palmer

Description: Tree with swollen base and open crown of spreading to horizontal or slightly drooping branches.
Height: 60–100′ (18–30 m).
Diameter: 1–3′ (0.3–0.9 m).
Leaves: 4–8″ (10–20 cm) long, 2–5″ (5–13 cm) wide. *Elliptical;* deeply divided into 7 *or* 5 *narrow long-pointed lobes* ending in a few bristle-tipped teeth; wide sinuses between lobes. *Dull dark green* above, paler with tufts of hairs in vein angles along midrib beneath; turning brown in fall and shedding gradually in early winter.
Bark: gray or brown, smooth; becoming black and furrowed into flat, scaly ridges.
Acorns: ¾–1¼″ (2–3 cm) long; *oblong,* usually dark-striped, ¼–½ enclosed by *deep thick cup* tapering to broad stalklike base; maturing second year.

Habitat: Wet, poorly drained, clay soils of flood plains; sometimes in pure stands.

Range: Alabama west to extreme E. Texas, and north to SE. Missouri; to 500′ (152 m).

Not distinguished as a species until 1927, when it was named for Thomas Nuttall (1786–1859), British-American botanist and ornithologist. The foliage resembles Pin Oak; the ranges overlap in Arkansas, but Pin Oak has smaller rounded acorns with a shallow cup.

49 **Oglethorpe Oak**
Quercus oglethorpensis Duncan

Description: Straight-trunked tree with crown of crooked branches.
Height: 60–80′ (18–24 m).
Diameter: 1–2′ (0.3–0.6 m).
Leaves: 2–5″ (5–13 cm) long, ¾–1½″ (2–4 cm) wide. *Narrowly elliptical to obovate, blunt at both ends; edges straight* or sometimes slightly wavy; thin; very short-stalked. Green and usually hairless above, paler and *velvety* with *light yellow star-shaped hairs beneath;* turning scarlet early in autumn.
Bark: light gray or brown; scaly or becoming furrowed.
Twigs: purplish-tinged, hairless.
Acorns: ⅜″ (10 mm) long; *egg-shaped,* ⅓ enclosed by *short cup;* becoming brown; stalkless or short-stalked; maturing first year.

Habitat: Poorly drained soils of level areas or "flatwoods" and adjacent slopes of Piedmont.

Range: W. South Carolina and NE. Georgia; at about 700′ (213 m).

This local oak was named in 1940 for Oglethorpe County, Georgia, where it is most abundant. It indirectly honors James Edward Oglethorpe (1695–1785), English general and founder of the colony of Georgia.

288, 517, 586 **Pin Oak**
"Swamp Oak" "Spanish Oak"
Quercus palustris Muenchh.

Description: Straight-trunked tree with spreading to *horizontal branches,* very slender *pinlike twigs,* and a broadly conical crown.
Height: 50–90′ (15–27 m).
Diameter: 1–2½′ (0.3–0.8 m).
Leaves: 3–5″ (7.5–13 cm) long, 2–4″ (5–10 cm) wide. *Elliptical; 5–7 deep lobes* nearly to midvein with few bristle-

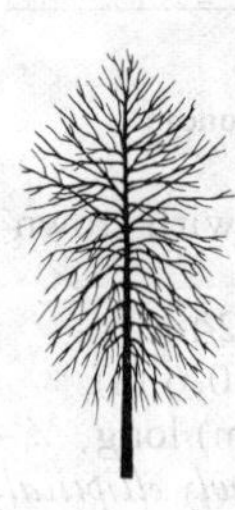

tipped teeth and wide rounded sinuses; base short-pointed. *Shiny dark green* above, light green and slightly shiny with tufts of hairs in vein angles along midvein beneath; turning red or brown in fall.

Bark: dark gray; hard; smooth, becoming fissured into short, broad, scaly ridges.

Acorns: ½″ (12 mm) long and broad; *nearly round;* becoming brown; ¼–⅓ enclosed by thin *saucer-shaped cup* tapering to base; maturing second year.

Habitat: In nearly pure stands on poorly drained, wet sites, including clay soils on level uplands; less common on deep, well-drained bottomland soils.

Range: Extreme S. Ontario to Vermont, south to central North Carolina, west to NE. Oklahoma, and north to S. Iowa; to 1000′ (305 m).

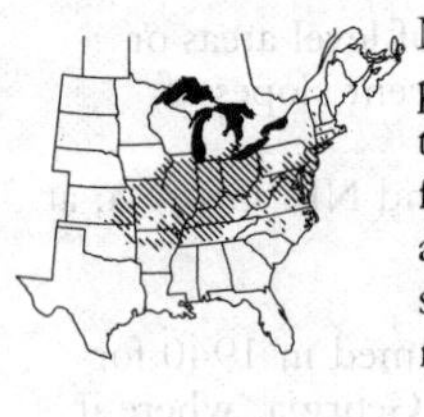

Named for the many short side twigs or pinlike spurs. A popular, graceful lawn tree with regular compact form and fine-textured foliage, Pin Oak is hardy and easily transplanted because the shallow fibrous root system lacks tap roots.

101 Willow Oak
"Pin Oak" "Peach Oak"
Quercus phellos L.

Description: Tree with conical or rounded crown of many slender branches ending in very slender, *pinlike twigs* with *willowlike foliage.*

Height: 50–80′ (15–24 m).

Diameter: 1–2½′ (0.3–0.8 m).

Leaves: 2–4½″ (5–11 cm) long, ⅜–¾″ (10–19 mm) wide. *Narrowly oblong* or *lance-shaped,* with tiny *bristle-tip; edges straight* or slightly wavy. *Light green* and slightly shiny above, dull light green and sometimes with fine gray hairs beneath; turning pale yellow in fall.

Bark: dark gray, smooth, and hard; becoming blackish, rough, and fissured into irregular narrow ridges and plates.
Acorns: ⅜–½" (10–12 mm) long and broad; *nearly round,* with *shallow saucer-shaped cup;* becoming brown; maturing second year.

Habitat: Moist alluvial soils of lowlands, chiefly flood plains or bottomlands of streams; sometimes in pure stands.

Range: New Jersey south to NW. Florida, west to E. Texas, and north to S. Illinois; to 1000′ (305 m).

A popular street and shade tree with fine-textured foliage, widely planted in Washington, D.C., and southward. Its disadvantage, however, is that it becomes too large to be grown around houses. Readily transplanted because of shallow roots. Easily distinguishable from most other oaks by the narrow leaves without lobes or teeth. While superficially the foliage resembles that of willows, it is recognized as an oak by the acorns and the tiny bristle-tip. City squirrels as well as wildlife consume and spread the acorns.

276, 604 **Chestnut Oak**
"Rock Chestnut Oak" "Rock Oak"
Quercus prinus L.

Description: Large tree with broad, open, irregular crown of *chestnutlike foliage.*
Height: 60–80′ (18–24 m).
Diameter: 2–3′ (0.6–0.9 m).
Leaves: 4–8" (10–20 cm) long, 2–4" (5–10 cm) wide. *Elliptical* or obovate, broadest beyond middle, short-pointed at tip; edges *wavy* with 10–16 *rounded teeth* on each side; gradually narrowed to base. Shiny green above, dull gray-green and sparsely hairy beneath; turning yellow in fall.
Bark: gray; becoming thick and deeply furrowed into broad or narrow ridges.

Acorns: ¾–1¼″ (2–3 cm) long; *egg-shaped,* ⅓ or more enclosed by *deep, thin cup* narrowed at base, composed of short, *warty,* hairy *scales not overlapping;* becoming brown; short-stalked; maturing first year.

Habitat: Sandy, gravelly, and rocky dry upland soils, but reaches greatest size on well-drained lowland sites; often in pure stands on dry rocky ridges.

Range: Extreme S. Ontario to SW. Maine, south to Georgia, west to NE. Mississippi, and north to SE. Michigan; at 1500–5000′ (457–1524 m).

Because of its high tannin content, the bark formerly served for tanning leather. The wood is marketed as White Oak. As a shade tree, it is adapted to dry rocky soil.

279, 375 **English Oak**
Quercus robur L.

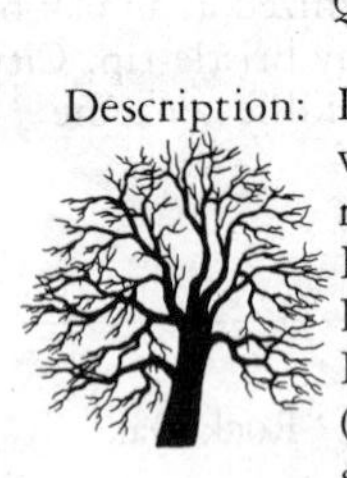

Description: Introduced tree with short, stout trunk; widespreading branches; and broad, rounded, open crown.
Height: 80′ (24 m), taller with age.
Diameter: 2–3′ (0.6–0.9 m).
Leaves: 2–5″ (5–13 cm) long, 1¼–2½″ (3–6 cm) wide. *Oblong;* with 6–14 shallow rounded lobes, including *2 small ear-shaped lobes* at very *short-stalked base.* Dark green above, pale blue-green beneath.
Bark: dark gray; deeply and irregularly furrowed; becoming thick.
Acorns: ⅝–1″ (1.5–2.5 cm) long; *egg-shaped,* about ⅓ enclosed by *half-round cup;* becoming brown; 1–5 on long slender stalk; maturing first year.

Habitat: Spreading from cultivation in moist soil, along roadsides and forest edges.

Range: Native of Europe, N. Africa, and W. Asia. Naturalized locally in SE. Canada and NE. United States; also planted in southeastern and Pacific states.

This noble oak is one of the most characteristic British trees, attaining very large size with age. It supplied timbers for wooden ships of the Royal Navy and oak paneling for famous buildings. The bark formerly was a source of tannin. Many horticultural varieties are distinguished by crown shape and leaf shape and color.

292, 370, 523 **Northern Red Oak**
"Red Oak" "Gray Oak"
Quercus rubra L.

Description: Large tree with rounded crown of stout, spreading branches.
Height: 60–90′ (18–27 m).
Diameter: 1–2½′ (0.3–0.8 m).
Leaves: 4–9″ (10–23 cm) long, 3–6″ (7.5–15 cm) wide. *Elliptical;* usually divided less than halfway to midvein into *7–11 shallow wavy lobes* with a few irregular bristle-tipped teeth. Usually *dull green* above, dull light green beneath with tufts of hairs in angles along midvein; turning brown or dark red in fall.
Bark: dark gray or blackish; rough, furrowed into scaly ridges; inner bark reddish.
Acorns: ⅝–1⅛″ (1.5–2.8 cm) long; *egg-shaped,* less than ⅓ enclosed by *broad cup* of reddish-brown, blunt, tightly overlapping scales; maturing second year.

Habitat: Moist, loamy, sandy, rocky, and clay soils; often forming pure stands.

Range: W. Ontario to Cape Breton Island, south to Georgia, west to E. Oklahoma, and north to Minnesota; to 5500′ (1676 m) in south.

The northernmost eastern oak, it is also the most important lumber species of red oak. Most are used for flooring, furniture, millwork, railroad cross-ties, mine timbers, fenceposts, pilings, and pulpwood. A popular handsome shade and street tree, with good form and dense foliage. One of the most rapid-growing oaks, it transplants easily, is hardy in city conditions, and endures cold.

293 Shumard Oak
"Spotted Oak" "Swamp Oak"
Quercus shumardii Buckl.

Description: Large tree with straight axis and broad, rounded, open crown.

Height: 60–90′ (18–27 m).
Diameter: 1–2½′ (0.3–0.8 m).
Leaves: 3–7″ (7.5–18 cm) long, 2½–5″ (6–13 cm) wide. *Elliptical;* usually *deeply divided* nearly to midvein into 5–9 *lobes* becoming broadest toward tip, with several spreading bristle-tipped teeth; large rounded sinuses between lobes, sometimes nearly closed. Slightly *shiny dark green* and hairy above, slightly shiny or dull green beneath with tufts of hairs in vein angles; turning red or brown in fall.
Bark: gray and smooth; becoming dark gray and slightly furrowed into ridges.
Acorns: ⅝–1⅛″ (1.5–2.8 cm) long; *egg-shaped,* ¼–⅓ enclosed by *shallow cup* of tightly overlapping *blunt scales;* green becoming brown; usually hairless; maturing second year.

Habitat: Moist well-drained soils including flood plains along streams, also on dry ridges and limestone hills.
Range: North Carolina to N. Florida, west to central Texas, and north to E. Kansas; local north to S. Michigan and S. Pennsylvania; to 2500′ (762 m).

A handsome shade tree, suggested as a substitute for Scarlet Oak, though not so hardy northward. Named for Benjamin Franklin Shumard (1820–69), state geologist of Texas. Texas Oak (var. *texana* (Buckl.) Ashe), a variety in central Texas and southern Oklahoma, has small, usually 5-lobed leaves, small acorns, and hairy red buds (instead of hairless brown).

270, 514 **Post Oak**
"Iron Oak"
Quercus stellata Wangenh.

Description: Tree with dense, rounded crown and distinctive *leaves suggesting a Maltese cross;* sometimes a shrub.

Height: 30–70′ (9–21 m).
Diameter: 1–2′ (0.3–0.6 m).
Leaves: 3¼–6″ (8–15 cm) long, 2–4″ (5–10 cm) wide. *Obovate;* with 5–7 *deep* broad rounded *lobes,* 2 middle lobes largest; with short-pointed base and rounded tip; slightly thickened. Shiny dark green and slightly rough with scattered hairs above, *gray-green* with tiny star-shaped hairs beneath; turning brown in fall.
Bark: light gray; fissured into scaly ridges.
Acorns: ½–1″ (1.2–2.5 cm) long; *elliptical,* ⅓–½ enclosed by *deep cup;* green becoming brown; usually stalkless or short-stalked; maturing first year.

Habitat: Sandy, gravelly, and rocky ridges, also moist loamy soils of flood plains along streams; sometimes in pure stands.

Range: SE. Massachusetts south to central Florida, west to NW. Texas, and north to SE. Iowa; to 3000′ (914 m).

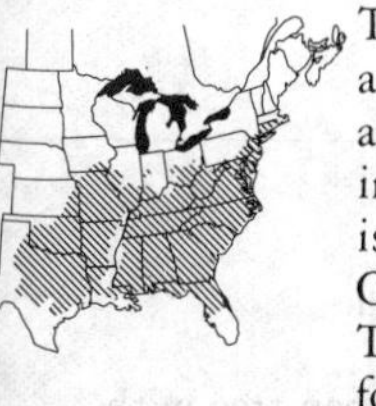

The wood is marketed as White Oak and used for railroad cross-ties, posts, and construction timbers. Of large size in the lower Mississippi Valley where it is known as "Delta Post Oak." Post Oak and Blackjack Oak form the Cross Timbers in Texas and Oklahoma, the forest border of small trees and transition zone to prairie grassland.

285, 516, 587 **Black Oak**
"Yellow Oak" "Quercitron Oak"
Quercus velutina Lam.

Description: Medium-sized to large tree with open, spreading crown.
Height: 50–80′ (15–24 m).
Diameter: 1–2½′ (0.3–0.8 m).
Leaves: 4–9″ (10–23 cm) long, 3–6″ (7.5–15 cm) wide. *Elliptical;* usually with 7–9 *lobes,* either *shallow or deep* and narrow, ending in a few bristle-tipped teeth; slightly thickened. *Shiny green* above, *yellow-green* and usually with *brown hairs* beneath; turning dull red or brown in fall.
Bark: gray and smooth on small trunks; becoming blackish, thick and rough, deeply furrowed into ridges; *inner bark yellow* or orange, very bitter.
Acorns: ⅝–¾″ (15–19 mm) long; *elliptical,* ½ enclosed by *deep thick top-shaped cup* narrowed at base, with fringed border of loose *rust-brown hairy scales;* maturing second year.

Habitat: Dry upland sandy and rocky ridges and slopes, also on clay hillsides; sometimes in pure stands.

Range: Extreme S. Ontario and SW. Maine, south to NW. Florida, west to central Texas, and north to SE. Minnesota; to 5000′ (1524 m).

Easily distinguishable by the yellow or orange inner bark, formerly a source of tannin, of medicine, and of a yellow dye for cloth. Peeled bark was dried, pounded to powder, and the dye sifted out.

47, 521 **Live Oak**
"Virginia Live Oak"
Quercus virginiana Mill.

Description: Medium-sized evergreen tree with short, broad trunk buttressed at the base forking into a few nearly

horizontal, long branches, and very broad, spreading, dense crown.
Height: 40–50′ (12–15 m).
Diameter: 2–4′ (0.6–1.2 m).
Leaves: *evergreen;* 1½–4″ (4–10 cm) long, ⅜–2″ (1–5 cm) wide. Elliptical or *oblong; thick;* rounded tip sometimes ending in tiny tooth; base short-pointed; *edges usually straight* and slightly *rolled under,* rarely, with few spiny teeth. Shiny dark green above, gray-green and densely *hairy beneath;* shedding after new leaves appear in spring.
Bark: dark brown; rough, deeply furrowed into scaly ridges.
Acorns: ⅝–1″ (1.5–2.5 cm) long; narrow and *oblong,* ¼–½ enclosed by *deep cup;* green becoming brown; long-stalked; maturing first year.

Habitat: Sandy soils including coastal dunes and ridges near marshes; often in pure stands.

Range: SE. Virginia south to S. Florida and west to S. and central Texas; local in SW. Oklahoma and NE. Mexico; to 300′ (91 m) and in Texas to 2000′ (610 m).

Live Oak timber was once important for building ships. The nation's first publicly owned timber lands were purchased as early as 1799 to preserve these trees for this purpose. Called Live Oak because of the evergreen foliage. The very broad branches are usually draped with Spanish-moss. A handsome shade tree, popular in the Southeast, where it attains very large size. Texas Live Oak (var. *fusiformis* (Small) Sarg.), a western variety in central Texas and local in southwestern Oklahoma and northeastern Mexico, has slightly smaller leaves, broadest toward the base, and acorns with cups narrowed at the base.

ELM FAMILY
(Ulmaceae)

Trees and shrubs, sometimes woody vines, including hackberries (*Celtis*) and elms (*Ulmus*). About 200 species nearly worldwide; 14 native and 1 naturalized tree species, and 2 native shrub species in North America.

Leaves: alternate in 2 rows, asymmetrical or unequal at base, often with 3 main veins, generally toothed, with paired stipules.

Flowers: tiny, inconspicuous, greenish; usually 1 to many, along twigs; male and female (bisexual in *Ulmus*), with calyx of 4–8 persistent sepals or lobes, no corolla, 4–8 stamens opposite sepals, and 1 pistil.

Fruit: a drupe or winged key (samara).

139 Sugarberry
"Sugar Hackberry" "Hackberry"
Celtis laevigata Willd.

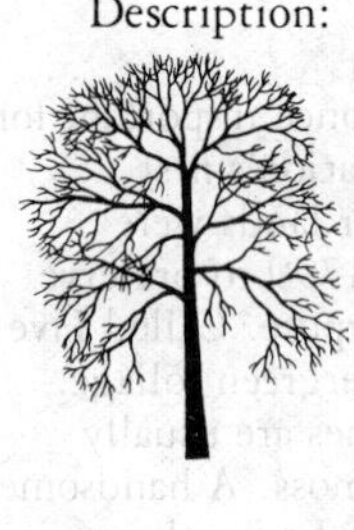

Description: Tree with broad, rounded, open crown of spreading or slightly drooping branches.

Height: 80′ (24 m).

Diameter: 1½′ (0.5 m).

Leaves: in 2 rows, 2½–4″ (6–10 cm) long, ¾–1¼″ (2–3 cm) wide. *Broadly lance-shaped,* long-pointed, often curved; 2 sides unequal; *without teeth,* sometimes with a few; *3 main veins* from base; *thin. Dark green and usually smooth above,* paler and usually hairless beneath.

Bark: light gray; thin, smooth, with prominent *corky warts.*

Twigs: greenish, slender, mostly hairless.

Flowers: ⅛″ (3 mm) wide; greenish; male and female at base of young leaves in early spring.

Fruit: ¼″ (6 mm) in diameter; orange-red or *purple* 1-seeded drupes; dry and sweet; slender-stalked at leaf bases.

Habitat: Moist soils, especially clay, on river flood plains; sometimes in pure stands but usually with other hardwoods.

Range: SE. Virginia south to S. Florida, west to central and SW. Texas, and north to central Illinois; also NE. Mexico; to 2000′ (610 m).

Robins, mockingbirds, and other songbirds eat the sweetish fruits. Principal uses of the wood are for furniture, athletic goods, and plywood.

96 Lindheimer Hackberry

"Palo Blanco"

Celtis lindheimeri Engelm.

Description: Small much-branched tree with broad, irregular crown.
Height: 30′ (9 m).
Diameter: 6″ (15 cm).
Leaves: in 2 rows; 1½–2¾″ (4–7 cm) long, ¾–1½″ (2–4 cm) wide. Narrowly ovate, with *3 main veins* from unequal-sided base; *without teeth* (with wavy teeth on young twigs); *thick;* with short hairy stalks. *Gray-green* and *rough* above, *pale and with soft hairs* and network of raised veins beneath.
Bark: brown; thick, smooth, with many *corky warts* and ridges.
Twigs: brown, slender, with soft hairs.
Flowers: ⅛″ (3 mm) wide; greenish; male and female at base of young leaves in early spring.
Fruit: ¼–5⁄16″ (6–8 mm) in diameter; shiny *light brown or reddish-brown,* 1-seeded drupes; slender-stalked at leaf bases; maturing in autumn and remaining until spring.

Habitat: Along streams in limestone hills.

Range: Edwards Plateau of central Texas, also NE. Coahuila, Mexico; at 1000–2000′ (305–610 m).

A local species closely related to Netleaf Hackberry, which has a wide range in western United States. The Spanish common name, "Palo Blanco," meaning "white tree," is applied in the Southwest to this and other hackberries, which have smooth light-colored bark. Named for its discoverer, Ferdinand Lindheimer (1801–79), a German-born botanical collector and Texas newspaper editor.

169, 551 **Hackberry**
"Sugarberry" "Nettletree"
Celtis occidentalis L.

Description: Tree with rounded crown of spreading or slightly drooping branches, often deformed as bushy growths called witches'-brooms.
Height: 50–90′ (15–27 m).
Diameter: 1½–3′ (0.5–0.9 m).
Leaves: in 2 rows; 2–5″ (5–13 cm) long, 1½–2½″ (4–6 cm) wide, *Ovate, long-pointed;* usually *sharply toothed* except toward *unequal-sided, rounded base; 3 main veins. Shiny green* and smooth (sometimes rough) above, paler and often hairy on veins beneath; turning yellow in autumn.
Bark: gray or light brown; smooth with *corky warts* or ridges, becoming scaly.
Twigs: light brown, slender, mostly hairy, slightly zigzag.
Flowers: ⅛″ (3 mm) wide; greenish; male and female at base of young leaves in early spring.
Fruit: ¼–⅜″ (6–10 mm) in diameter; orange-red to dark *purple* 1-seeded drupes; dry and sweet; slender-stalked at leaf bases; maturing in autumn.

Habitat: Mainly in river valleys, also on upland slopes and bluffs in mixed hardwood forests.

Range: Extreme S. Ontario east to New England, south to N. Georgia, west to NW. Oklahoma, north to North Dakota; local in S. Quebec and S. Manitoba; to 5000′ (1524 m).

Used for furniture, athletic goods, boxes and crates, and plywood. The common name apparently was derived from "hagberry," meaning "marsh berry," a name used in Scotland for a cherry. Many birds, including quail, pheasants, woodpeckers, and cedar waxwings, consume the sweetish fruits. Branches of this and other hackberries may become deformed bushy growths called witches'-brooms produced by mites and fungi. The leaves often bear rounded galls caused by tiny jumping plant lice.

---

170 Netleaf Hackberry

"Western Hackberry" "Sugarberry"
Celtis reticulata Torr.

Description: Shrub or small tree with short trunk and open, spreading crown.
Height: 20–30′ (6–9 m).
Diameter: 1′ (0.3 m).
Leaves: in 2 rows; 1–2½″ (2.5–6 cm) long, ¾–1½″ (2–4 cm) wide. Shape very variable, mostly *ovate;* short- or long-pointed; with *3 main veins* from unequal-sided, rounded, or slightly notched base; *without teeth* or sometimes coarsely saw-toothed; usually *thick.* Dark green and *rough above,* yellow-green with *prominent network of raised veins* and slightly hairy beneath; shedding in late autumn or winter.
Bark: *gray; smooth* or becoming rough and fissured, with *large corky warts.*
Twigs: light brown, slender, slightly zigzag, hairy.
Flowers: ⅛″ (3 mm) wide; greenish; male and female at base of young leaves in early spring.
Fruit: ¼–⅜″ (6–10 mm) in diameter; *orange-red* 1-seeded drupes; sweet, slender-stalked at leaf bases; maturing in autumn.

Habitat: Moist soils usually along streams, in canyons and on hillsides in desert, grassland, and woodland zones.

Range: Central Kansas south to Texas, west to S. California, and north to E. Washington; also N. and central Mexico; usually at 1500–6000′ (457–1829 m).

The native hackberry of western United States, mainly in the Southwest, but extending eastward into the prairie states. The sweetish fruits are eaten by wildlife and were a food source for Indians. The branches often have deformed bushy growths called witches'-brooms, produced by mites and fungi. The leaves bear rounded swollen galls caused by tiny jumping plant lice. Mostly confined to areas with constant water supply.

---

171 Georgia Hackberry

"Dwarf Hackberry" "Upland Hackberry"
Celtis tenuifolia Nutt.

Description: Shrub or small tree with a scattered distribution on exposed rocky uplands.
Height: 25′ (7.6 m).
Diameter: 4″ (10 cm).
Leaves: in 2 rows; ¾–3″ (2–7.5 cm) long, ⅜–1½″ (1–4 cm) wide. *Broadly ovate; 3 main veins* from rounded base; unequal sides; *without teeth* (sharply toothed on young twigs). Gray-green (or darker) and smooth to rough above, hairless or hairy beneath.
Bark: gray; with corky ridges.
Twigs: brownish, slender, often hairy.
Flowers: ⅛″ (3 mm) wide; greenish; male and female at base of young leaves in early spring.
Fruit: ¼–5/16″ (6–8 mm) in diameter; orange to *brown* or red 1-seeded drupes; dry and sweet; slender-stalked at leaf bases; maturing in late summer or autumn.

Habitat: Rocky uplands and especially limestone bluffs and ridges; in hardwood forests.

Range: Maryland south to N. Florida, west to extreme E. Texas, and north to SE. Kansas; local north to extreme S. Ontario; to 1500′ (457 m).

Although the Latin species name means "thin leaf," the leaves vary in thickness. The fruit is consumed by songbirds and other wildlife.

---

167 Water-elm
"Planertree"
Planera aquatica J. F. Gmel.

Description: Elmlike tree with broad crown of spreading branches.

Height: 40′ (12 m).
Diameter: 1′ (0.3 m).
Leaves: in 2 rows; 2–2½″ (5–6 cm) long, ¾–1″ (2–2.5 cm) wide. *Ovate,* short-pointed, base rounded with unequal sides; *wavy-toothed* with *blunt gland-tipped teeth; many straight side veins;* slightly thickened. Dull dark green and rough above, paler beneath.
Bark: light brown or gray; thin, shedding in *large scales* and *exposing red-brown inner layers.*
Flowers: ⅛″ (3 mm) wide; greenish; in early spring. 1–3 male flowers from bud; 1–3 female or bisexual flowers at leaf base of same twig.
Fruit: ⅜″ (10 mm) long; an elliptical drupe; *warty or tubercled,* dry, light brown, 1-seeded, short-stalked; maturing in early spring and not opening.

Habitat: Wet soil of riverbanks and swamps, especially where flooded annually.

Range: SE. North Carolina to N. Florida, west to E. Texas, north to extreme S. Illinois; to 500′ (152 m).

This distinctive and uncommon small tree is the only species of its genus; however, fossil relatives have been found in Eurasia. Dedicated to Johann

Jakob Planer (1743–89), German botanist and professor of medicine.

165, 497 **Winged Elm**
"Cork Elm" "Wahoo"
Ulmus alata Michx.

Description: Tree with short trunk and open rounded crown.
Height: 40–80′ (12–24 m).
Diameter: 1½′ (0.5 m).
Leaves: in *2 rows;* 1¼–2½″ (3–6 cm) long. Elliptical; often slightly curved with sides unequal; doubly saw-toothed; with yellow midvein and many straight side veins; thick and firm. *Dark green* and *hairless above,* with *soft hairs beneath;* turning yellow in autumn.
Bark: light brown; thin, irregularly furrowed.
Twigs: brownish, slender, often with 2 *broad corky wings.*
Flowers: ⅛″ (3 mm) wide; greenish; clustered along twigs in early spring.
Fruit: ⅜″ (10 mm) long; *elliptical reddish flat* 1-seeded *keys* (samaras); *hairy,* with *narrow wing* having 2 curved points at tip; maturing in early spring.

Habitat: Dry uplands including abandoned fields, also in moist valleys; in hardwood forests.

Range: S. Virginia, south to central Florida, west to central Texas, and north to central Missouri; to 2000′ (610 m).

In the 18th and 19th centuries, the fibrous inner bark was made into rope for fastening covers of cotton bales. The common and Latin species names refer to the distinctive broad, corky wings present on some twigs; "Wahoo" was the Creek Indian name.

162, 373, 498 **American Elm**
"White Elm" "Soft Elm"
Ulmus americana L.

Description: Large, handsome, graceful tree, often with enlarged buttresses at base, usually forked into *many spreading branches, drooping at ends,* forming a very broad, rounded, flat-topped or vaselike crown, often wider than high.
Height: 100′ (30 m).
Diameter: 4′ (1.2 m), sometimes much larger.
Leaves: in *2 rows;* 3–6″ (7.5–15 cm) long, 1–3″ (2.5–7.5 cm) wide. Elliptical, abruptly long-pointed, base rounded with sides unequal; doubly saw-toothed; with many straight parallel side veins; thin. *Dark green* and *usually hairless* or slightly rough above, paler and usually with soft hairs beneath; turning bright yellow in autumn.
Bark: light gray; deeply furrowed into broad, forking, scaly ridges.
Twigs: brownish, slender, hairless.
Flowers: ⅛″ (3 mm) wide; greenish; clustered along twigs in early spring.
Fruit: ⅜–½″ (10–12 mm) long; *elliptical flat* 1-seeded *keys* (samaras), with *wing hairy on edges,* deeply notched with points curved inward; long-stalked; maturing in early spring.

Habitat: Moist soils, especially valleys and flood plains; in mixed hardwood forests.

Range: SE. Saskatchewan east to Cape Breton Island, south to central Florida, and west to central Texas; to 2500′ (762 m).

This well-known, once abundant species, familiar on lawns and city streets, has been ravaged by the Dutch Elm disease, caused by a fungus introduced accidentally about 1930 and spread by European and native elm bark beetles. The wood is used for containers, furniture, and paneling.

205 Cedar Elm
"Basket Elm" "Southern Rock Elm"
Ulmus crassifolia Nutt.

Description: Tree with rounded crown of drooping branches and the *smallest leaves of any native elm.*
Height: 80′ (24 m).
Diameter: 2′ (0.6 m).
Leaves: in *2 rows;* 1–2″ (2.5–5 cm) long, ½–1″ (1.2–2.5 cm) wide. Elliptical or lance-shaped, blunt or short-pointed at tip, base rounded with unequal sides; coarsely *saw-toothed* with rounded teeth; *thick and slightly leathery;* with yellow midvein and straight side veins. Shiny dark green and *rough above,* with *soft hairs beneath;* turning bright yellow in late autumn.
Bark: light brown; furrowed into broad, scaly ridges.
Twigs: brownish, slender, hairy, often with *corky wings.*
Flowers: ⅛″ (3 mm) wide; greenish; short-stalked; at leaf bases in *late summer.*
Fruit: ⅜–½″ (10–12 mm) long; *elliptical flat* green 1-seeded *keys* (samaras), with narrow wing deeply notched at tip, covered with *soft white hairs;* short-stalked at leaf bases; maturing in *autumn.*

Habitat: Moist soils along streams and also upland limestone hills; with various hardwoods.

Range: Extreme SW. Tennessee south to Mississippi, west to S. Texas and extreme NE. Mexico, and north to S. Oklahoma; local in N. Florida; to 1500′ (457 m).

The common native elm in east Texas where it is planted for shade. Called Cedar Elm because it is often found with Ashe Juniper, which is locally called "cedar." The Latin species name means "thick leaf."

168 **Chinese Elm**
Ulmus parvifolia Jacq.

Description: Introduced tree with dense, broad, rounded crown of spreading branches and *small leaves.*
Height: 50′ (15 m).
Diameter: 1½′ (0.5 m).
Leaves: in *2 rows;* ¾–2″ (2–5 cm) long, ⅜–¾″ (10–19 mm) wide. Elliptical; *saw-toothed; slightly thickened. Shiny dark green* above, paler and hairy when young and in vein angles beneath; turning reddish or purplish in autumn or remaining nearly evergreen in warm climates.
Bark: *mottled brown;* smooth; shedding in irregular, *thin flakes* and exposing reddish-brown inner bark.
Twigs: slender, slightly zigzag, hairy.
Flowers: ⅛″ (3 mm) wide; greenish; clustered at base of leaves in *autumn.*
Fruit: ⅜″ (10 mm) long; *elliptical flat* 1-seeded *keys,* with *broad pale yellow wing;* maturing in *autumn.*

Habitat: Moist soils in humid, temperate regions.

Range: Native of China, Korea, and Japan. Planted across the United States, especially in Gulf and Pacific regions.

Fast-growing and hardy, Chinese Elm is a handsome ornamental with showy bark and a compact crown and is also cultivated for shade and shelterbelts. It should not be confused with Siberian Elm (*Ulmus pumila* L.), sometimes erroneously called Chinese Elm.

---

163 **English Elm**
Ulmus procera Salisb.

Description: Large, introduced shade tree with tall, straight trunk and dense, broad, rounded crown of spreading and nearly upright branches.
Height: 80′ (24 m).

Diameter: 3′ (0.9 m).
Leaves: in *2 rows;* 2–3¼″ (5–8 cm) long, 1¼–2″ (3–5 cm) wide. Broadly elliptical, abruptly long-pointed at tip, base with *very unequal* sides; *doubly saw-toothed;* with 10–12 straight veins on each side. Dark green and *rough* above, paler and covered with *soft hairs* and with tufts in vein angles beneath; remaining green late, turning yellow and shedding late in autumn.
Bark: gray; deeply furrowed into rectangular plates.
Twigs: brown, slender, densely covered with hairs when young; sometimes with corky wings.
Flowers: ⅛″ (3 mm) wide; dark red; clustered along twigs in early spring.
Fruit: ½″ (12 mm) long; *rounded flat* greenish *keys* (samaras), *hairless,* with 1 seed near narrow notch at tip; short-stalked; maturing in spring.

Habitat: Scattered in moist soil, roadsides, thickets, and forest borders, spreading from cultivation.

Range: Native of England and W. Europe. Widely planted since colonial times and escaping in northeastern and Pacific states.

English Elm is propagated by suckers from roots, which often encircle the trunk or appear at a distance. Several cultivated varieties differ in habit and leaf color (including white-striped, dark purple, and yellowish forms). Widely planted in England, where trees reach great size with age.

166 Siberian Elm
"Asiatic Elm" "Dwarf Elm"
Ulmus pumila L.

Description: Small to medium-sized, introduced tree with open, rounded crown of slender, spreading branches.
Height: 60′ (18 m).

Diameter: 1½′ (0.5 m), usually smaller.
Leaves: ¾–2″ (2–5 cm) long, ½–1″ (1.2–2.5 cm) wide. *Narrowly elliptical,* blunt-based; *saw-toothed;* with many straight side veins; slightly thickened. Dark green above, paler and nearly hairless beneath; turning yellow in autumn.
Bark: gray or brown; rough, furrowed.
Flowers: ⅛″ (3 mm) wide; greenish; in clusters; in early spring.
Fruit: ⅜–⅝″ (10–15 mm) long; several clustered *rounded flat* 1-seeded *keys* (samaras), bordered with *broad notched wing;* in early spring.

Habitat: Dry regions, tolerant of poor soils and city smoke; also scattered in moist soils along streams.

Range: Native from Turkestan to E. Siberia and N. China. Naturalized from Minnesota south to Kansas and west to Utah; at 1000–5000′ (305–1524 m).

A fast-growing tree in dry regions such as the Great Plains, but less suited to moist regions. Hardy and resistant to the Dutch Elm disease. This species has been known erroneously as Chinese Elm. However, Chinese Elm (*Ulmus parvifolia* Jacq.), a cultivated species in this country, has brown, smoothish, mottled bark, small leaves turning purplish in autumn, and elliptical flattened fruits borne in autumn.

160, 500 **Slippery Elm**
"Red Elm" "Soft Elm"
Ulmus rubra Muhl.

Description: Tree with broad, open, flat-topped crown of spreading branches and large rough leaves.
Height: 70′ (21 m).
Diameter: 2–3′ (0.6–0.9 m).
Leaves: in *2 rows;* 4–7″ (10–18 cm) long, 2–3″ (5–7.5 cm) wide.

Elliptical, abruptly long-pointed, base rounded with *sides very unequal;* doubly saw-toothed with many straight parallel side veins; *thick.* Green to dark green and *very rough above, densely covered with soft hairs beneath;* turning dull yellow in autumn.
Bark: dark brown; deeply furrowed; *inner bark mucilaginous.*
Twigs: brownish, stout, *hairy.*
Flowers: ⅛″ (3 mm) wide; greenish; numerous; short-stalked along twigs in early spring.
Fruit: ½–¾″ (12–19 mm) long; *nearly round flat* 1-seeded *keys* (samaras); with *light green broad hairless wing,* slightly notched at tip; maturing in spring.

Habitat: Moist soils, especially lower slopes and flood plains, but often on dry uplands; in hardwood forests.

Range: S. Ontario east to extreme S. Quebec and SW. Maine, south to NW. Florida, west to central Texas, and north to SE. North Dakota; to 2000′ (610 m).

The thick, slightly fragrant, edible, gluelike inner bark is dried and afterwards moistened for use as a cough medicine or as a poultice. This "slippery" inner bark (found by chewing through the outer bark of a twig) is helpful in identification. The Latin species name refers to the large brown buds covered with rust-colored hairs.

---

164 **September Elm**
"Red Elm"
Ulmus serotina Sarg.

Description: Medium-sized tree with broad crown of spreading and drooping branches.
Height: 70′ (21 m).
Diameter: 2′ (0.6 m).
Leaves: in *2 rows;* 2–3½″ (5–9 cm) long, 1–1¾″ (2.5–4.5 cm) wide.

Elliptical, long-pointed, base rounded with sides unequal; coarsely, doubly saw-toothed; with prominent yellow midvein and many straight side veins; thin. *Shiny green* and hairless above, pale and finely hairy on veins beneath; turning orange-yellow in autumn.
Bark: light brown; thin, fissured into broad scaly ridges.
Twigs: brown, slender, often with corky wings.
Flowers: ⅛″ (3 mm) wide; greenish; in long drooping clusters in *autumn.*
Fruit: ⅜–½″ (10–12 mm) long; *elliptical flat* 1-seeded *keys* (samaras); hairy, with narrow *wing fringed with long whitish hairs,* deeply notched, short-stalked at leaf bases; maturing in *autumn.*

Habitat: Moist soils along streams and on limestone hills; scattered with hardwoods.

Range: Local in Kentucky south to NW. Georgia and west to E. Oklahoma; at 500–1500′ (152–457 m).

Although similar to American Elm, it is distinguishable by its flowers, which appear in September. Both the Latin species name, meaning "late," and the common name refer to this late blooming; most elms flower in early spring.

161, 499 **Rock Elm**
"Cork Elm"
Ulmus thomasii Sarg.

Description: Tree with usually straight trunk and *narrow, cylindrical crown* of short, drooping branches.
Height: 50–100′ (15–30 m).
Diameter: 1½–3′ (0.5–0.9 m).
Leaves: 2–3½″ (5–9 cm) long, ¾–2″ (2–5 cm) wide. Elliptical, abruptly long-pointed, base mostly rounded with sides unequal; doubly saw-toothed;

with many straight, parallel side veins; thick. *Shiny dark green and hairless above, pale and with soft hairs beneath;* turning bright yellow in autumn.
Bark: gray; deeply furrowed into broad, scaly ridges.
Twigs: often with *3–4 irregular corky wings.*
Flowers: ⅛″ (3 mm) wide; greenish; in long drooping clusters in early spring.
Fruit: ⅜–¾″ (10–19 mm) long; *elliptical flat* 1-seeded *keys* (samaras); finely *hairy,* with *broad wing* notched at tip; long-stalked in drooping clusters at leaf bases; maturing in spring.

Habitat: Moist to dry uplands, especially rocky ridges and limestone bluffs, also flatlands; in hardwood forests.

Range: S. Ontario, extreme S. Quebec, and W. New England south to Tennessee, west to extreme NE. Kansas, and north to Minnesota; at 200–2500′ (61–762 m).

The valuable wood is hard, tough, and difficult to split, making it particularly suitable for agricultural implements and tool handles. In the 19th century, this durable timber was exported to England for construction of wooden battleships and sailing vessels.

---

153 Japanese Zelkova
Zelkova serrata (Thunb.) Mak.

Description: Large, introduced tree with short trunk and broad, rounded crown of many spreading branches.

Height: 70′ (21 m).
Diameter: 2′ (0.6 m).
Leaves: in *2 rows;* 1–3½″ (2.5–9 cm) long, ¾–1¾″ (2–4.5 cm) wide. *Ovate* or elliptical; *sharply saw-toothed;* with *8–14 straight parallel veins* on each side of midvein; short-stalked. *Dark green* and *rough* above, paler and usually hairless beneath; turning yellow to reddish-yellow in autumn.

Bark: gray; smooth, becoming scaly.
Twigs: slender, mostly hairless.
Flowers: ⅛″ (3 mm) long; greenish; male clustered at base of new lower leaves, 1 or few female at upper leaves; almost stalkless; in early spring.
Fruit: 3/16″ (5 mm) long and wide; *egg-shaped* drupes; oblique, almost stalkless; maturing in autumn.

Habitat: Moist soils in humid temperate regions.

Range: Native of Japan. Planted across the United States.

Japanese Zelkova resembles its relatives, the elms, which have the doubly saw-toothed leaf edges and winged fruits. It has been suggested as a substitute for American Elm, as it is resistant to the Dutch Elm disease. Propagated by seeds, layers, and grafts, it grows rapidly. In Japan, the wood is an important timber and valued for making furniture, lacquerware, and trays; the plants are often used for bonsai.

MULBERRY FAMILY
(Moraceae)

Trees and shrubs, including mulberries (*Morus*), sometimes herbs, with white sap or latex usually present and often abundant. About 1400 species mostly in tropical and subtropical areas, a few in temperate regions; 5 native and 3 naturalized tree species.
Leaves: alternate, often in 2 rows, simple; without teeth, toothed, or lobed; pinnate- or palmate-veined.
Twigs: with 1–2 large stipules forming long-pointed bud, soon falling and leaving scars or rings at nodes.
Flowers: tiny, often greenish, male and female on the same plant or separate plants, usually numerous and crowded, often in spikes or heads. Calyx usually with 4 sepals or lobes and no petals; male with 4–1 opposite stamens, and female with 1 pistil.
Fruit: a drupe or achene, often multiple and fleshy, sometimes edible.

---

236 Paper-mulberry
Broussonetia papyrifera (L.) Vent.

Description: Introduced, ornamental and shade tree with an irregular trunk, broad, spreading, rounded crown of gray-green foliage, and with *milky sap.*

Height: 50′ (15 m).
Diameter: 1½′ (0.5 m).
Leaves: alternate or *sometimes opposite;* 3–8″ (7.5–20 cm) long and nearly as wide. *Broadly ovate; 3 main veins* from notched base; *coarsely saw-toothed;* often *with 3 or more deep lobes;* long-stalked. *Gray-green* and with *rough hairs* above, paler and with soft hairs beneath.
Bark: *light gray; smoothish* or fissured.
Twigs: gray-green, hairy.
Flowers: tiny; greenish; male and female on separate trees in spring; male

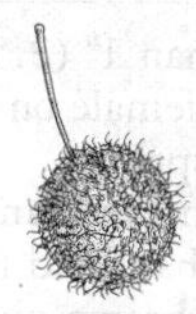

in cylindrical catkins, female clustered in balls.
Fruit: ¾″ (19 mm) in diameter; a *round* mulberry; showy, *orange,* with many individual pointed *red fruits;* maturing in early summer.

Habitat: Hardy in cities, tolerating dust, smoke, heat, and poor soil.

Range: Native of E. Asia. Planted and naturalized locally from S. New England south to Florida, west to Texas and Missouri.

The common and Latin species names both refer to the manufacture in the Orient of paper from the fibrous inner bark. Cloth was made from the bark in the South Pacific and Hawaii. Paper-mulberry spreads from root sprouts, often forming thickets along roadsides. Although seldom produced, the fruit is showy.

71, 391, 573 **Osage-orange**
"Bodark" "Hedge-apple"
Maclura pomifera (Raf.) Schneid.

Description: Medium-sized, spiny tree with short, often crooked trunk, broad rounded or irregular crown of spreading branches, single *straight stout spines* at base of some leaves, and *milky sap.*

Height: 50′ (15 m).
Diameter: 2′ (0.6 m).
Leaves: 2½–5″ (6–13 cm) long, 1½–3″ (4–7.5 cm) wide. Narrowly *ovate,* long-pointed; *not toothed;* hairless. Shiny dark green above, paler beneath; turning yellow in autumn.
Bark: gray or brown; thick, deeply furrowed into narrow forking ridges; *inner bark of roots orange,* separating into thin papery scales.
Twigs: brown, stout, with single spine ¼–1″ (0.6–2.5 cm) long at some nodes and short twigs or spurs.
Flowers: tiny; greenish; crowded in

rounded clusters less than 1″ (2.5 cm) in diameter; male and female on separate trees in early spring.
Fruit: 3½–5″ (9–13 cm) in diameter; a *heavy yellow-green ball,* hard and fleshy, containing many light brown nutlets; maturing in autumn and soon falling.

Habitat: Moist soils of river valleys.

Range: The native range uncertain. SW. Arkansas to E. Oklahoma and Texas; widely planted and naturalized in eastern and northwestern states.

Rows of these spiny plants served as fences in the grassland plains before the introduction of barbed wire. The name "Bodark" is from the French *bois d'arc,* meaning "bow wood," referring to the Indians' use of the wood for archery bows. It is also used for fenceposts. Early settlers extracted a yellow dye for cloth from the root bark. The fruit is eaten by livestock, which has given rise to yet another common name, "Horse-apple."

154, 237, 546, 607

White Mulberry
"Silkworm Mulberry" "Russian Mulberry"
Morus alba L.

Description: Naturalized small tree with rounded crown of spreading branches, *milky sap,* and edible mulberries.

Height: 40′ (12 m).
Diameter: 1′ (0.3 m).
Leaves: in 2 rows; 2½–7″ (6–18 cm) long, 2–5″ (5–13 cm) wide. Broadly ovate but variable in shape; with 3 main veins from rounded or notched base; *coarsely toothed; often divided into 3 or 5 lobes;* long-stalked. Shiny green above, paler and slightly hairy beneath.
Bark: light brown; smoothish, becoming furrowed into scaly ridges.
Twigs: light brown, slender.
Flowers: tiny; greenish; crowded in

short clusters; male and female on same or separate trees in spring.
Fruit: ⅜–¾″ (10–19 mm) long; a cylindrical *mulberry;* purplish, pinkish, or white; composed of many tiny beadlike 1-seeded fruits, sweet and juicy, edible; in late spring.

Habitat: Hardy in cities, drought-resistant, and adapted to dry, warm areas.

Range: Native of China. Widely cultivated across the United States; naturalized in the East and in the Pacific states.

White Mulberry has been cultivated for centuries, the leaves serving as the main food of silkworms. Introduced long ago in the southeastern United States, where silk production was not successful. It grows rapidly and produces abundant berries that are enjoyed by birds as well as by many people. The trees spread like weeds in cities, where the berries litter sidewalks. Varieties include a hardy one for shelterbelts, another with drooping or weeping foliage, and a fruitless form.

---

183 Texas Mulberry

"Mexican Mulberry" "Mountain Mulberry"
Morus microphylla Buckl.

Description: Small tree or clump-forming shrub, with variable leaf shape, milky sap and edible mulberries.

Height: 20′ (6 m).
Diameter: 8″ (20 cm).
Leaves: in 2 rows; 1–2½″ (2.5–6 cm) long, ¾–1¼″ (2–3 cm) wide. *Ovate* but variable in shape; sometimes 3-lobed; pointed at tip; with *3 main veins* from rounded or notched base; coarsely saw-toothed. Dark green and rough above, paler and with soft hairs beneath; turning yellow in autumn.
Bark: light gray; smooth, becoming furrowed and scaly.

Twigs: brown, slender; hairy when young.
Flowers: tiny; greenish; crowded in short clusters ⅜–¾″ (10–19 mm) long; male and female on separate trees in spring with leaves.
Fruit: ½″ (12 mm) long; a cylindrical *mulberry;* red to purple or black; composed of *many tiny beadlike 1-seeded fruits,* sweet and juicy, edible; in late spring.

Habitat: Moist soils, mostly along streams, canyons, washes, and rocky slopes; in foothills and mountains, woodland, upper desert, and grassland zones.

Range: S. Oklahoma and Texas west to Arizona; also N. Mexico; at 1000–6000′ (305–1829 m).

The small fruit is eaten by wildlife and was a food of southwestern Indians. It was introduced at the bottom of the Grand Canyon by Havasupai Indians, probably in prehistoric times from wild trees further south.

---

157 Red Mulberry
"Moral"
Morus rubra L.

Description: Medium-sized tree with short trunk, broad rounded crown, and milky sap.
Height: 60′ (18 m).
Diameter: 2′ (0.6 m).
Leaves: in 2 rows; 4–7″ (10–18 cm) long, 2½–5″ (6–13 cm) wide. Ovate; abruptly long-pointed; with *3 main veins* from often unequal base, coarsely *saw-toothed; often with 2 or 3 lobes* on young twigs. Dull dark green and rough above, with soft hairs beneath; turning yellow in autumn.
Bark: brown; fissured into scaly plates.
Twigs: brown, slender.
Flowers: *tiny,* about ⅛″ (3 mm) long; crowded in narrow clusters; male and

female on same or separate trees; in spring when leaves appear.

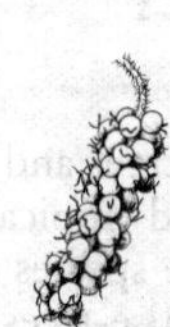

Fruit: 1–1¼″ (2.5–3 cm) long; a cylindrical *mulberry; red to dark purple;* composed of *many tiny beadlike 1-seeded fruits,* sweet and juicy, edible; in late spring.

Habitat: Moist soils in hardwood forests.

Range: S. Ontario east to Massachusetts, south to S. Florida, west to central Texas and north to SE. Minnesota; to 2000′ (610 m).

The wood is used locally for fenceposts, furniture, interior finish, and agricultural implements. People, domestic animals, and wildlife (especially songbirds) eat the berries. Choctaw Indians wove cloaks from the fibrous inner bark of young mulberry shoots.

MAGNOLIA FAMILY
(Magnoliaceae)

About 200 species of trees and shrubs in warm temperate and tropical regions. 11 native tree species in North America, including anise-trees (*Illicium*) and Yellow-poplar (*Liriodendron*) as well as magnolias (*Magnolia*).
Leaves: alternate, simple, not toothed, mostly with large stipules that form the bud and leave ring scars at nodes.
Flowers: often large and showy, frequently solitary, bisexual, regular, with 3 to many sepals, 6 to many commonly white petals, many stamens in spiral on elongated axis, and many simple pistils.
Fruit: many follicles or berries often united like a cone.

59, 365 **Florida Anise-tree**
"Stinkbush" "Star Anise"
Illicium floridanum Ellis

Description: Evergreen shrub or small tree, aromatic; *flowers with strong, unpleasant, fishy odor;* slender, crooked trunks, often leaning; small, open, rounded crown; and curious, *wheel-shaped fruit.*
Height: 10–20′ (3–6 m).
Diameter: 3″ (7.5 cm).
Leaves: *evergreen;* 2½–6″ (6–15 cm) long, 1–2″ (2.5–5 cm) wide. *Elliptical* to lance-shaped; long-pointed at ends; not toothed; *leathery;* hairless; crowded at twig ends. Dark green above, paler with gland-dots beneath.
Bark: gray or brown; smooth, becoming fissured.
Twigs: brown, slender, upright, hairless.
Flowers: nearly 2″ (5 cm) wide; *20–30 dark red or purple very narrow sepals,* with strong fishy odor; showy but often hidden; solitary on long, curved-down stalk at leaf base; in spring.

Fruit: 1–1¼″ (2.5–3 cm) wide; half-round, spreading like a *wheel or star;* composed of *11–15 individual dry fruits,* narrow and pointed, opening on top, and 1-seeded; maturing in summer.

Habitat: Wet soils of streams, ravines, and swamps.

Range: NW. Florida to central Alabama and SE. Louisiana; to 500′ (152 m).

This southeastern species has been planted along the coast north to Philadelphia, where it is a deciduous shrub. Florida Anise-tree is of no economic use; however, the unripe star-shaped fruit of the related Chinese species, Star Anise (*Illicium verum* Hook. f.), is used as a spice in cooking, especially in the Orient. The oil distilled from the fruit and foliage of that species is used in medicine and for flavoring alcoholic drinks.

60 Yellow Anise-tree

"Small-flower Anise-tree" "Star Anise"
Illicium parviflorum Michx. ex Vent.

Description: Evergreen, aromatic, large spreading shrub or small tree with *wheel-shaped fruits.*

Height: 23′ (7 m), in cultivation to 40′ (12 m).

Diameter: 6″ (15 cm).

Leaves: *evergreen;* alternate or in 3's or 4's; 2½–5″ (6–13 cm) long, 1–2″ (2.5–5 cm) wide. Elliptical, *rounded or blunt at tip,* short-pointed at base; not toothed; thin; dull green.

Bark: gray; smooth, becoming slightly fissured.

Twigs: brown, slender, hairless.

Flowers: ½″ (12 mm) wide; *12–15 broad yellow sepals;* 1–3 on slender stalks at leaf base; in late spring.

Fruit: ¾–1″ (2–2.5 cm) wide; half-round, spreading like a *wheel or star;*

composed of *10–13 individual dry fruits,* narrow and pointed, opening on top, and 1-seeded.

Habitat: Wet soils along streams.

Range: Local in 5 counties of central Florida at headwaters of St. Johns River; below 100′ (30 m).

Rare in the wild, this species is protected within the Ocala National Forest in Florida. Sometimes planted as an ornamental in southeastern gardens.

267, 390, 490, 618

Yellow-poplar

"Tuliptree" "Tulip-poplar"

Liriodendron tulipifera L.

Description: One of the tallest and most beautiful eastern hardwoods, with a long, straight trunk, a narrow crown that spreads with age, and large showy flowers resembling tulips or lilies.

Height: 80–120′ (24–37 m).
Diameter: 2–3′ (0.6–0.9 m), sometimes much larger.
Leaves: 3–6″ (7.5–15 cm) long and wide. Blades of unusual shape, with *broad tip* and base nearly straight *like a square,* and with *4 or sometimes 6 short-pointed paired lobes;* hairless; long-stalked. Shiny dark green above, paler beneath; turning yellow in autumn.
Bark: dark gray; becoming thick and deeply furrowed.
Twigs: brown, stout, hairless, with *ring scars at nodes.*
Flowers: 1½–2″ (4–5 cm) long and wide; *cup-shaped,* with 6 rounded green petals (orange at base); solitary and upright at end of leafy twig; in spring.
Fruit: 2½–3″ (6–7.5 cm) long; *conelike;* light brown; composed of many overlapping 1- or 2-seeded nutlets 1–1½″ (2.5–4 cm) long (including *narrow wing*); shedding from upright axis in autumn; the axis persistent in winter.

Habitat: Moist well-drained soils, especially

valleys and slopes; often in pure stands.

Range: Extreme S. Ontario east to Vermont and Rhode Island, south to N. Florida, west to Louisiana, and north to S. Michigan; to 1000′ (305 m) in north and to 4500′ (1372 m) in southern Appalachians.

Introduced into Europe from Virginia by the earliest colonists and grown also on the Pacific Coast. Very tall trees with massive trunks existed in the primeval forests but were cut for the valuable soft wood. Pioneers hollowed out a single log to make a long, lightweight canoe. One of the chief commercial hardwoods, Yellow-poplar is used for furniture, as well as for crates, toys, musical instruments, and pulpwood.

93, 389 **Cucumbertree**
"Cucumber Magnolia"
Magnolia acuminata L.

Description: Tree with straight trunk and narrow crown of short upright to spreading branches.
Height: 60–80′ (18–24 m).
Diameter: 2′ (0.6 m).
Leaves: 5–10″ (13–25 cm) long, often larger; 3–6″ (7.5–15 cm) wide. *Elliptical* or ovate; abruptly short-pointed; edges straight or wavy. Green and becoming hairless above, paler and *often with soft hairs* beneath; turning dull yellow or brown in autumn.
Bark: dark brown; furrowed into narrow scaly forking ridges.
Twigs: stout; with *ring scars at nodes;* young twigs and buds densely hairy.
Flowers: 2½–3½″ (6–9 cm) wide; *bell-shaped,* with 6 large *greenish-yellow* or bright yellow petals; solitary at end of twig; in spring.
Fruit: 2½–3″ (6–7.5 cm) long; *conelike;* oblong; dark red; composed of many

pointed fruits that split open, each with 2 seeds that hang down on threads; maturing in late summer.

Habitat: Moist soils of mountain slopes and valleys in mixed forests.

Range: Extreme S. Ontario and W. New York south to NW. Florida, west to Louisiana, and north to Missouri; at 100–4000′ (30–1219 m).

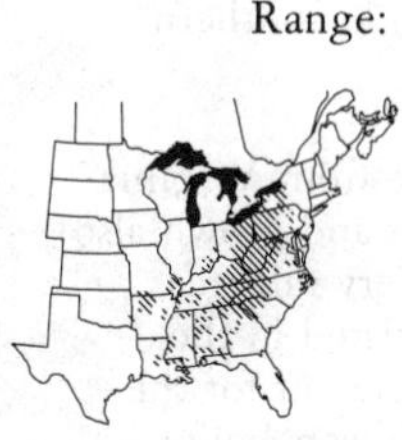

The common name refers to the shape of the fruit, and the Latin species name to the pointed leaves.

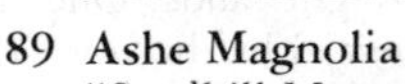

89 **Ashe Magnolia**
"Sandhill Magnolia"
Magnolia ashei Weatherby

Description: Shrub or small tree with broadly conical crown of many, slender, upright branches, and with *very large leaves and flowers.*

Height: 13–30′ (4–9 m).
Diameter: 4″ (10 cm).
Leaves: 8–24″ (20–61 cm) long, 4–11″ (10–28 cm) wide. Reverse ovate, blunt or short-pointed at tip, *broadest beyond middle,* tapering to *notched 2-lobed base;* edges straight or wavy; thin. Shiny light green above, *whitish or silvery* and often hairy beneath. Leafstalks 2–4″ (5–10 cm) long.
Bark: gray-brown; smooth or slightly rough.
Twigs: stout; with *silvery hairs* when young; with *ring scars at nodes.*
Flowers: 12″ (30 cm) wide; *cup-shaped* corolla of 6 curved *creamy white petals;* fragrant; in late spring.
Fruit: 1¾–2¾″ (4.5–7 cm) long; *conelike; narrowly cylindrical;* rose-red to brown; composed of many separate pointed 2-seeded fruits that split open in autumn.

Habitat: Upland bluffs in hardwood forests.

Range: Only in NW. Florida at 100–200′ (30–61 m).

This rare species of very local distribution is closely related to Bigleaf Magnolia, which is taller and has larger leaves and flowers. Named for its discoverer, William Willard Ashe (1872–1932), pioneer forester of the United States Forest Service.

91, 434 **Fraser Magnolia**
"Mountain Magnolia" "Umbrella-Tree"
Magnolia fraseri Walt.

Description: A tree often branched near the base, with an open crown of spreading branches, *large leaves,* and *very large flowers.*

Height: 30–70′ (9–21 m).
Diameter: 1–2′ (0.3–0.6 m).
Leaves: crowded; 8–18″ (20–46 cm) long, 5–8″ (13–20 cm) wide. Usually reverse ovate (sometimes ear-shaped); *broadest beyond middle,* short-pointed at tip, with *2 large pointed lobes at narrow base;* not toothed; hairless. Bright green above, pale and whitish beneath.
Bark: light gray; smooth or becoming scaly; thin.
Twigs: brown, stout, with *ring scars at nodes.*
Flowers: 8–10″ (20–25 cm) wide; 6–9 cream-colored petals; fragrant; solitary at end of twig; in spring.
Fruit: 4–5″ (10–13 cm) long; *conelike;* oblong; rose-red; composed of many long-pointed hairless 2-seeded fruits that split open in early autumn.

Habitat: Moist soils of mountain valleys in hardwood forests.

Range: W. Virginia and West Virginia south to N. Georgia; at 800–5000′ (244–1524 m).

Named for John Fraser (1750–1811), Scottish botanist, who introduced many North American plants to Europe. This species, of scattered distribution, is fairly common in the Great Smoky

Mountains National Park. Planted as an ornamental for the large flowers and coarse foliage.

58, 432, 565

Southern Magnolia

"Evergreen Magnolia" "Bull-bay"
Magnolia grandiflora L.

Description: One of the most beautiful native trees, evergreen with straight trunk, conical crown, and very fragrant, *very large, white flowers.*

Height: 60–80′ (18–24 m).

Diameter: 2–3′ (0.6–0.9 m).

Leaves: *evergreen;* 5–8″ (13–20 cm) long, 2–3″ (5–7.5 cm) wide. *Oblong or elliptical; thick and firm* with edges slightly turned under. *Shiny bright green above, pale and with rust-colored hairs beneath.* Stout leafstalks with rust-colored hairs.

Bark: dark gray; smooth, becoming furrowed and scaly.

Twigs: covered with *rust-colored hairs* when young; with *ring scars at nodes;* ending in *buds* also covered with *rust-colored hairs.*

Flowers: 6–8″ (15–20 cm) wide; *cup-shaped;* 3 *white* sepals and 6 or more petals; very fragrant; solitary at end of twig; in late spring and summer.

Fruit: 3–4″ (7.5–10 cm) long; *conelike;* oblong; pink to brown; covered with rust-colored hairs; composed of many separate short-pointed 2-seeded fruits that split open in early autumn.

Habitat: Moist soils of valleys and low uplands with various other hardwoods.

Range: E. North Carolina to central Florida and west to E. Texas; to 400′ (122 m).

Planted around the world in warm temperate and subtropical regions, it is a popular ornamental and shade tree, hardy north to Philadelphia. Several horticultural varieties have been developed. Principal uses of the wood are furniture, boxes, cabinetwork, and doors. The dried leaves are used by florists in decorations.

90, 430 **Bigleaf Magnolia**
"Silverleaf Magnolia" "Umbrella-tree"
Magnolia macrophylla Michx.

Description: The tree with the *largest flowers* and the *largest leaves of all native North American species* (except for tropical palms) and a broad, rounded crown of stout, spreading branches.

Height: 30–40′ (9–12 m).
Diameter: 1½′ (0.5 m).
Leaves: 15–30″ (38–76 cm) long, 6–10″ (15–25 cm) wide. Reverse ovate, *broadest beyond middle,* mostly blunt at tip; notched with *2 rounded lobes at base;* not toothed. Bright green above, with *silvery hairs beneath.* Stout, hairy leafstalks, 3–4″ (7.5–10 cm) long.
Bark: light gray; smooth, thin.
Twigs: stout, hairy; with *large leaf-scars at nodes* and ending in *large buds covered with white hairs.*
Flowers: 10–12″ (25–30 cm) wide; *cup-shaped* with 6 white petals with spot at base; fragrant; in late spring and early summer.
Fruit: 2½–3″ (6–7.5 cm) long; *conelike; elliptical* or nearly round; rose-red; composed of many separate short-pointed 2-seeded hairy fruits; maturing in autumn.

Habitat: Moist soil of valleys, especially ravines; in understory of hardwood forests.
Range: Central North Carolina south to W. Georgia and west to Louisiana; local in S. Ohio, NE. Arkansas, and SE. South Carolina.

Planted as an ornamental north to Massachusetts. However, in windy places the giant leaves become torn and unsightly. The "queenliest of all the deciduous magnolias" was named by the French naturalist and explorer André Michaux (1746–1802), who discovered this rare local tree near Charlotte, North Carolina, in 1789.

92 **Pyramid Magnolia**
"Southern Cucumbertree"
"Mountain Magnolia"
Magnolia pyramidata Bartr.

Description: Tree with *large leaves and flowers* and pyramid-shaped crown of upright branches.
Height: 30–40′ (9–12 m).
Diameter: 1′ (0.3 m).
Leaves: 3½–4½″ (9–11 cm) wide, 6–9″ (15–23 cm) long. Crowded; *reverse ovate; broadest beyond middle, blunt at tip,* with *2 large pointed lobes at narrow base;* not toothed; short-stalked. Bright green above, pale and whitish beneath.
Bark: dark brown; smoothish or becoming scaly; thin.
Twigs: stout; with *ring scars at nodes.*
Flowers: 4″ (10 cm) wide; *6–9 creamy white petals;* solitary at end of twig; in spring.
Fruit: 2–2½″ (5–6 cm) long; *conelike;* oblong; rose-red; composed of many separate 2-seeded fruits ending in *short points curved outward;* maturing in fall.
Habitat: Moist valley soils.
Range: Scattered from South Carolina to E. Texas; to 400′ (122 m).

This rare and local species of the Coastal Plain is closely related to Fraser Magnolia, a mountain species which has larger leaves and larger flowers with pale yellow petals. Discovered by William Bartram (1739–1823), naturalist of Pennsylvania, and named for its crown shape in his historical book, *Travels through North and South Carolina, Georgia, East and West Florida* (1791).

86 Saucer Magnolia
"Chinese Magnolia"
Magnolia ×soulangiana Soul.-Bod.

Description: Ornamental shrub or small tree usually with several trunks, and with widely spreading crown of coarse foliage and abundant, large flowers in early spring before the leaves.
Height: 25′ (7.6 m).
Diameter: 6″ (15 cm).
Leaves: 5–8″ (13–20 cm) long, 2½–4½″ (6–11 cm) wide. Reverse ovate or *elliptical,* broadest toward abruptly short-pointed tip, blunt at base; not toothed. Dull green above, hairy beneath.
Bark: light gray; smooth.
Twigs: light gray; stout; hairless; with *ring scars at nodes.*
Flowers: 5–10″ (13–25 cm) wide; 6 spreading petals; showy; *like large tulips; bell-shaped* but opening widely like *saucers;* pink, purple, or white; white within; often fragrant; borne singly at ends of twigs.

Fruit: 2½–4″ (6–10 cm) long; *conelike; cylindrical;* curved; composed of many separate pointed fruits, splitting open to expose red seeds; in early autumn.
Habitat: Humid temperate regions.
Range: Widely planted across the United States.

This popular magnolia with gorgeous early spring flowers is a hybrid of two Chinese species, Yulan Magnolia (*Magnolia heptapeta* (Buc'hoz) Dandy) and Lily Magnolia (*M. quinquepeta* (Buc'hoz) Dandy). It originated in 1820 as a chance seedling in the garden of Étienne Soulange-Bodin, a French nurseryman. It flowers when still a small shrub. Several cultivated varieties differ mainly in color, size, time of flowering, and in overall shape.

88, 431 **Umbrella Magnolia**
"Umbrella-tree" "Elkwood"
Magnolia tripetala L.

Description: Tree with *large leaves, very large flowers,* and a broad, open crown of spreading branches; often with sprouts at base.
Height: 30–40′ (9–12 m).
Diameter: 1′ (0.3 m).
Leaves: 10–20″ (25–51 cm) long, 5–10″ (13–25 cm) wide. Reverse ovate; *broadest beyond middle;* not toothed; crowded; short-stalked. Green above, with silky hairs beneath when young.
Bark: light gray; smooth, thin.
Twigs: *stout;* with *ring scars at nodes.*
Flowers: 7–10″ (18–25 cm) wide; 3 cup-shaped, light green sepals with 6 or 9 shorter white petals; disagreeable odor; at end of twig; in spring.
Fruit: 2½–4″ (6–10 cm) long; conelike; *oblong;* rose-red; composed of many separate 2-seeded short-pointed fruits; maturing in autumn.

Habitat: Moist soils of mountain valleys; in hardwood forests.

Range: S. Pennsylvania south to Georgia, west to SE. Mississippi, north to S. Indiana; local in Arkansas and SE. Oklahoma.

Fairly common at low altitudes in the Great Smoky Mountains National Park, North Carolina–Tennessee. The arrangement of spreading leaves somewhat resembles the ribs of an umbrella, hence the common name. The Latin species name, meaning "3 petals," probably refers to the 3 sepals, which are longer than the more numerous petals.

85, 433 **Sweetbay**
"Swampbay" "Swamp Magnolia"
Magnolia virginiana L.

Description: Tree with narrow, rounded crown that sheds its leaves in winter or is *almost*

evergreen southward, and with aromatic spicy foliage and twigs.
Height: 20–60′ (6–18 m).
Diameter: 1½′ (0.5 m).
Leaves: 3–6″ (7.5–15 cm) long, 1¼–2½″ (3–6 cm) wide. *Oblong,* blunt at tip, without teeth, slightly thickened; short-stalked, becoming shiny green above, *whitish and finely hairy beneath.*
Bark: gray; smooth, thin, aromatic.
Twigs: with *ring scars at nodes;* ending in *buds covered with whitish hairs.*
Flowers: 2–2½″ (5–6 cm) wide; *cup-shaped,* with *9–12 white petals;* fragrant; in late spring and early summer.

Fruit: 1½–2″ (4–5 cm) long; *conelike; elliptical;* dark red; composed of many separate pointed fruits, each with 2 red seeds; maturing in early autumn.

Habitat: Wet soils of coastal swamps and borders of streams and ponds.

Range: Long Island south to S. Florida and west to SE. Texas; local in NE. Massachusetts; to 500′ (152 m).

This attractive, native ornamental is popular for its fragrant flowers borne over a long period, showy conelike fruit, handsome foliage of contrasting colors, and smooth bark. Introduced into European gardens as early as 1688. Called "Beavertree" by colonists who caught beavers in traps baited with the fleshy roots.

ANNONA (CUSTARD APPLE) FAMILY
(Annonaceae)

Small to medium-sized trees and shrubs. About 2000 species in tropical and warm temperate regions; 4 native and 1 naturalized tree and 3 native shrub species in North America.
Leaves: alternate, commonly in 2 rows, simple, without teeth, with gland-dots, without stipules, and sometimes aromatic.
Flowers: generally solitary, large, bisexual, regular, with 3 sepals, 6 petals in 2 series of unequal size, many stamens in a spiral and on the axis, and few to many separate pistils.
Fruit: berries, follicles, or aggregate; often stalked, sometimes edible.

87, 364, 574 **Pawpaw**
"Pawpaw-apple" "False-banana"
Asimina triloba (L.) Dunal

Description: Shrub or small tree that forms colonies from root sprouts, with straight trunk, spreading branches, and large leaves.

Height: 30′ (9 m).
Diameter: 8″ (20 cm).
Leaves: 7–10″ (18–25 cm) long, 3–5″ (7.5–13 cm) wide. *Spreading in 2 rows* on long twigs; reverse ovate, *broadest beyond middle,* short-pointed at tip, tapering to base and short leafstalk; covered with *rust-colored hairs when young.* Green above, paler beneath; turning yellow in autumn. Bruised foliage has disagreeable odor.
Bark: dark brown, warty, thin.
Twigs: brown; often with rust-colored hairs; ending in small hairy buds.
Flowers: 1½″ (4 cm) wide; *3 triangular green to brown or purple outer petals,* hairy with prominent veins; nodding singly on slender stalks; in early spring.
Fruit: 3–5″ (7.5–13 cm) long, 1–1½″

(2.5–4 cm) in diameter; berrylike; brownish; cylindrical; slightly curved, *suggesting a small banana;* edible soft yellowish pulp has flavor of custard. Several shiny brown oblong seeds.

Habitat: Moist soils, especially flood plains; in understory of hardwood forests.

Range: S. Ontario and W. New York, south to NW. Florida, west to E. Texas, and north to SE. Nebraska; to 2600′ (792 m) in southern Appalachians.

Pawpaw is the northernmost New World representative of a chiefly tropical family, which includes the popular tropical fruits Annona, Custard-apple, Sugar-apple, and Soursop. The wild fruit was once harvested, but the supply has now decreased greatly due to the clearing of forests. The small crop is generally consumed only by wildlife, such as opossums, squirrels, raccoons, and birds. Attempts have been made to cultivate the Pawpaw as a fruit tree. First recorded by the DeSoto expedition in the lower Mississippi Valley in 1541. The name Pawpaw is from the Arawakan name of Papaya, an unrelated tropical American fruit.

LAUREL FAMILY
(Lauraceae)

Mostly large trees and a few shrubs, with aromatic bark, wood, and leaves. About 2000 species in tropical and warm temperate regions; 5 native and 2 naturalized tree, 3 native shrub, and 1 native herbaceous vine species in North America.
Leaves: mostly alternate, sometimes opposite or whorled; simple, commonly elliptical, without teeth, pinnately-veined with long curved side veins, often leathery with tiny gland-dots, without stipules.
Flowers: mostly small, many in branched clusters along twigs, bisexual or sometimes male and female on separate plants, regular, often with short cup, 3 sepals and 3 similar petals (or 6 sepals), 9–12 stamens (or some reduced to staminodes), the anthers opening by 2 or 4 pores with lids, and 1 pistil.
Fruit: a berry or drupe with 1 large seed, mostly with cup or tube from calyx or corolla persistent at base.

---

70 **Camphor-tree**
Cinnamomum camphora (L.) Karst.

Description: Introduced, aromatic, evergreen tree with rounded, *dense crown, wider than high,* and *odor of camphor in crushed foliage.*
Height: 40′ (12 m).
Diameter: 2′ (0.6 m).
Leaves: *evergreen; partly opposite;* 2½–4″ (6–10 cm) long, 1–2¼″ (2.5–6 cm) wide. *Elliptical,* pointed, with *3 main veins* from rounded base; not toothed; slightly thickened; long-stalked. Pinkish when young, becoming *shiny green above, dull whitish beneath.*
Bark: gray; smoothish, becoming rough, thick, and furrowed.

Twigs: green, slender, hairless.
Flowers: ⅛″ (3 mm) long; yellowish; in clusters 1½–3″ (4–7.5 cm) long; in spring.
Fruit: ⅜″ (10 mm) in diameter; a black 1-seeded berry with greenish cup and spicy taste of camphor.

Habitat: Moist soils, roadsides and waste places in humid subtropical regions.

Range: Native of tropical Asia from E. China to Vietnam, Taiwan, and Japan. Extensively cultivated and naturalized locally from Florida to S. Texas; also in S. California.

Popular as a street tree in the deep South, Camphor-tree produces dense shade. In Asia, camphor oil and gum for medicine and industry are obtained by steam distillation of leaf clippings and wood. The insect-repellent wood is used for cabinetry and chests.

---

69 Redbay
"Shorebay"
Persea borbonia (L.) Spreng.

Description: Handsome, aromatic, evergreen tree, with dense crown.
Height: 60′ (18 m).
Diameter: 2′ (0.6 m).
Leaves: *evergreen;* 3–6″ (7.5–15 cm) long, ¾–1½″ (2–4 cm) wide. *Elliptical* or lance-shaped; short-stalked; thick and leathery, with edges slightly rolled under. *Shiny green above, pale with whitish or rust-colored hairs beneath.*
Bark: dark or reddish-brown; furrowed into broad scaly ridges.
Flowers: 3⁄16″ (5 mm) long; light yellow; several in long-stalked cluster at leaf base; in spring.

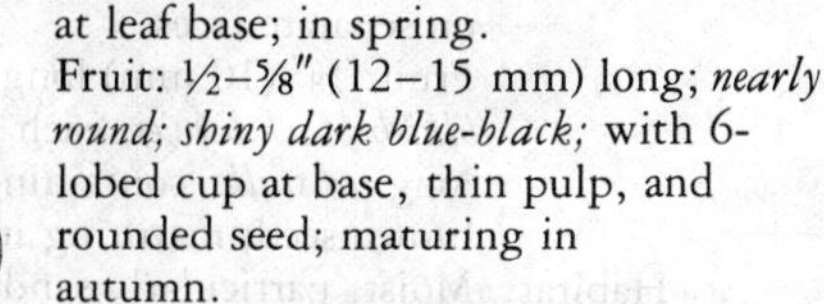

Fruit: ½–⅝″ (12–15 mm) long; *nearly round; shiny dark blue-black;* with 6-lobed cup at base, thin pulp, and rounded seed; maturing in autumn.

Habitat: Wet soils of valleys and swamps, also sandy uplands and dunes, in mixed forests.

Range: S. Delaware south to S. Florida and west to S. Texas; to 400′ (122 m).

The wood, which takes a beautiful polish, is used for fine cabinetwork and also for lumber. The spicy leaves can be used to flavor soups and meats. Birds eat the bitter fruit. Swampbay (var. *pubescens* (Pursh) Little) is a variety found in coastal swamps and characterized by twigs and lower leaf surfaces covered with rust-colored hairs.

239, 394, 544, 589, 621 **Sassafras**
Sassafras albidum (Nutt.) Nees

Description: Aromatic tree or thicket-forming shrub with variously shaped leaves and narrow, spreading crown of short, stout branches.

Height: 30–60′ (9–18 m).
Diameter: 1½′ (0.5 m), sometimes larger.
Leaves: 3–5″ (7.5–13 cm) long, 1½–4″ (4–10 cm) wide. *Elliptical, often with 2* mitten-shaped *lobes or 3* broad and blunt *lobes;* not toothed; base short-pointed; long slender leafstalks. Shiny green above, paler and often hairy beneath; turning yellow, orange, or red in autumn.
Bark: gray-brown; becoming thick and deeply furrowed.
Twigs: *greenish,* slender, sometimes hairy.
Flowers: ⅜″ (10 mm) long; *yellow-green;* several clustered at end of leafless twigs in early spring; male and female usually on separate trees.
Fruit: ⅜″ (10 mm) long; *elliptical* shiny *bluish-black berries;* each in *red cup* on long *red stalk,* containing 1 shiny brown seed; maturing in autumn.

Habitat: Moist, particularly sandy, soils of

uplands and valleys, often in old fields, clearings, and forest openings.

Range: Extreme S. Ontario east to SW. Maine, south to central Florida, west to E. Texas, and north to central Michigan; to 5000′ (1524 m) in southern Appalachians.

The roots and root bark supply oil of sassafras (used to perfume soap) and sassafras tea, and have been used to flavor root beer. Explorers and colonists thought the aromatic root bark was a panacea, or cure-all, for diseases and shipped quantities to Europe. The greenish twigs and leafstalks have a pleasant, spicy, slightly gummy taste. Sassafras apparently is the American Indian name used by the Spanish and French settlers in Florida in the middle of the 16th century. This is the northernmost New World representative of an important family of tropical timbers.

WITCH-HAZEL FAMILY
(Hamamelidaceae)

Trees, often large, and shrubs; about 100 species in subtropical and warm temperate regions; 2 native tree and 5 shrub species in North America.
Leaves: deciduous or in warm climates evergreen; alternate; simple; with gland-teeth or palmately lobed; often with star-shaped hairs; with paired stipules.
Flowers: bisexual or male and female; mostly tiny; in heads or racemes; composed of calyx of usually 4–5 sepals united at base and to ovary (sometimes no sepals), corolla of 4–5 petals borne on calyx or none, stamens 2 to many, separate, and 1 pistil.
Fruit: a capsule, often hard with 2-layered wall, splitting open across top into 2 parts; seeds few, sometimes winged.

230, 395, 529, 629 **Witch-hazel**
"Southern Witch-hazel"
Hamamelis virginiana L.

Description: Slightly aromatic shrub or small tree with a broad, open crown of spreading branches and small yellow flowers present in autumn or winter.

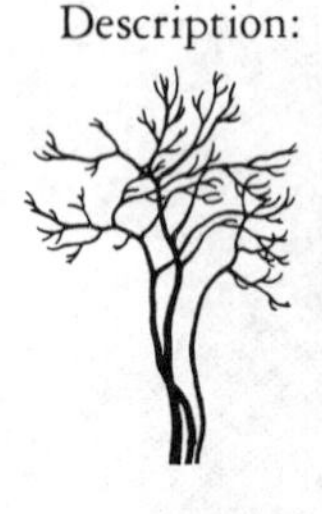

Height: 20–30′ (6–9 m).
Diameter: 4–8″ (10–20 cm).
Leaves: 3–5″ (7.5–13 cm) long, 2–3″ (5–7.5 cm) wide. *Broadly elliptical,* pointed or rounded at tip, blunt to notched and unequal at base, *broadest and wavy-lobed beyond middle;* with 5–7 straight veins on each side; hairy when young. Dull dark green above, paler below; turning yellow in autumn.
Bark: light brown; smooth or scaly.
Twigs: slender, zigzag, with gray or rust-colored hairs.
Flowers: 1″ (2.5 cm) wide; with *4 bright yellow petals, threadlike* and

twisted; few, short-stalked, along leafless twigs in *autumn or winter*.
Fruit: ½″ (12 mm) long; a hard *elliptical capsule* ending in *4 sharp curved points,* light brown, opening in 2 parts; maturing in autumn; with 1 or 2 shiny blackish seeds ¼″ (6 mm) long; ejected with force by contracting capsule walls.

Habitat: Moist soil in understory of hardwood forests.

Range: S. Ontario east to Nova Scotia, south to central Florida, west to E. Texas, and north to central Wisconsin; local in NE. Mexico; to 5000′ (1524 m), sometimes higher in southern Appalachians.

The aromatic extract of leaves, twigs, and bark is used in mildly astringent lotions and toilet water. A myth of witchcraft held that a forked branch of Witch-hazel could be used to locate underground water. The foliage and fruits slightly resemble those of the shrub hazel (*Corylus*). Upon drying, the contracting capsule can eject its small seed as far as 30′ (9 m).

266, 509, 594, 624 **Sweetgum**
"Redgum" "Sapgum"
Liquidambar styraciflua L.

Description: Large, aromatic tree with straight trunk and conical crown that becomes round and spreading.

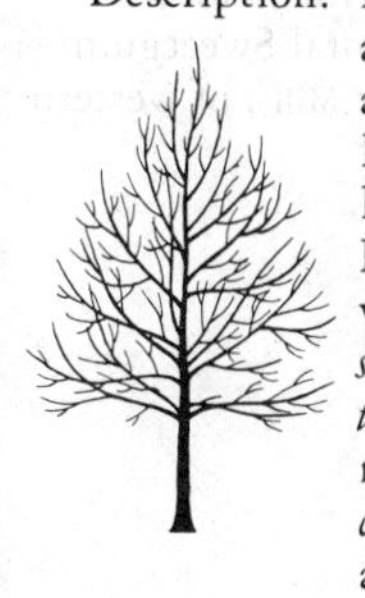

Height: 60–100′ (18–30 m).
Diameter: 1½–3′ (0.5–0.9 m).
Leaves: 3–6″ (7.5–15 cm) long and wide. *Star-shaped* or maplelike, with 5, *sometimes 7, long-pointed, finely saw-toothed lobes* and 5 main veins from notched base; with *resinous odor when crushed;* leafstalks slender, nearly as long as blades. Shiny dark green above, turning reddish in autumn.
Bark: gray; deeply furrowed into narrow scaly ridges.

Twigs: green to brown, stout, *often forming corky wings.*
Flowers: tiny; in *greenish ball-like clusters* in spring; male in several clusters along a stalk; female in drooping cluster on same tree.
Fruit: 1–1¼″ (2.5–3 cm) in diameter; a long-stalked drooping brown *ball* composed of many individual fruits, each ending in *2 long curved prickly points* and each with 1–2 long-winged seeds; maturing in autumn and persistent into winter.

Habitat: Moist soils of valleys and lower slopes; in mixed woodlands. Often a pioneer after logging, clearing, and in old fields.

Range: Extreme SW. Connecticut south to central Florida, west to E. Texas, and north to S. Illinois; also a variety in E. Mexico; to 3000′ (914 m) in southern Appalachians.

An important timber tree, Sweetgum is second in production only to oaks among hardwoods. It is a leading furniture wood, used for cabinetwork, veneer, plywood, pulpwood, barrels, and boxes. In pioneer days, a gum was obtained from the trunks by peeling the bark and scraping off the resinlike solid. This gum was used medicinally as well as for chewing gum. Commercial storax, a fragrant resin used in perfumes and medicines, is from the related Oriental Sweetgum (*Liquidambar orientalis* Mill.) of western Asia.

SYCAMORE FAMILY
(Platanaceae)

Large tree with massive trunk; 1 genus (*Platanus*), with 9 or fewer species in north temperate regions, 3 in North America.
Leaves: deciduous, alternate, simple, long-stalked, about as wide as long, palmately 3- to 9-lobed, often toothed, with star-shaped hairs and large paired toothed stipules united at base.
Bark: smooth, light-colored, peeling in large thin flakes and becoming mottled.
Twigs: slender, zigzag with ring scars; conical buds covered by 1 hairless scale and hidden inside enlarged base of leafstalk.
Flowers: male and female on same tree in spring with leaves; numerous, tiny, greenish tinged with red, crowded in 1–6 balls or dense round heads on long drooping stalk, with cup-shaped calyx, no or few petals. Male with 3–7 stamens; female with 5–9 pistils.
Fruit: many 1-seeded narrow 4-angled nutlets surrounded by long stiff hairs, crowded in ball or head.

---

257 London Planetree
Platanus ×acerifolia (Ait.) Willd.

Description: Large, introduced, shade tree with straight, stout trunk and broad, open crown of spreading to slightly drooping branches and coarse foliage.
Height: 70′ (21 m).
Diameter: 2′ (0.6 m).
Leaves: 5–10″ (13–25 cm) long and wide. *Palmately 3- or 5-lobed;* shallowly short-pointed and lobed, with few large teeth or none; *3 or 5 main veins* from notched base; becoming hairless or nearly so. Shiny green above, pale beneath. Leafstalk long, stout, covering side bud at enlarged base.
Bark: *smooth;* with patches of brown,

green, and gray; *peeling off in large flakes.*
Twigs: greenish, slender, zigzag, hairy, with ring scars at nodes.
Flowers: tiny; in *greenish ball-like* drooping *clusters;* male and female on separate twigs; in spring.
Fruit: 1″ (2.5 cm) in diameter; usually *2 bristly brown balls* hanging on long stalk, composed of many narrow *nutlets* with hair tufts; maturing in autumn, separating in winter.

Habitat: Moist soils in humid temperate regions; hardy and tolerant of city conditions.

Range: Planted across the United States.

London Planetree is a hybrid between Sycamore of eastern United States and Oriental Planetree (*Platanus orientalis* L.) of southeastern Europe and Asia Minor. Also popular as a street tree in Europe, where it originated probably before 1700. The plants can be clipped into screens and arbors. The genus name *Platanus* is the classical Latin and Greek name of Oriental Planetree, from the Greek word for "broad," which describes the leaves. The Latin species name means "maple leaf," referring to the resemblance of London Planetree's leaves to those of maples.

256, 508 Sycamore
"American Sycamore" "American Planetree"
Platanus occidentalis L.

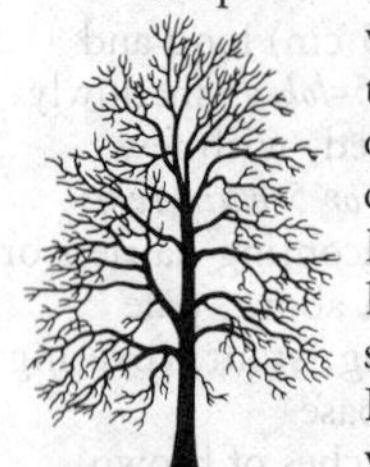

Description: One of the largest eastern hardwoods, with an enlarged base, massive, straight trunk, and large, spreading, often crooked branches forming a broad open crown.
Height: 60–100′ (18–30 m).
Diameter: 2–4′ (0.6–1.2 m), sometimes much larger.
Leaves: 4–8″ (10–20 cm) long and wide (larger on shoots). *Broadly ovate,*

with *3 or 5 shallow broad short-pointed lobes;* wavy edges with scattered large teeth; 5 or 3 main veins from notched base. Bright green above, paler beneath and becoming hairless except on veins; turning brown in autumn. Leafstalk long, stout, covering side bud at enlarged base.

Bark: *smooth, whitish and mottled; peeling off* in large thin flakes, exposing patches of brown, green, and gray; base of large trunks dark brown, deeply furrowed into broad scaly ridges.

Twigs: greenish, slender, zigzag, with ring scars at nodes.

Flowers: tiny; *greenish;* in 1–2 *ball-like* drooping *clusters;* male and female clusters on separate twigs; in spring.

Fruit: 1″ (2.5 cm) in diameter; usually 1 *brown ball* hanging on long stalk, composed of many narrow *nutlets* with *hair tufts;* maturing in autumn, separating in winter.

Habitat: Wet soils of stream banks, flood plains, and edges of lakes and swamps; dominant in mixed forests.

Range: SW. Maine, south to NW. Florida, west to S. central Texas, north to E. Nebraska; also NE. Mexico; to 3200′ (975 m).

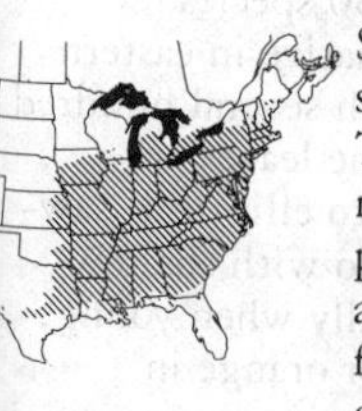

Sycamore pioneers on exposed upland sites such as old fields and strip mines. The wood is used for furniture parts, millwork, flooring, and specialty products such as butcher blocks, as well as pulpwood, particleboard, and fiberboard. A shade tree, Sycamore grows to a larger trunk diameter than any other native hardwood. The present champion's trunk is about 11′ (3.4 m) in diameter; an earlier giant's was nearly 15′ (4.6 m). The hollow trunks of old, giant trees were homes for chimney swifts in earlier times.

ROSE FAMILY
(Rosaceae)

About 2000 species of trees, shrubs, and herbs worldwide; approximately 77 native and 9 naturalized tree species and many species of shrubs and herbs in North America; including serviceberries (*Amelanchier*), hawthorns (*Crataegus*), apples (*Malus*), plums and cherries (*Prunus*), and mountain-ashes (*Sorbus*).
Leaves: alternate, generally simple, with paired stipules.
Flowers: small to large, bisexual, usually regular or sometimes slightly irregular, generally with cuplike base that bears 5 sepals, 5 petals, and many separate stamens, and with 1 to many simple pistils or 1 compound pistil.
Fruit: a drupe, pome, achene, or follicle.

Crataegus, commonly known as Hawthorn, Haw, Thornapple, Red Haw, or Hog-apple, is a large genus of many difficult-to-distinguish species. Much-branched shrubs and small spreading trees, they commonly have a dense rounded or flattened crown and long thorns. About 30 species (described here) are native in eastern United States, though several hundred have been named. The leaves are usually small, ovate to elliptical, saw-toothed and often also with shallow lobes. Hairy, especially when young, they often turn red or orange in autumn. On young twigs, leaves are larger and lobed. Leafstalks are short and slender. Bark is commonly gray or brown, and scaly. Twigs are slender, often much-branched, and slightly zigzag, with long slender spines or thorns. Flowers have a cuplike base with 5 narrow calyx lobes, 5 small rounded white petals and 5–20 stamens with pale yellow or pink anthers, enclosing pistil with 1–5 styles.

Flowers are usually in clusters at end of twigs in spring. The commonly red fruits are mostly small and round in clusters, with 5 calyx lobes usually persisting at tip, and are often sweet and edible; 1–5 nutlets are each 1-seeded. Fruit matures in late summer or autumn and sometimes remains attached in winter.

---

200 **Western Serviceberry**
"Saskatoon" "Western Shadbush"
Amelanchier alnifolia (Nutt.) Nutt.

Description: Shrub or small tree, usually with several trunks, and with star-shaped white flowers.
Height: 30′ (9 m).
Diameter: 8″ (20 cm).
Leaves: ¾–2″ (2–5 cm) long and almost as broad. *Broadly elliptical to nearly round;* rounded at both ends; *coarsely toothed above middle;* usually with 7–9 *straight veins on each side.* Dark green and becoming hairless above, paler and hairy when young beneath.
Bark: gray or brown; thin, smooth or slightly fissured.
Twigs: red-brown, slender, hairless.
Flowers: ¾–1¼″ (2–3 cm) wide; with 5 narrow white petals; in small terminal clusters; in spring with leaves.
Fruit: ½″ (12 mm) in diameter; like a *small apple; purple* or blackish, edible, juicy and sweet, with several seeds; in early summer.

Habitat: Moist soils in forests and openings.

Range: Central Alaska southeast to Manitoba, W. Minnesota, and Colorado, and west to N. California; local east to SE. Quebec; to 6000′ (1829 m).

The fruit of this and related species are eaten fresh, prepared in puddings, pies, and muffins, and dried like raisins and currants. They are also an important food for wildlife from songbirds to squirrels and bears. Deer and livestock browse the foliage.

175, 418, 595 **Downy Serviceberry**
"Shadbush" "Juneberry" "Shadblow"
Amelanchier arborea (Michx. f.) Fern.

Description: Tree with narrow, rounded crown, or an irregularly branched shrub, with star-shaped, white flowers.

Height: 40′ (12 m).
Diameter: 1′ (0.3 m).
Leaves: 1½–4″ (4–10 cm) long, 1–2″ (2.5–5 cm) wide. *Ovate* or elliptical, pointed at tip, notched at base; *finely saw-toothed;* with *soft hairs when young,* especially beneath; with *11–17 straight veins on each side.* Dull green above, paler beneath; turning yellow to red in autumn.
Bark: light gray; smooth, becoming furrowed into narrow ridges.
Twigs: red-brown, slender, often covered with white hairs when young.
Flowers: 1¼″ (3 cm) wide; with 5 *narrow white* petals; on slender stalks in terminal clusters; in spring before leaves.
Fruit: ¼–⅜″ (6–10 mm) in diameter; like a *small apple; purple,* edible, nearly dry or juicy and sweet, with several seeds; in early summer.

Habitat: Moist soils in hardwood forests.

Range: S. Newfoundland and Nova Scotia south to NW. Florida, west to Louisiana and E. Oklahoma, and north to Minnesota; to 6000′ (1829 m) in southern Appalachians.

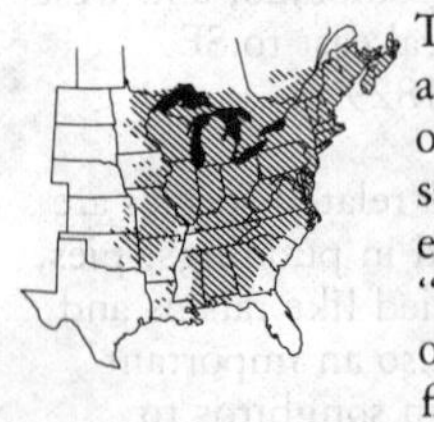

The names "Shadbush" and "Shadblow" allude to the fact that the showy masses of white flowers tend to occur at the same time that shad ascend the rivers in early spring to spawn. An older name is "Sarvis." Sometimes planted as an ornamental for the showy clusters of flowers.

199 Roundleaf Serviceberry
"Roundleaf Juneberry" "Shore Shadbush"
Amelanchier sanguinea (Pursh) DC.

Description: Shrub or small tree with 1 or several slender trunks, upright branches, and star-shaped, white flowers.
Height: 20′ (6 m).
Diameter: 4″ (10 cm).
Leaves: 1¼–2¾″ (3–7 cm) long. *Elliptical* to nearly round; *rounded or blunt at tip; coarsely toothed;* with *12–15 straight veins* on each side; becoming hairless or nearly so. Dull green above, paler beneath.
Bark: gray; smooth, becoming slightly fissured.
Twigs: *reddish when young,* slender.
Flowers: nearly 1″ (2.5 cm) wide; with *5 narrow white petals* on slender stalks in terminal clusters; in spring before leaves.
Fruit: 5/16″ (8 mm) in diameter; like a *small apple; blackish,* edible, juicy and sweet, with several seeds; in summer.

Habitat: Rocky slopes and stream banks in open forests.

Range: SW. Quebec and Maine south to N. New Jersey and west to N. Iowa and N. Minnesota; also south in mountains to W. North Carolina and E. Tennessee; to 2000′ (610 m).

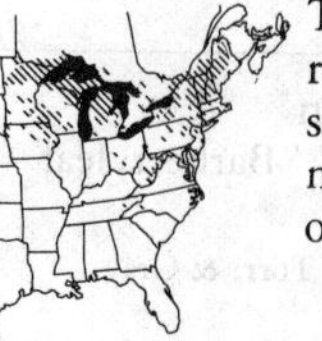

This species is recognized by the nearly round leaves, as the common name suggests. The Latin species name, meaning "blood red," refers to the color of the young twigs.

241 May Hawthorn
"Apple Hawthorn" "Shining Hawthorn"
Crataegus aestivalis (Walt.) Torr. & Gray

Description: Large shrub or small tree with rounded, compact crown.

Height: 30′ (9 m).
Diameter: 6″ (15 cm).
Leaves: 1–2″ (2.5–5 cm) long, ⅜–¾″ (10–19 mm) wide. *Elliptical* to ovate, blunt or short-pointed at tip, gradually narrowed to base; *coarsely saw-toothed beyond middle,* with gland-tipped teeth (on young twigs often with few small lobes). *Shiny* dark green and *hairless above, yellow-green* with tufts of *hairs in vein angles* beneath.
Bark: gray with red-brown inner layers; fissured and scaly.
Twigs: brown, slender, hairless, often with straight spines.
Flowers: ¾″ (19 mm) wide; with 5 white petals, *15–20 pink stamens,* and usually 3 styles; short-stalked; with or *before leaves* in early spring.
Fruit: about ⅜″ (10 mm) in diameter; shiny *red;* sour juicy pulp; usually 3 nutlets; maturing in summer.

Habitat: Wet soils of riverbanks and borders of swamps.

Range: S. North Carolina southwest to N. Florida and S. Mississippi; to 200′ (61 m).

The name May Hawthorn is misleading, because the flowers are borne earlier and the fruits ripen later, as the Latin species name, meaning "summer," indicates.

---

216 **Barberry Hawthorn**
"Bigtree Hawthorn" "Barberryleaf Hawthorn"
Crataegus berberifolia Torr. & Gray

Description: Small tree with broad, flat crown of spreading branches.
Height: 30′ (9 m).
Diameter: 8″ (20 cm).
Leaves: 1¼–2½″ (3–6 cm) long, ¾–1¼″ (2–3 cm) wide. *Obovate to elliptical, rounded or blunt at tip,* widest beyond middle; saw-toothed except near narrow

base; slightly thickened and *usually with coat of whitish hairs*. Shiny dark green above, paler beneath.
Bark: gray; scaly.
Twigs: covered with *white hairs* when young; sometimes with long straight spines.
Flowers: to ¾″ (19 mm) wide; with 5 white petals, about 20 (sometimes 10) pink or yellow stamens, and 2–3 styles; 5–10 flowers on short hairy stalks in compact clusters; in early spring.
Fruit: ⅜–½″ (10–12 mm) in diameter; *rounded; orange or red;* thin mealy pulp; 2–3 nutlets; *few,* in drooping clusters; maturing in autumn.

Habitat: Moist soils along streams and uplands in open areas and borders of oak and pine forests.

Range: S. Illinois south to Mississippi, west to E. Texas, and north to SE. Kansas; at 200–1000′ (61–305 m).

The common and Latin species names both refer to the resemblance of the leaves to those of barberries (*Berberis*), native and introduced shrubs.

Blueberry Hawthorn
"Blue Haw"
Crataegus brachyacantha Sarg. & Engelm.

Description: A handsome tree with broad, rounded, dense crown of stout, light gray branches, and round, blue fruit.
Height: 40′ (12 m).
Diameter: 1½′ (0.5 m).
Leaves: 1–2″ (2.5–5 cm) long, ½–1″ (1.2–2.5 cm) wide. *Elliptical,* blunt or rounded at tip, broadest beyond middle and gradually narrowed to base; *wavy saw-toothed with rounded gland-tipped teeth;* slightly thickened. Hairy above when young, becoming *shiny dark green,* paler beneath. Leaves on young twigs larger, ovate, deeply lobed, with *veins extending to notches* as well as to lobes.

Bark: dark brown; thick, deeply furrowed and scaly.
Twigs: slightly hairy when young; with many stout *short curved spines.*
Flowers: ⅝" (15 mm) wide; with 5 white petals, fading to orange; 15–20 yellow stamens and 3–5 styles; many, crowded in leafy clusters; in spring.

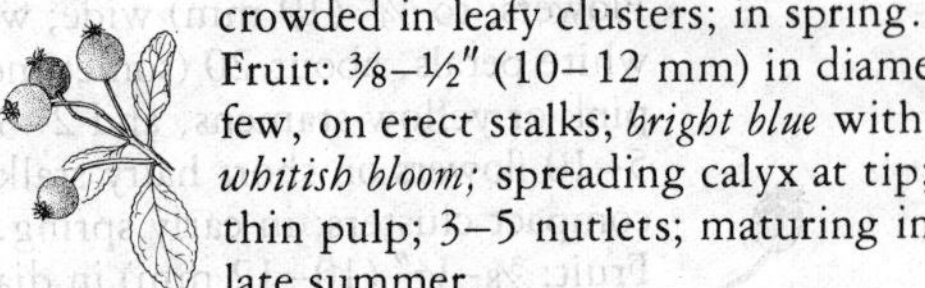

Fruit: ⅜–½" (10–12 mm) in diameter; few, on erect stalks; *bright blue* with *whitish bloom;* spreading calyx at tip; thin pulp; 3–5 nutlets; maturing in late summer.

Habitat: Moist soils along streams in prairies.

Range: SW. Georgia west to SW. Arkansas and E. Texas; to 300' (91 m).

The scientific name, meaning "short-spined," refers to the twigs.

224 Brainerd Hawthorn
Crataegus brainerdii Sarg.

Description: Much-branched shrub or small tree with broad, rounded crown of spreading branches.
Height: 20' (6 m).
Diameter: 6" (15 cm).
Leaves: 2–3" (5–7.5 cm) long, 1½–2" (4–5 cm) wide. *Elliptical* or ovate; long-pointed; *many shallow lobes irregularly saw-toothed;* slightly thickened; hairy when young. *Dark green and rough* above, paler and hairless beneath.
Bark: gray; becoming rough and scaly.
Twigs: stout, hairless, with long slender straight or curved spines.
Flowers: ¾" (19 mm) wide; with 5 white petals, 5–20 mostly pink stamens, and 2–5 styles; 4–12 flowers in broad, loose, hairless clusters at end of leafy twigs; in spring.

Fruit: ⅜–½" (10–12 mm) in diameter; *red;* thick dry mealy pulp; 2–5 nutlets; in drooping clusters; maturing in early autumn.

Habitat: Moist valleys and slopes.

Range: S. Quebec to Nova Scotia south to Connecticut, west to Ohio, and north to Michigan; to 500′ (152 m) or above.

Named for its discover, Ezra Brainerd (1844–1924), botanist of Vermont and president of Middlebury College.

225 **Pear Hawthorn**
Crataegus calpodendron (Ehrh.) Medic.

Description: Shrub or small tree with broad, flattened crown of slender, spreading, nearly horizontal branches, pear-shaped fruit, and brilliant autumn foliage; mostly without spines.
Height: 20′ (6 m).
Diameter: 6″ (15 cm).
Leaves: 2–4½″ (5–11 cm) long, 1–3″ (2.5–7.5 cm) wide. *Elliptical to ovate;* short- or long-pointed; usually *doubly saw-toothed* with gland-tipped teeth; *often with shallow lobes;* dotlike glands on leafstalks. Dull yellow-green *with sunken veins* (and hairy when young) above, *finely hairy beneath; turning orange and scarlet* in autumn.
Bark: dark gray; scaly, becoming thick and furrowed.
Twigs: densely hairy when young; *mostly without spines,* sometimes with few short spines.
Flowers: more than ½″ (12 mm) wide; with 5 white petals, 15–20 pink stamens, and 2–3 styles; *many* flowers in broad hairy clusters; in spring.
Fruit: ⅜″ (10 mm) in diameter; *elliptical or pear-shaped; orange-red* or red; *thin sweet juicy pulp;* 2–3 nutlets; many, in upright clusters; maturing in autumn and remaining attached.

Habitat: Moist soils of valleys; especially rocky stream banks and uplands.

Range: S. Ontario and New York, south to Georgia, west to E. Texas, north to Minnesota; to 2000′ (610 m) or above.

The Latin species name, meaning "urn tree," and the common name both describe the fruit, which instead of being round is longer than it is broad. Introduced in European gardens more than 2 centuries ago.

247 **Fireberry Hawthorn**
"Roundleaf Hawthorn" "Golden-fruit Hawthorn"
Crataegus chrysocarpa Ashe

Description: Much-branched shrub or small tree with a broad, rounded crown.
Height: 20′ (6 m).
Diameter: 8″ (20 cm).
Leaves: 1½–2″ (4–5 cm) long, 1–1½″ (2.5–4 cm) wide. *Elliptical or nearly round;* finely and often doubly *saw-toothed;* lower teeth gland-tipped; *shallow-lobed* beyond middle; slightly thickened. *Shiny green with sunken veins* above, paler beneath.
Bark: dark brown; scaly.
Twigs: mostly hairless, with *many* spines.
Flowers: ⅝″ (15 mm) wide; with 5 white petals, usually 5–10 pale yellow stamens and 3–4 styles; *many,* in broad loose clusters; in spring.
Fruit: ⅜–½″ (10–12 mm) in diameter; *rounded; dark red or rarely yellow;* thin, yellow, dry, sweet pulp; 3–4 nutlets; *many,* in drooping clusters; maturing in autumn.
Habitat: Moist valleys.
Range: Alberta east to Newfoundland, south to Virginia, west to Missouri, and north to South Dakota; south in western mountains to Colorado; to 2000′ (610 m), higher in the West.

The Latin species name means "golden fruit," although the fruit is commonly dark red and rarely yellow.

251 Scarlet Hawthorn

Crataegus coccinea L.

Description: Small tree with rounded crown of long, slender branches.
Height: 20′ (6 m).
Diameter: 1′ (0.3 m).
Leaves: 2–4″ (5–10 cm) long, 1½–3″ (4–7.5 cm) wide. *Broadly ovate,* short-pointed at tip, blunt or rounded at base; *coarsely and often doubly saw-toothed,* with *4–5 short-pointed shallow lobes* beyond middle on each side. *Dark green and rough* above, paler and nearly hairless beneath.
Bark: red-brown; scaly.
Twigs: hairy when young; with straight or slightly curved spines.
Flowers: ⅝–¾″ (15–19 mm) wide; with 5 white petals, 10 *pink or red stamens,* and 4–5 styles; long-stalked; in spreading loose clusters; in spring.
Fruit: ½–⅝″ (12–15 mm) in diameter; *elliptical,* shiny *red* with many dark dots; gland-toothed calyx at tip; thin dry mealy pulp; 4–5 nutlets; in drooping clusters; maturing in early autumn.

Habitat: Moist soils of valleys and hillsides.

Range: S. Quebec south to Maine and New York, west to Iowa, and north to SE. Minnesota; to 1000′ (305 m); local south in mountains to W. North Carolina.

One of the three native hawthorns described by Linnaeus in 1753. Its common and Latin species names both refer to the color of the fruit.

Kansas Hawthorn

Crataegus coccinioides Ashe

Description: Small tree with broad, rounded, dense crown of spreading branches, large flowers, large shiny dark red fruit, and showy autumn foliage; or a large

thicket-forming shrub.
Height: 20′ (6 m).
Diameter: 8″ (20 cm).
Leaves: 2½–3″ (6–7.5 cm) long, 2–2½″ (5–6 cm) wide. *Broadly ovate,* short-pointed at tip, rounded or straight at base; sharply and often *doubly saw-toothed with several shallow lobes beyond middle.* Reddish-tinged when young, turning dull dark green above, paler and slightly hairy beneath; becoming orange and red in autumn. Leafstalks often reddish with red gland-dots.
Bark: dark brown; scaly.
Twigs: hairless, with many stout spines.
Flowers: ¾–1″ (2–2.5 cm) wide; with 5 white petals, 20 pink or red stamens, and 5 styles; 4–7 flowers in compact, usually hairless clusters; in spring.
Fruit: to ¾″ (19 mm) in diameter; rounded but flattened at ends; often angled; shiny *dark red* with many pale dots; *reddish* gland-toothed *calyx* with *enlarged lobes;* thick juicy pulp; 5 nutlets; many in spreading clusters; maturing in autumn.

Habitat: Dry upland slopes, especially limestone hills, at edges of woods.

Range: S. Illinois south to NE. Oklahoma and N. Arkansas and west to SE. Kansas; at 400–1500′ (122–457 m).

The Latin species name refers to this tree's resemblance to Scarlet Hawthorn. The latter has smaller flowers and smaller fruit with thin, dry pulp and calyx lobes not enlarged.

219, 412 **Cockspur Hawthorn**
"Hog-apple" "Newcastle-thorn"
Crataegus crus-galli L.

Description: Small, spiny, thicket-forming tree with short, stout trunk and broad, dense crown of spreading and horizontal

branches; *hairless throughout*.
Height: 30′ (9 m).
Diameter: 1′ (0.3 m).
Leaves: 1–4″ (2.5–10 cm) long, ⅜–2″ (1–5 cm) wide. *Spoon-shaped or narrowly elliptical; short-pointed or rounded* at tip, widest beyond middle, tapering to narrow base; sharply saw-toothed beyond middle with gland-tipped teeth; slightly *thick and leathery. Shiny dark green* above, pale with prominent network of veins beneath; turning orange and scarlet in autumn.
Bark: dark gray or brown; scaly, with *branched spines*.
Twigs: stout; usually with *many very long slender brown spines*.
Flowers: ½–⅝″ (12–15 mm) wide; with 5 white petals, 10 to sometimes 20 pink or pale yellow stamens, and 2–3 styles; many, in large clusters; in late spring or early summer.
Fruit: ⅜–½″ (10–12 mm) in diameter; *rounded;* greenish or *dull dark red; thin hard pulp;* usually 2–3 nutlets; several, in drooping clusters; maturing in late autumn and *persisting* until spring.

Habitat: Moist soils of valleys and low upland slopes.

Range: S. Ontario and S. Quebec south to N. Florida, west to E. Texas, and north to Iowa; to 2000′ (610 m).

The common and Latin species names both describe the numerous and extremely long spines, which are used locally as pins. The long spines and shiny dark green spoon-shaped leaves make this one of the most easily recognized hawthorns. Common and widespread, it has been planted for ornament and as a hedge since colonial times.

246, 449 Broadleaf Hawthorn
"Apple-leaf Hawthorn"
Crataegus dilatata Sarg.

Description: Much-branched shrub or small tree with straight trunk, broad, rounded crown of spreading branches, large flowers, and large, scarlet fruit with dark dots.
Height: 20′ (6 m).
Diameter: 4″ (10 cm).
Leaves: 2–2½″ (5–6 cm) long, and almost as wide. *Broadly ovate,* short-pointed at tip, notched or rounded at base; coarsely doubly *saw-toothed and usually with several shallow lobes;* hairy when young. Dark green and becoming hairless above, paler beneath.
Bark: gray-brown; scaly.
Twigs: short, *hairless,* with few spines.
Flowers: 1″ (2.5 cm) wide; with 5 white petals, 20 *rose-red stamens,* and 5 styles; 8–12 flowers on long hairy stalks in broad clusters; in spring.
Fruit: *large,* to ¾″ (19 mm) in diameter; *scarlet with many dark dots;* enlarged reddish gland-toothed calyx; thick juicy sweet pulp; 5 nutlets; many in drooping clusters; maturing and falling in early autumn.

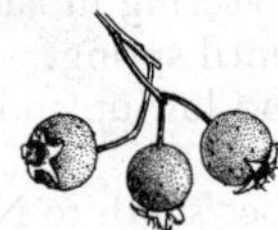

Habitat: From borders of salt marshes and streams to moist hillsides.
Range: S. Ontario, S. Quebec, New England, and New York; to 1000′ (305 m).

The Latin species name, meaning "dilated" or "spread out," describes the broad leaves.

245 Fanleaf Hawthorn
Crataegus flabellata (Bosc) K. Koch

Description: Much-branched shrub or small tree with irregular crown.
Height: 20′ (6 m).
Diameter: 6″ (15 cm).
Leaves: 1½–3″ (4–7.5 cm) long, 1¼–

2½″ (3–6 cm) wide. Broadly ovate or *fan-shaped;* short- or long-pointed at tip, blunt or *rounded at base;* sharply and doubly saw-toothed with 4–6 shallow lobes. Green, becoming nearly hairless above, paler beneath; turning yellow and brown in autumn.
Bark: dark gray or brown; scaly.
Twigs: stout, hairless, with many long slender curved spines.
Flowers: ⅝″ (15 mm) or more wide; with 5 white petals, 5–20 *dark red stamens,* and 3–5 styles; 5–15 flowers on slender stalks in compact hairless clusters; in spring.
Fruit: ½″ (12 mm) in diameter; *rounded;* red; soft sweet pulp; 3–5 nutlets; few in drooping clusters; maturing in autumn.

Habitat: Rocky upland soils in thickets, borders of forests, pastures, and old fields.

Range: Newfoundland and Nova Scotia south to Georgia, west to Louisiana, and north to Minnesota; to 3000′ (914 m), locally in southern Appalachians to 6000′ (1829 m).

One of the common hawthorns, known also by the scientific name *Crataegus macrosperma* Ashe. It is widely distributed in mountains and is the most common hawthorn in the Great Smoky Mountains National Park. The Latin species name, meaning "fanlike," refers to the shape of the broad leaves.

218 Yellow Hawthorn
"Summer Haw"
Crataegus flava Ait.

Description: Thicket-forming shrub or small tree with broad, open crown of stout branches and yellow or orange fruit.
Height: 20′ (6 m).
Diameter: 8″ (20 cm).
Leaves: 2–2½″ (5–6 cm) long, 1½–2″ (4–5 cm) wide. *Elliptical to broadly*

obovate, short-pointed at tip, *gradually narrowed to base;* coarsely *doubly saw-toothed* with *gland-tipped teeth;* sometimes with few shallow lobes. *Yellow-green* and becoming hairless with sunken veins above, paler and hairy on veins beneath. Leafstalks long and hairy, with *gland-dots.*

Bark: dark brown, becoming reddish-tinged; thick and furrowed into narrow ridges.

Twigs: hairy when young; with straight spines.

Flowers: ¾″ (19 mm) wide; with 5 white petals, 10–20 *rose-red stamens,* and 3–5 styles; 4–7 flowers in compact, hairy clusters; in spring.

Fruit: ½″ (12 mm) in diameter; *elliptical; orange, brown, or yellow; tubular gland-toothed calyx* often shedding early; thick dry mealy pulp; 3–5 nutlets; few, in drooping clusters; maturing and falling in autumn.

Habitat: Dry, sandy and gravelly soils, at woodland borders and in pine forests.

Range: Virginia southwest to N. Florida, Mississippi, and E. Tennessee; to about 1500′ (457 m).

The Latin species name, meaning "yellow," refers to the fruit, which is often that color. It is used locally to make jelly.

Gregg Hawthorn

Crataegus greggiana Eggl.

Description: Shrub or small tree with *small flowers* and *hairy red fruit.*

Height: 20′ (6 m).

Diameter: 6″ (15 cm).

Leaves: 1½–3″ (4–7.5 cm) long, 1¼–2¼″ (3–6 cm) wide. *Broadly ovate; blunt at ends;* doubly saw-toothed, sometimes lobed near tip; *slightly thickened;* hairy when young. *Dull green and rough* above, pale and nearly hairless beneath.

Bark: brown or gray; thin, scaly.
Twigs: hairy when young; with many straight spines.
Flowers: less than ½″ (12 mm) wide; with 5 white petals, 10 yellow stamens, and 3–5 styles; 5–12 flowers in compact hairy clusters; in early spring.

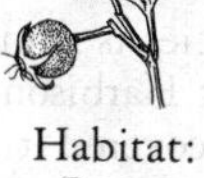

Fruit: to ½″ (12 mm) in diameter; few, on densely hairy stalks; *red, hairy;* calyx-lobes pressed against surface; *thick dry pulp;* 3–5 nutlets; maturing in autumn.

Habitat: Dry rocky soils in woodlands.

Range: Edwards Plateau in central Texas south to NE. Mexico; at 900–1600′ (274–488 m).

Preserves are prepared from the sweetish fruit. The common and Latin species names honor Josiah Gregg (1806–50), early American explorer.

---

231 Harbison Hawthorn

Crataegus harbisonii Beadle

Description: Small tree with open, rounded crown of widespreading branches.
Height: 20′ (6 m).
Diameter: 8″ (20 cm).
Leaves: 1½–3″ (4–7.5 cm) long, 1–2½″ (2.5–6 cm) wide. *Broadly ovate to elliptical, short-pointed to rounded at ends;* irregularly and coarsely saw-toothed; sometimes slightly lobed and larger; hairy when young; leafstalks with black-tipped glands. Shiny dark green and slightly *rough* with sunken veins above, paler and hairy beneath; turning red in autumn.
Bark: light gray; fissured and scaly; the *spines sometimes branched.*
Twigs: covered with white hairs when young, usually with stout straight spines.
Flowers: ¾″ (19 mm) wide; with 5 white petals, 10–20 yellow stamens, 3–5 styles, and *gland-toothed calyx;*

3–10 flowers in broad silky hairy clusters; in spring.

Fruit: ⅝″ (15 mm) in diameter; *round,* bright *red;* thick dry mealy pulp; 3–5 nutlets; maturing in autumn.

Habitat: Rocky upland slopes and ridges, old fields and woods.

Range: Tennessee, Alabama, and Mississippi; at 100–500′ (30–152 m).

This rare and local species is named for botanist Thomas Grant Harbison (1862–1936), who discovered it near Nashville, Tennessee.

250, 450 **Biltmore Hawthorn**
"Thicket Hawthorn" "Allegheny Hawthorn"
Crataegus intricata Lange

Description: Much-branched, thicket-forming shrub or small tree with irregular, open crown.
Height: 20′ (6 m).
Diameter: 6″ (15 cm).
Leaves: 1–2½″ (2.5–6 cm) long, 1–2″ (2.5–5 cm) wide. *Elliptical or broadly ovate;* mostly blunt at ends; sharply or *doubly saw-toothed,* with *gland-tipped teeth; often shallowly lobed;* hairless or nearly so. Green above, paler beneath. Stout leafstalks often with *gland-dots.*
Bark: brown or gray; scaly.
Twigs: hairless, with long spines.
Flowers: ⅝″ (15 mm) wide; with 5 white petals, 10 pale yellow or pink stamens, and 3–5 styles; short-stalked; usually 4–8 flowers in compact, slightly hairy clusters; in late spring.
Fruit: nearly ½″ (12 mm) in diameter; *rounded to elliptical, greenish or reddish-brown;* large gland-toothed calyx often shedding early; thick hard dry pulp; 3–5 nutlets; maturing and usually falling in autumn.

Habitat: Dry uplands and moist valleys, open areas, borders of woods, stream banks.

Range: S. Ontario east to New Hampshire, south to Georgia, west to SE. Oklahoma, and north to S. Michigan; to 3000′ (914 m) in southern Appalachians.

The common name honors the Biltmore Estate in North Carolina, where early studies of hawthorns were made. The Latin species name, meaning "entangled," describes the branches.

217 **Pensacola Hawthorn**
"Weeping Hawthorn" "Sandhill Hawthorn"
Crataegus lacrimata Small

Description: Much-branched shrub or small tree with rounded crown of *long, slender, drooping branches.*
Height: 20′ (6 m).
Diameter: 4″ (10 cm).
Leaves: *small,* ⅜–1″ (1–2.5 cm) long, ¼–¾″ (6–19 mm) wide. *Spoon-shaped* or obovate; saw-toothed toward rounded tip, with curved *gland-tipped teeth;* broadest beyond middle and gradually tapering to narrow short-stalked base; *3 main veins; slightly thickened. Shiny green and hairless* above, paler with hairs in vein angles beneath.
Bark: dark gray; thick, scaly or deeply furrowed.
Twigs: slender, *very zigzag,* hairless, with many slender *short spines.*
Flowers: ⅝″ (15 mm) wide; with 5 white petals, 20 light yellow stamens, and 3 styles; 2–5 flowers in short-stalked hairless clusters; in early spring.
Fruit: ⅜″ (10 mm) in diameter; rounded, dull *yellow or orange to reddish* with *dark dots;* tubular gland-toothed calyx often shedding early; thin dry mealy pulp; 3 nutlets; maturing in late summer.

Habitat: Dry or moist sandy soils with scrub oaks and on stream banks.

Range: NW. Florida (Walton to Escambia counties); to 300′ (91 m).

This very local but common species is named for its distribution around Pensacola. The Latin species name, meaning "of tears," was suggested by the "weeping" branches. Easily recognized also by the small, spoon-shaped leaves.

242, 451 **Parsley Hawthorn**
Crataegus marshallii Eggl.

Description: Small tree with wide-spreading, slender branches and broad, irregular, open crown of parsleylike foliage; or often, a low much-branched shrub.

Height: 20′ (6 m).
Diameter: 4″ (10 cm).
Leaves: ¾–2″ (2–5 cm) long, ¾–1½″ (2–4 cm) wide. Broadly ovate; *deeply divided nearly to midvein into 5–7 narrow short-pointed saw-toothed lobes; veins running to notches* as well as to points of lobes; hairless or nearly so; very long slender leafstalks. Shiny green above, paler beneath.
Bark: gray; thin, smooth, peeling off in patches and mottled with brown.
Twigs: light brown; hairy when young; with straight spines.
Flowers: ⅝″ (15 mm) wide; with 5 white petals, about 10 red stamens, and 1–3 styles; 3–12 flowers clustered on long slender hairy stalks; in spring.
Fruit: ⅜″ (10 mm) long, and half as wide; *oblong, bright red;* thin juicy pulp; 1–3 (usually 2) nutlets; maturing in autumn, persisting until winter.

Habitat: Moist valley soils.

Range: SE. Virginia south to central Florida, west to E. Texas, and north to SE. Missouri; to 600′ (183 m).

One of the easiest hawthorns to recognize, with its small, divided leaves and small, oblong fruit. The Latin species name honors Humphry Marshall (1722–1801), U.S. botanist.

220, 452 **Downy Hawthorn**
Crataegus mollis Scheele

Description: Handsome tree with tall trunk and compact, rounded crown of spreading branches, large broad hairy leaves, *many large flowers,* and large scarlet fruit.

Height: 40′ (12 m).
Diameter: 1′ (0.3 m).
Leaves: 3–4″ (7.5 – 10 cm) long and wide. *Broadly ovate,* short-pointed at tip, rounded or slightly notched at base; doubly saw-toothed; with *4–5 veins on each side* ending in shallow pointed lobes; *densely covered with white hairs when young.* Dark yellow-green above, pale and slightly hairy beneath.
Bark: brown to gray; fissured into scaly plates; becoming thick.
Twigs: *covered with white hairs when young;* with stout spines, though sometimes nearly thornless.
Flowers: to 1″ (2.5 cm) wide; with 5 white petals, 20 light yellow stamens, and 4–5 styles; in broad clusters; in spring.
Fruit: ¾″ (19 mm) in diameter; nearly round or short oblong; *scarlet or crimson* with dark dots; slightly hairy; thick juicy edible pulp; 4–5 nutlets; few, in drooping clusters; maturing in late summer or autumn.

Habitat: Moist soil of valleys and hillsides in open woods.

Range: S. Quebec and Nova Scotia, south to West Virginia and Alabama, west to S. central Texas, and north to SE. North Dakota; to 1500′ (457 m) or higher.

One of the largest trees of its genus, Downy Hawthorn was originally called "White Thorn." It was introduced into

European gardens as early as 1683. The common and Latin species names both refer to the soft hairy foliage.

---

210 **Riverflat Hawthorn**
"May Hawthorn" "Apple Haw"
Crataegus opaca Hook. & Arn.

Description: Small tree with relatively tall trunk and narrow, rounded crown; or a large shrub.

Height: 30′ (9 m).
Diameter: 8″ (20 cm).
Leaves: 2–2½″ (5–6 cm) long, ½–1″ (1.2–2.5 cm) wide. *Elliptical,* gradually narrowed from middle to short-pointed ends; *finely and wavy saw-toothed* beyond middle with tiny gland-tipped teeth; *covered with gray hairs when young.* Dull dark green and hairless or nearly so above, *with rust-brown hairs* beneath, especially on veins.
Bark: dark reddish-brown; fissured and scaly.
Twigs: hairy when young; sometimes with stout spines.
Flowers: 1″ (2.5 cm) wide; with 5 white petals, 20 pink stamens, and 3–5 styles; short-stalked; *before* or with *leaves* in early spring.

Fruit: ⅝″ (15 mm) in diameter; *round,* shiny *red,* with pale dots; *juicy, slightly sour* edible pulp; 3–5 nutlets; maturing in *late spring.*

Habitat: In pure stands in wet soil of riverbanks and borders of swamps.

Range: SW. Alabama west to S. Arkansas and E. Texas; to 300′ (91 m).

The large, edible fruit is made into preserves and jellies.

243, 448, 552 **Washington Hawthorn**
"Washington-thorn"
Crataegus phaenopyrum (L. f.) Medic.

Description: Shrub or small tree with short trunk and regular, rounded crown of upright branches, abundant small flowers in late spring, many small, round, red fruit, and brilliant autumn foliage; hairless throughout.

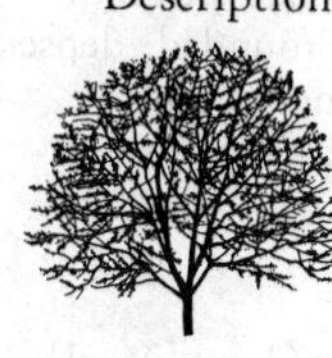

Height: 30′ (9 m).
Diameter: 1′ (0.3 m).
Leaves: 1½–2½″ (4–6 cm) long, 1–1¾″ (2.5–4.5 cm) wide. *Broadly ovate* to triangular or *3-lobed;* short-pointed at tip; nearly straight to slightly notched at base; *coarsely saw-toothed;* often with 5 shallow lobes; slightly hairy when young. Tinged with red, becoming shiny dark green above, paler beneath; turning scarlet and orange in autumn.
Bark: light brown; smooth; thin, becoming scaly.
Twigs: shiny brown, with slender spines.
Flowers: more than ½″ (12 mm) wide; with 5 white petals, 20 pale yellow stamens, and 3–5 styles; many flowers in compact hairless clusters; in *late spring.*
Fruit: ¼″ (6 mm) in diameter; *shiny red* or scarlet, with ring scar from *shed calyx;* thin dry pulp; 3–5 nutlets exposed at ends; maturing in autumn and *persisting* until spring.

Habitat: Moist soil of valleys.

Range: Virginia south to N. Florida, west to Arkansas, and north to S. Missouri; naturalized locally northeast to Massachusetts; to 2000′ (610 m).

One of the showiest and most desirable hawthorns for planting. In the early 19th century, it was introduced into Pennsylvania from Washington, D.C., as a hedge plant and is thus called "Washington-thorn." The Latin species name refers to the pearlike foliage.

244 Frosted Hawthorn
"Waxy-fruit Thorn"
Crataegus pruinosa (H. L. Wendl.) K. Koch

Description: Handsome, much-branched shrub or small tree, with broad, rounded, dense crown of spreading branches, large white flowers, and dark purplish-red fruit.
Height: 20′ (6 m).
Diameter: 4″ (10 cm).
Leaves: 1–2″ (2.5–5 cm) long, ¾–1½″ (2–4 cm) wide. *Elliptical* or ovate, short-pointed at tip, blunt at base; irregularly and *often doubly saw-toothed,* with 6–8 shallow lobes; reddish when unfolding, becoming hairless and slightly thickened; leafstalks often reddish. *Dark blue-green* above *(often with whitish bloom),* paler beneath; turning orange-red in autumn.
Bark: gray; scaly, thin, becoming rough and slightly furrowed.
Twigs: hairless, with many straight spines.
Flowers: nearly 1″ (2.5 cm) wide; with 5 white petals, 20 *rose-red or pink stamens,* and 5 styles; few flowers on long stalks in hairless clusters; in *late spring.*
Fruit: ⅜–⅝″ (10–15 mm) in diameter; *often slightly angled;* with *whitish bloom,* turning *dark purple-red* with many small dots; thick mealy sweet edible pulp; 5 nutlets; maturing in late autumn.

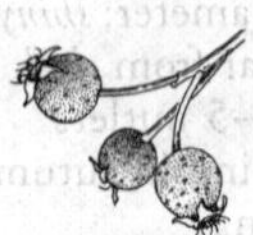

Habitat: Hillsides, especially in limestone soils.
Range: S. Quebec and Newfoundland south to North Carolina, west to E. Oklahoma, and north to Wisconsin; to 2500′ (762 m) in southern Appalachians.

The Latin species name, meaning "with a frostlike bloom," describes the fruit.

249 Beautiful Hawthorn
Crataegus pulcherrima Ashe

Description: Much-branched shrub or small tree with spreading, rounded crown and showy flowers.
Height: 20′ (6 m).
Diameter: 4″ (10 cm).
Leaves: 1–2″ (2.5–5 cm) long, ¾–1½″ (2–4 cm) wide. *Broadly ovate,* elliptical, or rounded; short-pointed at tip, gradually narrowed to base; finely saw-toothed; *4–6 veins on each side* ending in shallow lobes; becoming *hairless.* Shiny green above, paler beneath; turning orange, yellow, or brown in autumn.
Bark: gray or brown; scaly.
Twigs: brown, hairless, often with long slender spines.
Flowers: ¾″ (19 mm) wide; with 5 white petals, about 20 *purplish stamens,* and 3–5 styles; *3–10 flowers in compact hairless clusters;* in spring.

Fruit: *very small,* 5/16″ (8 mm) in diameter; *red, rounded;* thin dry mealy pulp; 3–5 nutlets; maturing in autumn.

Habitat: Moist soils of stream banks, upland forests, and old fields.

Range: SW. Georgia, N. Florida, and S. Alabama; to 500′ (152 m).

Common and Latin species names both refer to the beautiful flowers. This distinct local species was named in 1900 by William Willard Ashe (1872–1932), U.S. forester.

232 Dotted Hawthorn
"Large-fruit Thorn"
Crataegus punctata Jacq.

Description: Small, thicket-forming tree with round or flattened, dense crown of stout, spreading branches, often broader than high.
Height: 30′ (9 m).

Diameter: 1′ (0.3 m).
Leaves: 2–3″ (5–7.5 cm) long, ¾–2″ (2–5 cm) wide. *Obovate or elliptical,* blunt at tip, *broadest beyond middle* and tapering to base; sharply and often doubly saw-toothed and lobed; slightly thickened. *Dull green* with *sunken veins* and becoming hairless above, paler and hairy beneath; turning orange or scarlet in autumn.
Bark: gray or brown; fissured and scaly.
Twigs: hairy when young; usually with many spines.
Flowers: ½–¾″ (12–19 mm) wide; with 5 white petals, 20 pink or yellow stamens, and 2–5 styles; numerous flowers in compact hairy clusters; in late spring.
Fruit: ½–¾″ (12–19 mm) in diameter; *rounded* or slightly elliptical; *dull red* or sometimes yellow, with *whitish dots;* thick mellow pulp; 2–5 nutlets; many, in drooping clusters; maturing and falling in autumn.

Habitat: Moist soils of valleys and on rocky upland slopes, especially on limestone; in forest openings and borders.

Range: S. Quebec and Newfoundland south to Georgia, west to E. Oklahoma, and north to Minnesota; to nearly 6000′ (1829 m) in southern Appalachians.

The common and Latin species names refer to the small dots on the large fruit. Introduced into cultivation as an ornamental in Europe at an early date, and named in 1770 from plants grown in the Botanic Garden in Vienna.

---

252 Reverchon Hawthorn
Crataegus reverchonii Sarg.

Description: Much-branched, thicket-forming shrub or sometimes a small tree with rounded crown, becoming *hairless throughout.*
Height: 23′ (7 m).
Diameter: 4″ (10 cm).

Leaves: 1¼–1½″ (3–4 cm) long, ¾–1″ (2–2.5 cm) wide. *Elliptical to obovate, rounded at tip,* short-pointed at base; doubly saw-toothed; *slightly thick and leathery. Shiny dark green* above, dull and paler beneath.
Bark: gray to orange-brown; scaly.
Twigs: light brown, slender, with very long slender straight spines.
Flowers: ½″ (12 mm) wide; with 5 white petals, 10–15 yellow stamens, and 2–5 styles; in compact clusters; in spring.
Fruit: about ⅜″ (10 mm) in diameter; *rounded, shiny or dull red* with dark dots; thick dry or juicy sweet pulp; usually 2–4 nutlets; maturing in autumn.

Habitat: Moist soils along streams and on rocky hillsides in prairies and in open woods.

Range: S. Missouri and SE. Kansas south to central Texas and Arkansas; at 400–1500′ (122–457 m).

A southwestern relative of the widespread eastern species, Cockspur Hawthorn. Named for Julien Reverchon (1837–1905), a French-born Texan.

---

240 Littlehip Hawthorn

"Small-fruit Hawthorn" "Pasture Hawthorn"
Crataegus spathulata Michx.

Description: Much-branched shrub or small tree with broad, open crown of spreading branches, spoon-shaped leaves, small white flowers, and small red fruit.
Height: 25′ (7.6 m).
Diameter: 8″ (20 cm).
Leaves: ¾–1½″ (2–4 cm) long, ⅜–1″ (1–2.5 cm) wide. *Spoon-shaped;* usually *rounded at tip; wavy saw-toothed; sometimes 3-lobed; gradually narrowed* from middle to long-pointed base; fine, nearly parallel *veins running both to notches* and points of lobes; hairy only when young.

Shiny dark green above, paler beneath.
Bark: brown; thin, peeling off in patches and mottled with brown.
Twigs: hairless, usually with straight spines.
Flowers: ⅜″ (10 mm) wide; with 5 white petals, about 20 pale yellow stamens, and 2–5 styles; numerous flowers in short-stalked compact hairless clusters; in spring.
Fruit: ¼″ (6 mm) in diameter; *rounded, red;* thin dry mealy pulp; 3–5 nutlets; maturing in autumn.

Habitat: Wet or moist soils of stream banks and swamps bordering hardwood and pine forests and upland slopes in open woods and clearings.

Range: Virginia south to N. Florida, west to E. Texas, and north to S. Missouri; to about 1000′ (305 m).

The Latin species name means "like a spatula," referring to the spoon-shaped leaves. The common name "Littlehip" describes the small, roselike fruit.

---

226 Fleshy Hawthorn

"Long-spine Hawthorn" "Succulent Hawthorn"
Crataegus succulenta Schrad.

Description: Shrub or small tree with a short trunk and broad irregular, dense crown of stout branches.
Height: 20′ (6 m).
Diameter: 6″ (15 cm).
Leaves: 2–2½″ (5–6 cm) long 1–1½″ (2.5–4 cm) wide. *Elliptical,* gradually narrowed from middle to base; doubly saw-toothed, *shallow-lobed beyond middle;* with 4–7 fine *sunken veins* on each side; slightly thickened; hairy when young. Shiny dark green above, pale yellow-green with fine hairs on midvein beneath.
Bark: dark red-brown; scaly.
Twigs: stout, hairless, with many long,

stout, slightly curved spines.
Flowers: ⅝–¾″ (15–19 mm) wide; with 5 white petals, 10–20 *white or pink stamens,* and 2–4 styles; in broad hairy clusters; in late spring.

Fruit: ½–⅝″ (12–15 mm) in diameter; *bright red;* gland-toothed calyx at tip; *thick juicy sweet pulp;* 2–4 nutlets; in drooping clusters on long stalks; maturing in early autumn.

Habitat: Moist soils of valleys and open upland areas.

Range: S. Manitoba east to Nova Scotia, south to W. North Carolina, and west to Kansas and Nebraska; to 3000′ (914 m); locally to Colorado at 5000–7000′ (1524–2134 m).

This handsome ornamental sometimes flowers when a low shrub. The names both refer to the succulent, soft fruit.

---

248 Texas Hawthorn
Crataegus texana Buckl.

Description: Shrub or small tree with tall trunk and broad, irregular or rounded crown, many large clusters of flowers, and many large, edible, pale-dotted red fruit.
Height: 30′ (9 m).
Diameter: 1′ (0.3 m).
Leaves: 3–4″ (7.5–10 cm) long, 2½–3″ (6–7.5 cm) wide. *Broadly ovate,* short-pointed at ends; coarsely doubly saw-toothed, *usually with 8–10 shallow* lobes beyond middle; *thick.* Shiny dark green above and becoming hairless, paler and hairy with prominent veins beneath.
Bark: reddish-brown to gray; scaly.
Twigs: covered with white hairs when young, with long slender spines, or sometimes spineless.
Flowers: ¾″ (19 mm) wide; *hairy,* with 5 white petals, 20 red stamens, and 5 styles; on long hairy stalks; in large clusters; in spring.

Fruit: ½–¾″ (12–19 mm) in diameter; *rounded or slightly oblong, bright red with large pale dots,* hairy when young; gland-toothed calyx usually present; thick sweet edible pulp; 5 nutlets; many in drooping clusters; maturing in autumn.

Habitat: Moist soils of river flood plains and valleys.

Range: SE. and S. Texas, mostly near coast; to 600′ (183 m).

This species of limited range within Texas was first recognized and named in 1862 by Samuel Botsford Buckley (1809–84), a Texas naturalist.

Threeflower Hawthorn

Crataegus triflora Chapm.

Description: Large shrub or small tree with *large white flowers,* usually 3 in a cluster.
Height: 20′ (6 m).
Diameter: 4″ (10 cm).
Leaves: ¾–2¾″ (2–7 cm) long, ⅜–2″ (1–5 cm) wide. *Ovate* to elliptical, short-pointed at tip, rounded or blunt at base; saw-toothed, often with shallow short-pointed lobes; leafstalks with gland-hairs. Green above, covered with *soft hairs* beneath.
Bark: brown; scaly.
Twigs: hairy, with straight spines to 2″ (5 cm) long.
Flowers: 1″ (2.5 cm) or more wide; with 5 white petals, 20 pale yellow stamens, and 3–5 styles; *usually 3* (2–5) flowers in a cluster on hairy stalks; in spring.

Fruit: to ⅝″ (15 mm) in diameter; *red, hairy;* gland-toothed calyx at tip; 3–5 nutlets; maturing in autumn.

Habitat: Rocky banks of streams.

Range: Georgia, Alabama, and Mississippi; to 600′ (183 m).

Common and Latin species names both refer to the number of flowers usually found in a cluster. This distinct local species was named by Alvan Wentworth Chapman (1809–99) in his *Flora of the Southern United States*.

234 Oneflower Hawthorn

"Dwarf Hawthorn"
Crataegus uniflora Muenchh.

Description: A low spreading shrub or sometimes a small bushy tree, with short, stout trunks and dense, rounded crown of stout, crooked branches.
Height: 2–5′ (0.6–1.5 m), sometimes to 16′ (5 m).
Diameter: 6″ (15 cm).
Leaves: ¾–1½″ (2–4 cm) long, ½–1¼″ (1.2–3.2 cm) wide. *Spoon-shaped, broadest beyond middle* and tapering to almost stalkless base; coarsely and wavy saw-toothed, sometimes lobed; *slightly thick;* with sunken veins. Shiny dark green, rough, and becoming hairless above, paler and hairy beneath.
Bark: gray-brown; smooth, becoming fissured into scaly plates near base.
Twigs: stiff, hairy, with *many slender spines.*
Flowers: nearly ⅝″ (15 mm) wide; with 5 white petals, about 20 pale yellow or white stamens, and 3–5 styles; *usually solitary* (sometimes 2–3) on short stalks; in late spring after leaves.
Fruit: to ½″ (12 mm) in diameter; rounded; *yellow to dull red* to brown; *large gland-toothed calyx* at tip; *hard dry* mealy pulp; 3–5 nutlets; maturing in late summer or autumn.

Habitat: Dry sandy and rocky uplands in open woods, forest borders, and old fields.

Range: New York (Long Island) and New Jersey south to N. Florida, west to E. Texas, and northeast to S. Missouri; to 2000′ (610 m) or above.

This species occurs in tree form only in northern Florida. Sometimes planted as a curiosity in botanical gardens because of its dwarf size.

221, 453 **Green Hawthorn**
"Southern Hawthorn"
Crataegus viridis L.

Description: Thicket-forming tree with straight, often fluted, trunk and rounded, dense crown of spreading branches, shiny foliage, showy flowers, and small red to yellow fruit.
Height: 40′ (12 m).
Diameter: 1½′ (0.5 m).
Leaves: 1–2½″ (2.5–6 cm) long, ½–1½″ (1.2–4 cm) wide. *Elliptical or nearly 4-angled,* short-pointed at tip, *gradually narrowed to long-pointed base; finely saw-toothed;* sometimes slightly 3-lobed; *shiny dark green* and becoming hairless above, pale with tufts of hairs in vein angles beneath; often turning scarlet in autumn.
Bark: pale gray with orange-brown inner bark; scaly.
Twigs: gray, hairless, *usually without spines.*
Flowers: ⅝″ (15 mm) wide; with 5 white petals, about 20 pale yellow stamens, and usually 5 (2–5) styles; *many* flowers (mostly 8–20), on long slender stalks in branching clusters; in spring.

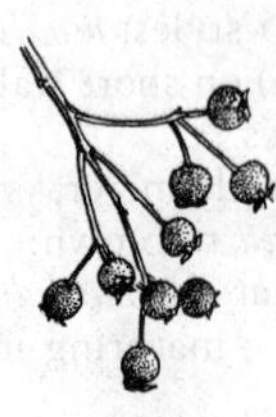

Fruit: ¼″ (6 mm) in diameter; *bright red,* orange-red, or yellow; thin juicy pulp; usually 5 nutlets; many, in drooping clusters; maturing in autumn and persisting into winter.

Habitat: Wet or moist soils of valleys and low upland slopes.

Range: Delaware south to N. Florida, west to E. Texas, and north to SW. Indiana; to about 500′ (152 m).

In 1753 Linnaeus gave this species its Latin name, meaning "green," from a specimen with shiny green foliage sent from Virginia.

121 Southern Crab Apple

"Narrow-leaf Crab Apple"
Malus angustifolia (Ait.) Michx.

Description: Small tree with short trunk, spreading branches, and broad, open crown.
Height: 30′ (9 m).
Diameter: 1′ (0.3 m).
Leaves: 1–2¾″ (2.5–7 cm) long, ½–¾″ (12–19 mm) wide. *Elliptical* or oblong, usually *blunt at tip; wavy saw-toothed;* hairy when young. Dull green above, paler beneath; turning brown in autumn. Leafstalks slender, ½–¾″ (12–19 mm) long, hairy when young.
Bark: gray or brown; furrowed into narrow scaly ridges.
Twigs: brown; densely covered with hairs when young.
Flowers: 1–1½″ (2.5–4 cm) wide; with 5 rounded pink petals; in clusters, on long stalks; fragrant; in spring.
Fruit: ¾–1″ (2–2.5 cm) in diameter; *like small apples;* yellow-green, sour, long-stalked; maturing in late summer.

Habitat: Moist soils of valleys and lower slopes, borders of forests, fence rows, and old fields.

Range: S. Virginia south to N. Florida, west to Louisiana, and north to Arkansas; local from S. New Jersey to S. Ohio, and in SE. Texas; to 2000′ (610 m).

This is the crab apple that grows at low altitudes in the Southeast, often forming thickets. Quantities of the fruit are consumed by bobwhites, grouse, pheasants, rabbits, squirrels, opossums, raccoons, skunks, and foxes. The hard, heavy wood has been used to make tool handles.

135, 458, 572 **Sweet Crab Apple**
"Garland-tree"
Malus coronaria (L.) Mill.

Description: A small tree with a short trunk and several stout branches forming broad, open crown.

Height: 30′ (9 m).
Diameter: 1′ (0.3 m).
Leaves: 2–4″ (5–10 cm) long, 1½″ (4 cm) wide. *Ovate; coarsely saw-toothed* beyond middle; *slightly lobed on young twigs;* both blades and leafstalks with fine reddish hairs when young. Yellow-green above, paler beneath; turning yellow in autumn.
Bark: red-brown; fissured and scaly.
Twigs: red-brown; covered with gray hairs when young.
Flowers: nearly 1½″ (4 cm) wide; with 5 rounded white or pink petals; in clusters, on long stalks; in spring.
Fruit: 1–1¼″ (2.5–3 cm) in diameter; *like a small apple;* yellow-green, long-stalked; maturing in late summer.

Habitat: Moist soils in openings and borders of forests.

Range: S. Ontario east to New York, south to extreme N. Georgia, west to NE. Arkansas, and north to N. Illinois; to 3300′ (1006 m) in southern Appalachians.

The common crab apple of the Ohio Valley, it is sometimes planted as an ornamental. Double-flowered varieties have a greater number of larger and deeper pink flowers. The fruit can be made into preserves and cider.

130, 457 **Prairie Crab Apple**
"Iowa Crab"
Malus ioensis (Wood) Britton

Description: A sometimes spiny shrub or small tree, with spreading branches and broad, open crown.

Height: 10–30′ (3–9 m).
Diameter: 1′ (0.3 m).
Leaves: 2½–4″ (6–10 cm) long, 1–1½″ (2.5–4 cm) wide. Elliptical; *wavy saw-toothed, often slightly lobed; with prominent veins; slightly thickened;* hairy when young; leafstalks slender and hairy. Shiny dark green above, *paler and hairy beneath;* turning yellow in autumn.
Bark: reddish-brown; scaly.
Twigs: reddish-brown; hairy when young.
Flowers: 1½–2″ (4–5 cm) wide; 5 *rounded pink or white petals;* in clusters, on long stalks; in spring.

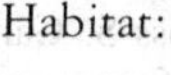

Fruit: 1–1¼″ (2.5–3 cm) in diameter; *like a small apple;* yellow-green, long-stalked; maturing in late summer.

Habitat: Moist soils along streams in prairie and forest borders.

Range: N. Indiana south to Arkansas and Oklahoma, and north to extreme SE. South Dakota; local in central Texas and Louisiana; at 500–1500′ (152–457 m).

This is the crab apple of the eastern prairie region in the upper Mississippi Valley. A handsome double-flowered variety is grown as an ornamental. Numerous species of birds, including bobwhites and pheasants, and squirrels, rabbits, and other mammals consume the fruit.

133, 456, 570 **Apple**
"Common Apple" "Wild Apple"
Malus sylvestris (L.) Mill.

Description: This familiar fruit tree naturalized locally has a short trunk, spreading rounded crown, showy pink-tinged blossoms, and delicious red fruit.
Height: 30–40′ (9–12 m).
Diameter: 1–2′ (0.3–0.6 m).
Leaves: 2–3½″ (5–9 cm) long, 1¼–2¼″ (3–6 cm) wide. *Ovate* or elliptical; *wavy saw-toothed;* hairy leafstalk. Green

above, *densely covered with gray hairs beneath.*
Bark: gray; fissured and scaly.
Twigs: greenish, turning brown; densely covered with white hairs when young.
Flowers: 1¼″ (3 cm) wide; with 5 *rounded petals, white tinged with pink;* in early spring.
Fruit: the familiar *edible apple;* 2–3½″ (5–9 cm) in diameter; shiny red or yellow; sunken at ends; thick sweet pulp; *star-shaped core* contains up to 10 seeds; matures in late summer.

Habitat: Moist soils near houses, fences, roadsides, and clearings.

Range: Native of Europe and W. Asia; naturalized locally across S. Canada, in E. continental United States, and in Pacific states.

The Apple has been cultivated since ancient times. Numerous improved varieties have been developed from this species and from hybrids with related species. Although well known, it is sometimes not recognized when growing wild. For nearly fifty years Jonathan Chapman (1774–1845), better known as Johnny Appleseed, traveling mostly on foot, distributed apple seeds to everybody he met. With seeds from cider presses, he helped to establish orchards from Pennsylvania to Illinois. Wildlife consume quantities of fallen fruit after harvest.

122, 445 **Allegheny Plum**
"Allegheny Sloe" "Sloe Plum"
Prunus alleghaniensis Porter

Description: An uncommon, thicket-forming shrub or small tree with many upright branches, profuse blooms in spring, and small plums.
Height: 20′ (6 m).
Diameter: 6″ (15 cm).

Leaves: 2–3½″ (5–9 cm) long, ¾–1¼″ (2–3 cm) wide. Narrowly elliptical or lance-shaped; *2 dotlike glands* at base; finely and *sharply saw-toothed;* slightly thick; covered with soft hairs when young. Dark green above, paler, often hairy beneath.
Bark: dark brown or gray; scaly.
Twigs: dark *reddish-brown;* hairy when young; short twigs sometimes ending in spine.
Flowers: ½″ (12 mm) wide; with 5 rounded petals, white turning pink; 2–4 flowers on slender stalk; in early spring with leaves.
Fruit: a *plum* ⅜–½″ (10–12 mm) in diameter; *reddish-purple* with whitish bloom; sour yellow pulp; large stone; maturing in late summer.

Habitat: Moist soils and dry rocky mountain slopes.

Range: Mountains from E. Pennsylvania south to E. West Virginia; local in NE. Tennessee and Connecticut; at 1200–2000′ (366–610 m).

The fruit can be made into preserves and jelly. Allegheny Plum's abundant flowers and showy fruit make it useful as an ornamental. Listed as a rare tree because of its restricted occurrence.

125, 420, 568 **American Plum**
"Red Plum" "River Plum"
Prunus americana Marsh.

Description: A thicket-forming shrub or small tree with short trunk, many spreading branches, broad crown, showy large white flowers, and red plums.

Height: 30′ (9 m).
Diameter: 1′ (0.3 m).
Leaves: 2½–4″ (6–10 cm) long, 1¼–1¾″ (3–4.5 cm) wide. *Elliptical, long-pointed* at tip; *sharply and often doubly saw-toothed;* slightly thickened. *Dull green* with *slightly sunken veins* above,

paler and often slightly hairy on veins beneath.
Bark: dark brown; scaly.
Twigs: light brown, slender, hairless; short twigs ending in spine.
Flowers: ¾–1″ (2–2.5 cm) wide; with 5 rounded white petals; in clusters of *2–5 on slender equal stalks;* slightly unpleasant odor; in early spring before leaves.
Fruit: a *plum* ¾–1″ (2–2.5 cm) in diameter; thick *red* skin; juicy sour edible pulp; large stone; maturing in summer.

Habitat: Moist soils of valleys and low upland slopes.

Range: SE. Saskatchewan east to New Hampshire, south to Florida, west to Oklahoma, and north to Montana; to 3000′ (914 m) in the South and to 6000′ (1829 m) in the Southwest.

The plums are eaten fresh and used in jellies and preserves, and are also consumed by many kinds of birds. Numerous cultivated varieties with improved fruit have been developed. A handsome ornamental with large flowers and relatively big fruit, American Plum is also grown for erosion control, spreading by root sprouts.

---

123 Chickasaw Plum
"Sand Plum"
Prunus angustifolia Marsh.

Description: Thicket-forming shrub or sometimes a small tree, with slender, spreading branches, small white flowers, and red plums.

Height: 5–15′ (1.5–4.6 m).
Diameter: 4″ (10 cm).
Leaves: 1–2½″ (2.5–6 cm) long, ⅜–1″ (1–2.5 cm) wide. *Lance-shaped,* with *sides curved up* from midvein; sharply saw-toothed with tiny gland teeth.

Shiny green above; dull, paler, and nearly hairless beneath. Leafstalks slender, *red,* often with 2 red gland-dots near tip.
Bark: dark reddish-brown; thin, slightly fissured and scaly.
Twigs: reddish, shiny, slender, slightly zigzag; often hairy when young; short twigs often ending in spine.
Flowers: ⅜″ (10 mm) wide; with 5 rounded white petals; *2–4 flowers clustered on short equal stalks;* in spring before leaves.
Fruit: a *plum* ½–¾″ (12–19 mm) in diameter; thin shiny *red or yellow skin;* thick juicy edible pulp; large stone; maturing in summer.

Habitat: Moist soils near houses, along roadsides, borders of fields, and waste places.

Range: Naturalized from New Jersey south to central Florida, west to Texas, and north to Kansas; native probably in Texas and Oklahoma; to 3000′ (914 m).

Apparently brought from the Southwest by the Chickasaw Indians and cultivated before the arrival of European colonists. This plum is eaten fresh and made into jellies and preserves. Improved varieties are grown in the Southeast.

129, 547 **Sweet Cherry**
"Mazzard" "Mazzard Cherry"
Prunus avium (L.) L.

Description: Introduced fruit tree with tall trunk and cylindrical crown of stout, gray branches, bearing sweet cherries.
Height: 50–70′ (15–21 m).
Diameter: 1½′ (0.5 m).
Leaves: 3–6″ (7.5–15 cm) long, 1½–2½″ (4–6 cm) wide. *Ovate or reverse ovate,* abruptly long-pointed; *coarsely and often doubly saw-toothed* with blunt teeth; often drooping. Dull green and

nearly hairless with slightly sunken veins above, paler beneath and with *soft hairs* at least on veins. Slightly hairy leafstalks usually with 2 gland-dots.
Bark: reddish-brown; smooth, often peeling in horizontal strips; becoming thick and furrowed.
Twigs: stout, hairless.
Flowers: 1–1¼″ (2.5–3 cm) wide; with 5 rounded white petals; 2–6 flowers on long stalks; with leaves in early spring.
Fruit: a *cherry* ¾–1″ (2–2.5 cm) in diameter; *round or egg-shaped; red* to purplish (sometimes yellow) skin; *sweet* edible pulp; smooth elliptical stone; maturing in summer.

Habitat: Along roadsides and borders of woods.

Range: Native of Europe and Asia; naturalized locally in SE. Canada, NE. United States, and in Pacific states.

Sweet Cherry is cultivated in many varieties. "Mazzard" is the common name of the wild form, used as a root stock in grafting improved varieties. The Latin species name, meaning "of birds," stresses the value of the fruit to wildlife.

138, 408 **Carolina Laurelcherry**
"Cherry-laurel" "Carolina Cherry"
Prunus caroliniana (Mill.) Ait.

Description: Evergreen tree with narrow or broad crown of spreading branches, glossy foliage, and black, inedible fruit.

Height: 40′ (12 m).
Diameter: 10″ (25 cm), larger in cultivation.
Leaves: *evergreen;* 2–4″ (5–10 cm) long, ¾–1½″ (2–4 cm) wide. Elliptical or narrowly elliptical; *thick and slightly turned* under at edges; sometimes with scattered teeth; hairless; aromatic when crushed, with bitter taste. *Shiny dark green with obscure veins* above, paler beneath.

Bark: gray; thin; smooth or fissured.
Twigs: greenish, slender, hairless.
Flowers: 3/16″ (5 mm) wide; with 5 *small rounded cream-colored petals;* in dense clusters along an axis at leaf base; in early spring.
Fruit: a *plum* 3/8–1/2″ (10–12 mm) long; elliptical, short-pointed; *shiny black* thick skin; thin dry inedible pulp; large stone; maturing in fall, remaining attached until spring.

Habitat: Moist soils of valleys and lowlands, forests and forest borders, often forming dense thickets.

Range: SE. North Carolina south to central Florida, and west to E. Texas; to 500′ (152 m).

A handsome ornamental and hedge plant in southeastern states and California, often escaping from cultivation. The partially wilted foliage contains hydrocyanic acid, which has been known to kill livestock that have browsed upon it. Birds consume the dry fruit.

---

126 **Sour Cherry**
"Pie Cherry" "Morello Cherry"
Prunus cerasus L.

Description: Small, introduced fruit tree or thicket-forming shrub with short trunk, broad rounded crown, and slender, spreading and drooping branches, bearing sour cherries.

Height: 30′ (9 m).
Diameter: 8″ (20 cm).
Leaves: 2–3 1/2″ (5–9 cm) long, 1–2 1/4″ (2.5–6 cm) wide. *Ovate or reverse ovate,* abruptly short-pointed; *finely, often doubly saw-toothed with rounded teeth.* Light green and slightly shiny above, paler and hairless beneath.
Bark: gray; scaly, becoming rough.
Twigs: gray, stout, hairless.
Flowers: 1″ (2.5 cm) wide; with 5

rounded white petals; *2–5 flowers on equal stalks* in many clusters; in spring with leaves.
Fruit: a *cherry* ⅝–¾″ (15–19 mm) in diameter; *shiny red;* soft juicy *sour* edible pulp; round stone; maturing in summer.

Habitat: Along roadsides, fences, and borders of woods.

Range: Long cultivated in W. Asia and SE. Europe. Naturalized locally in SE. Canada and E. and NW. United States.

The widely cultivated common "Pie Cherry" has several ornamental varieties; one is a double-flowered form. The scientific name is the classical Latin and Greek term for the cherry, introduced into Europe from Crimea (ancient Cerasus).

128, 421 **Garden Plum**
"Damson Plum" "Bullace Plum"
Prunus domestica L.

Description: Small, introduced fruit tree with a spreading, rounded, open crown.
Height: 30′ (9 m).
Diameter: 1′ (0.3 m).
Leaves: 2–4″ (5–10 cm) long, 1–2″ (2.5–5 cm) wide. *Elliptical* or reverse ovate; *coarsely wavy saw-toothed; thickened.* Dull green (and hairy when young) above, *hairy with prominent network of veins beneath.*
Bark: gray; smooth or fissured, thick.
Twigs: reddish, stout, often hairy; a few short twigs ending in spines.
Flowers: ¾–1″ (2–2.5 cm) wide; with 5 rounded white petals; 1–2 flowers on hairy stalks in clusters; in early spring.
Fruit: a *large plum* 1–2″ (2.5–5 cm) in diameter; *elliptical, bluish-black;* thick juicy sweetish edible pulp; smooth *stone, free* from pulp; maturing in summer.

Habitat: Along roadsides and fence rows.

Range: Native of W. Asia and Europe. Naturalized locally in SE. Canada and NE. and NW. United States.

The name "Damson" means "plum of Damascus," alluding to the original range of this species. As the Latin species name indicates, this is the common domesticated plum, long cultivated for its edible fruit. It was introduced by early British and French colonists. Numerous improved varieties have been developed for fruit; others, including a double-flowered form, are grown as ornamentals. The prune, a large plum with firm flesh readily dried, is derived from this species.

140 Hortulan Plum
"Miner Plum"
Prunus hortulana Bailey

Description: Small tree with short trunk, stout, spreading branches, and broad, rounded crown.
Height: 20′ (6 m).
Diameter: 8″ (20 cm).
Leaves: 2¾–4½″ (7–11 cm) long, ¾–1¼″ (2–3 cm) wide. *Narrowly ovate or lance-shaped; long narrow pointed tip; finely saw-toothed;* slightly thickened. Shiny dark green above, paler and hairy at least on veins beneath. Leafstalks slender, slightly hairy, with a *few gland-dots toward tip.*
Bark: dark brown; thin, scaly, curling and peeling off.
Twigs: dark brown, stout, sometimes spiny.
Flowers: ½–⅝″ (12–15 mm) wide; with 5 rounded white petals; *2–4 flowers clustered on slender equal stalks;* in spring with leaves.
Fruit: a *plum* ¾–1¼″ (2–3 cm) in diameter; thick *red* or sometimes yellow skin; thick edible pulp; pointed stone; maturing in early autumn.

Habitat: Moist soils along streams; not forming thickets from root sprouts.

Range: SW. Ohio west to E. Kansas and NE. Oklahoma; at 400–1200′ (122–366 m).

Perhaps the most important native plum in cultivation; many improved varieties have been developed. The common and Latin species names both mean "of gardens." This species became known through the work of horticulturists.

132, 422 **Mahaleb Cherry**
"Mahaleb" "Perfumed Cherry"
Prunus mahaleb L.

Description: Small introduced tree with a short, often crooked trunk, open crown of many spreading branches, and bitter black cherries; the foliage, flowers, seeds, bark, and wood are all *aromatic.*

Height: 20′ (6 m).
Diameter: 8″ (20 cm).
Leaves: 1¼–2¾″ (3–7 cm) long, ¾–2″ (2–5 cm) wide. *Rounded to broadly ovate; finely saw-toothed* with blunt teeth. Shiny dark green above, paler and hairy on midvein beneath. Leafstalks slender, with 1–2 glands.
Bark: dark gray or brown; rough, with low irregular ridges.
Twigs: stout; finely hairy when young.
Flowers: ⅝″ (15 mm) wide; with 5 rounded white petals; very fragrant; 4–10 flowers along end of leafy axis, with small leaflike scales at base; in spring.
Fruit: a *small cherry* ¼–⅜″ (6–10 mm) in diameter; *black* (sometimes dark red) skin; thin *bitter inedible pulp;* rounded stone; maturing in early summer.

Habitat: Along roadsides, fence rows, and borders of woods.

Range: Native of Europe and W. Asia. Naturalized locally in SE. Canada and NE. and NW. United States.

Long cultivated as a stock for grafting cherries and for hedges; several ornamental varieties are also grown. In Europe the dark red aromatic wood was once used in cabinetwork and woodenware; pipe stems and canes can be made from young branches. "Mahaleb" is the Arabic name.

---

144 **Mexican Plum**
"Inch Plum" "Bigtree Plum"
Prunus mexicana Wats.

Description: Small tree with single trunk and open crown.
Height: 20′ (6 m).
Diameter: 6″ (15 cm).
Leaves: 2–4½″ (5–11 cm) long, 1¼–2″ (3–5 cm) wide. *Elliptical;* abruptly long-pointed at tip; base often with glands; *finely doubly saw-toothed.* Yellow-green and usually hairless above; paler, with *soft hairs,* and *prominently veined* beneath. Leafstalks mostly with 1–3 gland-dots near tip.
Bark: gray; scaly, becoming rough and deeply furrowed.
Twigs: light brown, shiny, slender, hairless.
Flowers: ¾″ (19 mm) wide; with 5 rounded white petals; *2–4 flowers clustered on slender equal stalks;* in spring before leaves.
Fruit: a *plum* ½–1″ (1.2–2.5 cm) in diameter; skin *purplish-red* with whitish bloom; thick sweetish pulp; large stone; maturing in late summer.

Habitat: Moist soils of valleys and dry uplands, prairies, open areas, and oak forests; not forming thickets.

Range: S. Ohio south to Alabama, west to S. Texas, north to SE. South Dakota; also NE. Mexico; to 1500′ (457 m).

The common wild plum of the forest-prairie border from Missouri and eastern Kansas to Texas. The fruit is

eaten fresh and made into preserves and is also consumed by birds and mammals. This species has served as a stock for grafting cultivated varieties o plums.

142 Wildgoose Plum
"Munson Plum"
Prunus munsoniana Wight & Hedr.

Description: Small tree with rounded crown of spreading branches; sometimes formin thickets.
Height: 20′ (6 m).
Diameter: 6″ (15 cm).
Leaves: 2½–4″ (6–10 cm) long, ¾–1¼″ (2–3 cm) wide. *Elliptical or lance-shaped, long-pointed* at tip; finely saw-toothed. Shiny green above, paler and slightly hairy beneath. Leafstalks slender, slightly hairy, usually with 2 gland-dots near tip.
Bark: brown, smooth, thin; becoming gray and scaly.
Twigs: shiny red-brown, slender, hairless.
Flowers: ½–⅝″ (12–15 mm) wide; with 5 rounded white petals; *2–4 flowers clustered on slender equal stalks* along twigs and spurs; in early spring before or with leaves.
Fruit: a *plum* ½–¾″ (12–19 mm) in diameter; skin *bright red* with slight whitish bloom; juicy sweet aromatic pulp; large stone; maturing in late summer.

Habitat: Moist soils on slopes.

Range: SW. Ohio south to E. Tennessee, wes to central Texas; at 100–1000′ (30–305 m).

Improved cultivated varieties are grow commercially and are naturalized locally. The odd common name refers to the discovery of a seed in the craw a wild goose killed by a hunter in the early 19th century. The tree from thi seed grew rapidly and bore superior fruit. This species was introduced and propagated as a fruit tree long before it was given a scientific name.

146 Canada Plum
"Horse Plum" "Red Plum"
Prunus nigra Ait.

Description: Small tree with a short trunk, several upright branches, and narrow crown.
Height: 20′ (6 m).
Diameter: 8″ (20 cm).
Leaves: 2½–4″ (6–10 cm) long, 1–2″ (2.5–5 cm) wide. *Elliptical; long narrow pointed tip;* coarsely and *doubly saw-toothed* with small gland teeth. Dull dark green above, paler and usually hairy beneath. Leafstalks slender, with 2 gland-dots near tip.
Bark: light gray-brown; thin, scaly.
Twigs: green to dark brown, stout, slightly zigzag, hairless; stout spur twigs ending in spines.
Flowers: ¾–1″ (2–2.5 cm) wide; with 5 rounded petals, white fading to pink; *3–4 flowers clustered on slender equal stalks;* in spring with or before leaves.

Fruit: a *plum* 1–1¼″ (2.5–3 cm) long; *elliptical;* thick *red or yellowish-red skin;* yellow edible pulp; large stone; maturing in late summer.

Habitat: Moist soils of valleys and hillsides.

Range: SE. Manitoba east to S. Quebec, south to Connecticut, and west to Illinois; to 1500′ (457 m).

The northernmost native plum was first recorded along the St. Lawrence River in 1535 by Jacques Cartier (1491–1557), French explorer. Earlier, Indians had brought him the dried fruits. The Latin species name, meaning "black," refers to the dark branches. Thickets are formed from root sprouts. These plums are eaten fresh and made into preserves and jellies.

124, 423, 550, 584 **Pin Cherry**
"Fire Cherry" "Bird Cherry"
Prunus pensylvanica L. f.

Description: Small tree or shrub with horizontal branches; narrow, rounded, open crow shiny red twigs; bitter, aromatic bark and foliage; and *tiny red cherries.*
Height: 30′ (9 m).
Diameter: 1′ (0.3 m).
Leaves: 2½–4½″ (6–11 cm) long, ¾–1¼″ (2–3 cm) wide. Broadly lance-shaped, *long-pointed; finely and sharply saw-toothed;* becoming hairless. Shiny green above, paler beneath; turning bright yellow in autumn. Slender leafstalks often with 2 gland-dots near tip.
Bark: reddish-gray, smooth, thin; becoming gray and fissured into scaly plates.
Flowers: ½″ (12 mm) wide; with 5 rounded white petals; 3–5 flowers on long equal stalks; in spring with leav
Fruit: a *cherry* ¼″ (6 mm) in diameter *red* skin; thin sour pulp; large stone; i summer.

Habitat: Moist soil, often in pure stands on burned areas and clearings; with aspe Paper Birch, and Eastern White Pine

Range: British Columbia and S. Mackenzie e across Canada to Newfoundland, sout to N. Georgia, west to Colorado; to 6000′ (1829 m) in southern Appalachians.

This species is often called "Fire Cherry" because its seedlings come u after forest fires. The plants grow rapidly and can be used for fuel and pulpwood. It is also a "nurse" tree, providing cover and shade for the establishment of seedlings of the next generation of larger hardwoods. The cherries are made into jelly and are al consumed by wildlife.

141, 459, 569 **Peach**
Prunus persica Batsch

Description: A well-known, small fruit tree with a short trunk, spreading rounded crown, showy pink blossoms, long narrow leaves, and yellow to pink juicy fruit.
Height: 30′ (9 m).
Diameter: 1′ (0.3 m).
Leaves: 3½–6″ (9–15 cm) long, ¾–1″ (2–2.5 cm) wide. *Lance-shaped* or narrowly oblong; *finely saw-toothed; sides often curved up* from midvein; leafstalks short with glands near tip. *Shiny green above,* paler beneath. Crushed foliage has a strong odor and bitter taste.
Bark: dark reddish-brown; smooth, becoming rough; bitter.
Twigs: greenish turning reddish-brown; long, slender, hairless.
Flowers: 1–1¼″ (2.5–3 cm) wide; with 5 *rounded pink petals; usually single* and *nearly stalkless;* in early spring before leaves.
Fruit: a *peach* 2–3″ (5–7.5 cm) in diameter; *nearly round;* grooved; fine *velvety hairs* covering the yellow-to-pink skin; thick sweet edible pulp; large elliptical pitted stone; maturing in summer.

Habitat: Along roadsides and fence rows.

Range: Native of China. Naturalized locally in E. United States, S. Ontario, and California; mostly in the Southeast.

Peach has been grown as a fruit tree since ancient times. Numerous cultivated varieties include freestone peaches, with pulp separating from the stone; clingstones, with pulp adhering to the stone; and smaller, hairless fruits known as nectarines. Other varieties are ornamentals, some with double flowers and with white or red petals. Spanish colonists introduced the Peach into Florida, and American Indians then planted it widely.

136, 447, 548 **Black Cherry**
"Wild Cherry" "Rum Cherry"
Prunus serotina Ehrh.

Description: Aromatic tree with tall trunk, oblong crown, abundant small white flowers, and small black cherries; crushed foliage and bark have distinctive *cherrylike odor* and *bitter taste.*

Height: 80′ (24 m).
Diameter: 2′ (0.6 m).
Leaves: 2–5″ (5–13 cm) long, 1¼–2″ (3–5 cm) wide. *Elliptical; 1–2 dark re glands* at base; *finely saw-toothed with curved or blunt teeth;* slightly thickened Shiny dark green above, light green a often hairy along midvein beneath; turning yellow or reddish in autumn.
Bark: dark gray; smooth, with horizontal lines; becoming irregularly fissured and scaly, exposing reddish-brown inner bark; *bitter and aromatic.*
Twigs: red-brown, slender, hairless.
Flowers: ⅜″ (10 mm) wide; 5 rounde white petals; *many flowers along spreadi or drooping axis* of 4–6″ (10–15 cm) a end of leafy twig; in late spring.
Fruit: a *cherry* ⅜″ (10 mm) in diamet skin *dark red turning blackish;* slightly *bitter, juicy, edible* pulp; elliptical stor maturing in late summer.

Habitat: On many sites except very wet or very dry soils; sometimes in pure stands.

Range: S. Quebec to Nova Scotia, south to central Florida, west to E. Texas, and north to Minnesota; varieties from central Texas west to Arizona and sou to Mexico; to 5000′ (1524 m) in southern Appalachians and at 4500–7500′ (1372–2286 m) in the Southwest.

This widespread species is the largest and most important native cherry. Th valuable wood is used particularly for furniture, paneling, professional and scientific instruments, handles, and toys. Wild cherry syrup, a cough medicine, is obtained from the bark, and jelly and wine are prepared from the fruit. One of the first New World trees introduced into English gardens, it was recorded as early as 1629. As many as 5 geographical varieties have been distinguished.

127 Flatwoods Plum

"Hog Plum" "Black Sloe"
Prunus umbellata Ell.

Description: Small tree with short trunk and broad, compact, flattened crown of slender branches; sometimes a thicket-forming shrub.
Height: 20′ (6 m).
Diameter: 6″ (15 cm).
Leaves: 1½–2¾″ (4–7 cm) long, ¾–1½″ (2–4 cm) wide. *Elliptical;* finely saw-toothed; usually 2 dark glands at base. Dark green above, paler with hairy midvein beneath.
Bark: dark brown; scaly.
Twigs: shiny red, slender, hairy when young.
Flowers: ⅝″ (15 mm) wide; 5 rounded white petals; *2–5 flowers clustered on slender equal stalks;* in early spring before leaves.
Fruit: a *plum* ⅜–⅝″ (10–15 mm) in diameter; thick *black,* red, or yellow skin with whitish bloom; thick sour pulp; elliptical stone; maturing in late summer.

Habitat: Along streams and in sandy or rocky slopes, uplands, edges of forests, and open areas.

Range: S. North Carolina south to central Florida, west to S. Arkansas and E. Texas; to 1000′ (305 m).

The wild plums of this species are gathered for making jellies and jams. The scientific name describes the flower clusters, known as umbels, that are rounded, with equal stalks like an umbrella.

134, 446, 549, 605 **Common Chokecherry**
"Eastern Chokecherry" "Western Chokecherry"
Prunus virginiana L.

Description: Shrub or small tree, often forming dense thickets, with dark red or blackish chokecherries.

Height: 20′ (6 m).
Diameter: 6″ (15 cm).
Leaves: 1½–3¼″ (4–8 cm) long, ⅝–1½″ (1.5–4 cm) wide. *Elliptical;* finely and *sharply saw-toothed;* slightly thickened. Shiny dark green above, light green and sometimes slightly hairy beneath; turning yellow in autumn. Leafstalks slender, usually with 2 gland-dots.
Bark: brown or gray; smooth or becoming scaly.
Twigs: brown, slender, with disagreeable odor and bitter taste.
Flowers: ½″ (12 mm) wide; with 5 rounded white petals; in unbranched clusters to 4″ (10 cm) long; in late spring.
Fruit: a *chokecherry* ¼–⅜″ (6–10 mm) in diameter; shiny dark red or blackish skin; juicy, *astringent or bitter* pulp; large stone; maturing in summer.

Habitat: Moist soils, especially along streams in mountains, forest borders, clearings and roadsides.

Range: N. British Columbia east to Newfoundland, south to W. North Carolina, and west to S. California; to 8000′ (2438 m) in the Southwest.

As the common name suggests, chokecherries are astringent or puckery especially when immature or raw; but they can be made into preserves and jelly. However, the fruit stones are poisonous; also wilted foliage of this and other cherries occasionally contains hydrocyanic acid that can poison livestock. Sometimes divided into three geographic varieties based on minor differences of leaves and fruits.

Tent caterpillars (*Malacosoma*) often construct their silvery webs on the branches of this species.

131, 455, 575 **Pear**
"Common Pear"
Pyrus communis L.

Description: The well-known, naturalized fruit tree with broad crown of shiny green foliage, white flowers in early spring, and edible pears in autumn.

Height: 40′ (12 m).
Diameter: 1′ (0.3 m).
Leaves: 1½–3″ (4–7.5 cm) long, 1–2″ (2.5–5 cm) wide. *Broadly ovate to elliptical;* finely *wavy saw-toothed;* becoming nearly *hairless;* often crowded on spur twigs; leafstalks slender. *Shiny green* above, paler beneath.
Bark: gray-brown; smooth, becoming scaly.
Twigs: both long and short; *hairless;* the many stout spurs sometimes ending in spines.
Flowers: 1¼″ (3 cm) wide; with 5 *rounded white petals;* in long-stalked clusters; in early spring with leaves.
Fruit: a *pear* 2½–4″ (6–10 cm) long; green to brown skin; thick juicy sweet edible pulp; star-shaped *gritty core;* maturing in late summer.

Habitat: Moist soils near houses, fences, roadsides, clearings, and borders of forests.

Range: Native of Europe and W. Asia. Naturalized locally from Maine to Florida, Texas, and Missouri, and in the Northwest.

The Pear has been cultivated since ancient times. Numerous varieties have been developed from this species and from hybrids. French provincial cabinetmakers prized the light brown wood for dressers, armoires, and other furniture.

320, 416, 564 **American Mountain-ash**
"American Rowan-tree" "Roundwood"
Sorbus americana Marsh.

Description: Small tree with spreading crown or a shrub with many stems, and with showy white flowers, and bright red berries.
Height: 30′ (9 m).
Diameter: 8″ (20 cm).
Leaves: pinnately compound; 6–8″ (15–20 cm) long. *11–17* stalkless, *lance-shaped leaflets* 1½–4″ (4–10 cm) long, ½–1″ (1.2–2.5 cm) wide; *long-pointed; saw-toothed;* becoming hairless. Yellow-green above, paler beneath; turning yellow in autumn.
Bark: light gray; smooth or scaly, thin.
Twigs: reddish-brown, stout, hairy when young.
Flowers: ¼″ (6 mm) wide; with 5 white rounded petals; *numerous* flowers, *crowded* in upright clusters 3–5″ (7.5–13 cm) wide; in late spring.
Fruit: ¼″ (6 mm) in diameter; *like small apples; bright red* skin; bitter pulp; with few seeds; *many,* in clusters; maturing in autumn.

Habitat: Moist soils of valleys and slopes; in coniferous forests.

Range: W. Ontario to Newfoundland, south to N. Georgia, and northwest to N. Illinois; to 5000–6000′ (1524–1829 m) in southern Appalachians.

A handsome ornamental, its showy red fruit persists into winter. The berries are eaten by birds, especially grouse, grosbeaks, and cedar waxwings. Moose browse the foliage, winter twigs, and fragrant inner bark. According to one explanation, Roan Mountain, North Carolina–Tennessee, was named for the "Rowan-trees" along its summit.

321 European Mountain-ash
"Rowan-tree"
Sorbus aucuparia L.

Description: Introduced tree with open, spreading, rounded crown, showy white flowers, and many bright red berries.
Height: 20–40′ (6–12 m).
Diameter: 1′ (0.3 m).
Leaves: pinnately compound; 4–8″ (10–20 cm) long. 9–17 leaflets 1–2″ (2.5–5 cm) long, ½–¾″ (12–19 mm) wide; *oblong* or lance-shaped; *short-pointed; saw-toothed* except near base; stalkless. Dull green above, with *white hairs beneath;* turning reddish in autumn.
Bark: dark gray; smooth, with horizontal lines; aromatic.
Twigs: stout; *densely covered with white hairs when young.*
Flowers: ⅜″ (10 mm) wide; with 5 rounded white petals; in clusters 3–6″ (7.5–15 cm) wide of *75–100 flowers* on stalks with white hairs; in late spring.

Fruit: 5⁄16″ (8 mm) in diameter; *like small apples; bright red* skin; bitter pulp; numerous in clusters; maturing in late spring.
Habitat: Along roads, forming thickets.
Range: Native of Eurasia; naturalized from SE. Alaska and across S. Canada to Newfoundland, and from Maine to Minnesota and California.

Introduced in colonial times, this handsome ornamental has showy red fruit that remains on the tree into early winter, providing food for birds which then spread the seeds. The Latin species name, meaning "to catch birds," refers to the former use by fowlers of the sticky fruit to make bird lime, which they smeared on branches to trap birds. The only tree naturalized in Alaska, it was introduced in the southeastern part of the state and has escaped around towns. The word "rowan" is derived from an old Scandinavian word meaning red, and refers to the brightly colored berries.

319, 415 **Showy Mountain-ash**
Sorbus decora (Sarg.) Schneid.

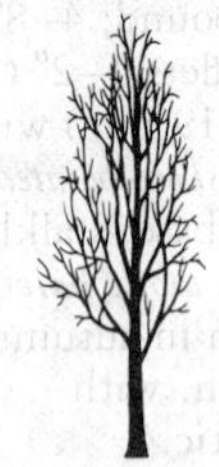

Description: Shrub or small tree with spreading rounded crown, showy white flowers, and large bright red berries.
Height: 30′ (9 m).
Diameter: 8″ (20 cm).
Leaves: pinnately compound; 4–6″ (10–15 cm) long. *15 stalkless leaflets* (sometimes 11–17) 1¼–2¾″ (3–7 cm) long, ⅝–1″ (1.5–2.5 cm) wide. *Elliptical to oblong; short-pointed* at ends; *saw-toothed.* Blue-green above, paler (and hairy when young) beneath; turning orange in autumn.
Bark: dark gray; smooth or scaly.
Twigs: stout; hairy when young.
Flowers: ⅜″ (10 mm) wide; with 5 white rounded petals; in terminal upright flattened clusters 2–4″ (5–10 cm) wide; in late spring.
Fruit: ⅜–½″ (10–12 mm) in diameter; *like small apples; bright red* skin; bitter pulp; with few seeds; many, in clusters; maturing in early autumn.

Habitat: Moist soils of valleys and slopes.

Range: W. Ontario east to Newfoundland and S. Greenland south to Connecticut and west to NE. Iowa; to 2000′ (610 m).

The common name and the Latin species name, meaning "showy" or "ornamental," both refer to the bright red fruit, which is larger than that of related species.

LEGUME FAMILY
(Leguminosae)

The third largest family of seed plants, characterized by pealike flowers and pods, with approximately 12,000 species of herbs, shrubs, and trees worldwide. About 44 native and 6 naturalized tree species and numerous species of shrubs and herbs in North America.
Leaves: alternate; compound (mostly pinnately compound; also bipinnate and with 3 leaflets), rarely simple; with paired stipules sometimes becoming spines.
Flowers: small to large and showy; often in racemes, spikes, and heads; bisexual; mostly irregular in shape of bean flower or butterfly; calyx usually tubular with 5 lobes, corolla of 5 unequal petals, 10 or more stamens distinct or united at base, and 1 pistil.
Fruit: generally a pod (legume), which opens on 2 lines and contains 1 or more elliptical bean-shaped seeds.

297, 396 **Huisache**
"Sweet Acacia" "Cassie"
Acacia farnesiana (L.) Willd.

Description: Spiny much-branched shrub or small tree with a widely spreading, flattened crown and fragrant, yellow balls of tiny flowers.

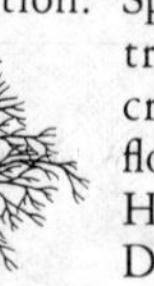

Height: 16′ (5 m).
Diameter: 4″ (10 cm).
Leaves: alternate or clustered; *bipinnately compound;* 2–4″ (5–10 cm) long; usually with 3–5 pairs of side axes. *10–20 pairs of leaflets* ⅛–¼″ (3–6 mm) long; oblong; mostly hairless; stalkless; gray-green.
Bark: grayish-brown; thin; smooth or scaly.
Twigs: slightly zigzag, slender, covered with fine hairs when young; with *straight*

slender paired white spines at nodes.
Flowers: 3⁄16″ (5 mm) long; *yellow or orange; very fragrant;* including many tiny stamens clustered in stalked *balls* ½″ (12 mm) in diameter; mainly in late winter and early spring.
Fruit: 1½–3″ (4–7.5 cm) long, ⅜–½″ (10–12 mm) in diameter; a *cylindrical* pod; short-pointed at ends, dark brown or black, hard; maturing in summer, remaining attached, often opening late; many elliptical flattened shiny brown seeds.

Habitat: Sandy and clay soils, especially in open areas, borders of woodlands, and roadsides.

Range: S. Texas and local in S. Arizona; cultivated and naturalized from Florida west to Texas and S. California; also in Mexico; to 5000′ (1524 m).

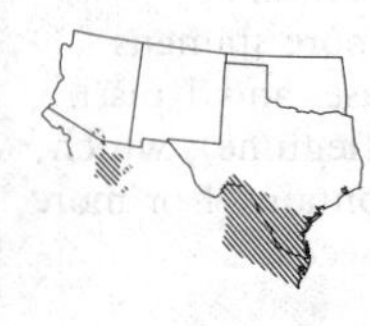

In southern Europe this species is extensively planted for the "cassie" flowers, which are a perfume ingredient. After drying in the shade, the flowers can be used in sachets to keep clothes smelling fragrant. The tender foliage and pods are browsed by livestock; it is also a honey plant. Mucilage is produced from the gum of the trunk, and tannin and dye from the pods and bark.

295, 397 **Gregg Catclaw**
"Devilsclaw" "Catclaw Acacia"
Acacia greggii Gray

Description: Spiny, much-branched, thicket-forming shrub; occasionally a small tree with a broad crown.
Height: 20′ (6 m).
Diameter: 6″ (15 cm).
Leaves: clustered; *bipinnately compound;* 1–3″ (2.5–7.5 cm) long; the slender axis usually with 2–3 pairs of side axes. 3–7 *pairs of leaflets* ⅛–¼″ (3–6 mm) *long;* oblong; rounded at ends; thick;

usually hairy; almost stalkless; dull green.

Bark: gray; thin, becoming deeply furrowed.

Twigs: brown, slender, angled, covered with fine hairs; with many *scattered stout spines* ¼" (6 mm) long, *hooked* or curved backward.

Flowers: ¼" (6 mm) long; light yellow; fragrant; stalkless; including many tiny stamens in long narrow clusters 1–2" (2.5–5 cm) long; in early spring and irregularly in summer.

Fruit: 2½–5" (6–13 cm) long, ½–¾" (12–19 mm) wide; thin, flat, ribbonlike, *oblong pod;* brown, *curved, much twisted,* often narrowed between seeds; maturing in summer and shedding in winter, remaining closed; several beanlike, nearly round, flat, brown seeds.

Habitat: Washes, slopes, and rocky canyons, especially limestone soils, desert, and desert grassland; forming impenetrable thickets.

Range: Central, S., and Trans-Pecos Texas west to S. Nevada and SE. California, south to N. Mexico; to 5000′ (1524 m).

One of the most despised southwestern shrubs. As indicated by the common names (including the Spanish, *uña de gato*), the sharp, stout, hooked spines, like a cat's claws, tear clothing and flesh. The hard, heavy wood with reddish-brown heartwood and yellow sapwood is used for souvenirs and locally for tool handles and fuel. Catclaw honey (also known as Uvalde honey, from the Texas county of that name) is made from the flowers of this and related species. Indians once made meal called "pinole" from the seeds.

300 Roemer Catclaw
"Roemer Acacia"
Acacia roemeriana Scheele

Description: Spiny, much-branched, irregularly spreading shrub with slender, weak stems and whitish-yellow balls of tiny flowers; sometimes a small tree with rounded crown.

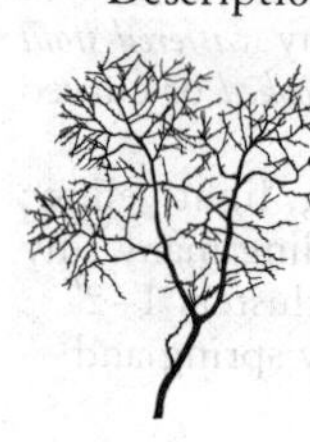

Height: 3–6′ (0.9–1.8 m), rarely 15′ (4.6 m).
Diameter: 1′ (0.3 m).
Leaves: *bipinnately compound;* 1½–4″ (4–10 cm) long; with 1–3 pairs of side axes, each with *4–12 pairs of stalkless leaflets* 3/16–3/8″ (5–10 mm) long; *elliptical to oblong,* rounded at ends, hairless, gray-green.
Bark: gray to brown; smooth but becoming scaly.
Twigs: gray or brown, slender, hairless, usually with *single or paired short curved spines.*
Flowers: 3/16″ (5 mm) long including many tiny stamens; *whitish-yellow;* fragrant; clustered in stalked *balls* 3/8″ (10 mm) in diameter, at leaf bases; from early spring to early summer.
Fruit: 2–4″ (5–10 cm) long, ¾–1¼″ (2–3 cm) wide; *flat oblong pods;* slightly curved; *reddish-brown;* leathery and flexible; several small flattened seeds; maturing in summer and splitting open.

Habitat: Dry rocky and gravelly slopes and bluffs, mainly limestone outcrops; forming thickets, or "brush," with junipers and evergreen oaks.

Range: Central, SW., and Trans-Pecos Texas, SE. New Mexico, and NE. Mexico; at 500–4500′ (152–1372 m).

Common to abundant as a thicket-forming shrub on uplands, Roemer Catclaw becomes a small tree in moist areas. Named for Karl Ferdinand Roemer (1818–91) of Germany, who collected plants in Texas in 1846–47.

296, 502 **Wright Catclaw**
"Texas Catclaw" "Wright Acacia"
Acacia wrightii Benth.

Description: Much-branched and often spiny, thicket-forming shrub or small tree, with broad, irregular crown.
Height: 20′ (6 m).
Diameter: 6″ (15 cm).
Leaves: alternate or clustered; *bipinnately compound;* 1–2″ (2.5–5 cm) long; with 1–2 pairs of side axes. *2–7 paired leaflets* ¼–½″ (6–12 mm) long; oblong; green above, paler beneath.
Bark: gray; thin, fissured into narrow scales.
Twigs: brown, often with *scattered hooked stout spines* ½″ (12 mm) long.
Flowers: ¼″ (6 mm) long; *light yellow;* fragrant; stalkless; including many tiny stamens in narrow clusters 1½″ (4 cm) long; in early spring and irregularly in summer.
Fruit: 2–6″ (5–15 cm) long, ⅝–1″ (1.5–2.5 cm) wide; a flat *oblong pod;* reddish-brown; *slightly curved;* maturing in summer and shedding in autumn; several beanlike *flat oblong* brown seeds.

Habitat: Along streams, in canyons, and on dry rocky slopes of plains and foothills.

Range: Central, S., and Trans-Pecos Texas and NE. Mexico; to 2000′ (610 m), sometimes higher.

The common names (including the Spanish, *uña de gato*) liken the hooked spines, capable of tearing clothing and flesh, to a cat's claws, although this species is much less thorny than its close relative, Gregg Catclaw.

310, 463 **Silktree**
"Mimosa-tree" "Powderpuff-tree"
Albizia julibrissin Durazzini

Description: Small ornamental with short trunk or several trunks and a very *broad, flattened*

crown of spreading branches and with showy pink flower clusters.
Height: 20′ (6 m).
Diameter: 8″ (20 cm).
Leaves: *bipinnately compound;* 6–15″ (15–38 cm) long; *fernlike;* with *5–12 pairs* of side axes covered with fine hairs. Each axis has *15–30 pairs* of oblong pale green leaflets, ⅜–⅝″ (10–15 mm) long.
Bark: blackish or gray; nearly smooth.
Twigs: brown or gray; often angled.
Flowers: more than 1″ (2.5 cm) long; with long *threadlike pink stamens* whitish toward base; crowded in long-stalked *ball-like clusters* 1½–2″ (4–5 cm) wide; grouped at ends of twigs; throughout summer.
Fruit: 5–8″ (13–20 cm) long; flat pointed *oblong pod;* yellow-brown; maturing in summer, remaining closed; several beanlike flattened shiny brown seeds.

Habitat: Open areas including wasteland and dry gravelly soils.

Range: Native from Iran to China; naturalized from Maryland to S. Florida, west to E. Texas, north to Indiana; to 2000′ (610 m).

The hardiest tree of its genus, Silktree has an unusually long flowering period. Often called "Mimosa-tree" because the flowers are similar to those of the related herbaceous sensitive-plants (genus *Mimosa*). Silktree leaflets fold up at night; those of sensitive-plants fold up when touched.

99, 466 **Eastern Redbud**
"Judas-tree"
Cercis canadensis L.

Description: Tree with a short trunk, rounded crown of spreading branches, and pink flowers that cover the twigs in spring.
Height: 40′ (12 m).

Diameter: 8″ (20 cm).
Leaves: 2½–4½″ (6–11 cm) long and broad. *Heart-shaped, with broad short point;* without teeth; with 5–9 main veins; long-stalked. Dull green above, paler and sometimes hairy beneath; turning yellow in autumn.
Bark: dark gray or brown; smooth, becoming furrowed into scaly plates.
Twigs: brown, slender, angled.
Flowers: ½″ (12 mm) long; pea-shaped, with 5 *slightly unequal purplish-pink petals,* rarely white; 4–8 flowers in a cluster on slender stalks; in early spring before leaves.

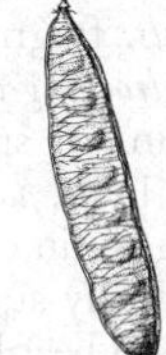

Fruit: 2½–3¼″ (6–8 cm) long; *flat* narrowly *oblong pods;* pointed at ends; pink, turning blackish; splitting open on 1 edge; falling in late autumn or winter. Several beanlike flat elliptical dark brown seeds.

Habitat: Moist soils of valleys and slopes and in hardwood forests.

Range: New Jersey south to central Florida, west to S. Texas, and north to SE. Nebraska; also N. Mexico; to 2200′ (671 m).

Very showy in early spring, when the leafless twigs are covered with masses of pink flowers, Eastern Redbud is often planted as an ornamental. The flowers can be eaten as a salad, or fried. According to a myth, Judas Iscariot hanged himself on the related Judas-tree (*Cercis siliquastrum* L.) of western Asia and southern Europe, after which the white flowers turned red with shame or blood.

350 Yellowwood
"Virgilia"
Cladrastis kentukea (Dum.-Cours.) Rudd

Description: Medium-sized tree with short trunk and broad, rounded crown of spreading branches.

Height: 50′ (15 m).
Diameter: 1½′ (0.5 m).
Leaves: pinnately compound; 8–12″ (20–30 cm) long; with base of leafstalk enclosing bud. *7–11 leaflets* 2½–4″ (6–10 cm) long, 1¼–2″ (3–5 cm) wide; *elliptical;* nearly in pairs except at end; short-stalked; not toothed; becoming hairless. Shiny green above, paler beneath; turning yellow in autumn.
Bark: *gray; smooth,* thin, resembling beech (*Fagus*).
Twigs: brown, slender, zigzag, hairless.
Flowers: 1–1¼″ (2.5–3 cm) long; *pea-shaped,* with 5 *white petals;* fragrant; long-stalked; many, in *drooping clusters* about 1′ (30 cm) long; in late spring.

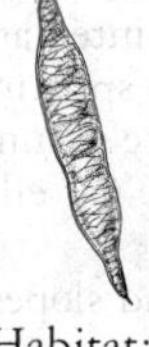

Fruit: 2–3¼″ (5–8 cm) long; *flat,* narrowly *oblong pod* hanging in clusters; maturing and falling in early autumn before splitting open; 2–6 beanlike seeds.

Habitat: Moist soils, especially limestone cliffs, stream banks, and rich rocky coves (or deep valleys of mountains); in hardwood forests.

Range: SW. Virginia, W. North Carolina, and NE. Georgia west to E. Oklahoma; local north to S. Indiana; at 300–3500′ (91–1067 m).

A handsome ornamental. Charles Sprague Sargent, (1841–1927) author of the classic 14-volume *Silva of North America,* called this rare species one of the most beautiful flowering trees of the American forests. The clear yellow heartwood, which turns light brown on exposure, has been used as a source of a yellow dye. The generic name is from Greek words meaning "branch" and "brittle." Scattered in the Great Smoky Mountains National Park, where wild trees bear abundant flowers at irregular intervals of 2–5 years.

354, 368 **Southeastern Coralbean**
"Cherokee-bean" "Red-cardinal"
Erythrina herbacea L.

Description: A spiny shrub with many slender stems, in south Florida becoming a small tree with crooked trunk, spreading brittle branches, and rounded crown.
Height: 20′ (6 m).
Diameter: 8″ (20 cm).
Leaves: pinnately compound; 6–8″ (15–20 cm) long; with slender stalks, sometimes prickly. *3 leaflets,* 1½–3″ (4–7.5 cm) long and almost as wide, sometimes larger; *triangular* or slightly 3-lobed; not toothed; midvein often prickly beneath; light green.
Bark: light gray; smooth, becoming thick and furrowed; sometimes spiny.
Twigs: light green, stout, brittle, with scattered short curved *spines* or prickles.
Flowers: 2″ (5 cm) long; *very narrow;* showy, with dark red tubular calyx and 5 *narrow unequal red or scarlet petals;* many, in upright long-stalked clusters 8–12″ (20–30 cm) long.
Fruit: 4–8″ (10–20 cm) long; *cylindrical pod;* dark brown or black, long-pointed at ends, narrowed between seeds; maturing in late summer and opening along 1 edge; several beanlike *shiny red* poisonous seeds.

Habitat: Moist sandy soils.

Range: SE. North Carolina to S. Florida, including Florida Keys, and west to E. and S. Texas; to 500′ (152 m).

This unusual tropical tree extends its range northward as a shrub or perennial herb, but is killed back to the ground each winter. Planted for the showy flowers and seeds, although the brittle branches are subject to damage by windstorms. In Mexico, the toxic seeds have been used for poisoning rats and fish. Although novelties and necklaces can be made from the seeds, they should be kept away from children.

309 Waterlocust
Gleditsia aquatica Marsh.

Description: Medium-sized, spiny tree with short trunk and broad, flattened crown of spreading branches.
Height: 50′ (15 m).
Diameter: 2′ (0.6 m).
Leaves: *pinnately and bipinnately* compound; 4–8″ (10–20 cm) long; the axis often with 3–4 pairs of side axes or forks. *Numerous* paired *leaflets* ½–1¼″ (1.2–3.2 cm) long; *oblong;* with *finely wavy edges;* nearly hairless; stalkless. Shiny dark green above, dull yellow-green beneath.
Bark: gray or brown; thin, smooth, fissured into small scaly plates; with large branched *spines.*
Twigs: brown, usually with *stout shiny brown spines* to 4″ (10 cm) long, often slightly *flattened* and curved, mostly unbranched.
Flowers: ¼″ (6 mm) wide; *greenish-yellow; bell-shaped,* with 5 spreading petals, covered with fine hairs; in short narrow clusters at leaf bases in late spring; usually male and female on separate twigs or trees.
Fruit: 1–2″ (2.5–5 cm) long, ¾″ (19 mm) wide; *flat elliptical pod;* shiny brown; with *thin papery walls* and without pulp; in drooping clusters; maturing in late summer, opening late shedding in autumn; usually with 1 rounded flattened brown seed.

Habitat: Wet soils of riverbanks, flood plains, and swamps, especially where submerged for long periods; in floodplain forests.

Range: South Carolina to central Florida, west to E. Texas, and north to S. Illinois an extreme SW. Indiana; to 400′ (122 m,

The common and Latin species names both refer to Waterlocust's wet habita The generic name honors Johann Gottlieb Gleditsch (1714–86), directc of the Berlin Botanical Gardens.

306, 506 Honeylocust
"Sweet-locust" "Thorny-locust"
Gleditsia triacanthos L.

Description: Large, spiny tree with open, flattened crown of spreading branches.

Height: 80′ (24 m).
Diameter: 2½′ (0.8 m).
Leaves: *pinnately and bipinnately compound;* 4–8″ (10–20 cm) long; the axis often with 3–6 pairs of side axes or forks; in late spring. *Many oblong leaflets* ⅜–1¼″ (1–3 cm) long; paired and stalkless; with *finely wavy edges.* Shiny dark green above, dull yellow-green and nearly hairless beneath; turning yellow in autumn.
Bark: gray-brown or black; fissured in long narrow scaly ridges; with *stout brown spines, usually branched,* sometimes 8″ (20 cm) long, with 3 to many points.
Twigs: shiny brown, stout, zigzag, with long spines.
Flowers: ⅜″ (10 mm) wide; *bell-shaped,* with 5 petals; *greenish-yellow,* covered with fine hairs; in short narrow clusters at leaf bases in late spring; usually male and female on separate twigs or trees.
Fruit: 6–16″ (15–41 cm) long, 1¼″ (3 cm) wide; *flat pod;* dark brown, hairy, *slightly curved and twisted,* thick-walled; shedding unopened in late autumn; many beanlike flattened dark brown seeds in *sweetish edible pulp.*

Habitat: Moist soils of river flood plains in mixed forests; sometimes on dry upland limestone hills; also in waste places.

Range: Extreme S. Ontario to central Pennsylvania, south to NW. Florida, west to SE. Texas, and north to SE. South Dakota; naturalized eastward; to 2000′ (610 m).

Livestock and wildlife consume the honeylike, sweet pulp of the pods. Honeylocust is easily recognized by the large, branched spines on the trunk; thornless forms, however, are common in cultivation and are sometimes found wild. The spines have been used as pins. This hardy species is popular for shade, hedges, and attracting wildlife.

311, 503 **Kentucky Coffeetree**
Gymnocladus dioicus (L.) K. Koch

Description: Usually a short-trunked tree with narrow, open crown of coarse branches and very large, twice-compound leaves.

Height: 70′ (21 m).
Diameter: 2′ (0.6 m).
Leaves: *bipinnately compound;* 12–30″ (30–76 cm) long; the axis with 3–8 pairs of side axes or forks. Upper axis with *6–14 mostly paired leaflets* (sometimes 1 at end) 1–3″ (2.5–7.5 cm) long, ¾–2″ (2–5 cm) wide; *ovate;* without teeth; pink when unfolding; becoming nearly hairless. Dull green above, paler beneath; turning yellow in autumn.
Bark: gray; thick, deeply furrowed into *narrow scaly ridges* often projecting toward one side.
Twigs: *brown, few, stout,* hairy when young, with thick brown pith.
Flowers: ⅝–¾″ (15–19 mm) long; *greenish-white* and hairy, with *narrow tube* and *4–5 spreading petals;* in large upright terminal clusters; usually male and female on separate trees in late spring.
Fruit: 4–7″ (10–18 cm) long and 1½–2″ (4–5 cm) wide; dark *red-brown pod;* thick-walled; hanging on stout stalks; falling unopened in winter. Several seeds ¾″ (19 mm) in diameter; beanlike, rounded, shiny dark brown, thick-walled.

Habitat: Moist soils of valleys with other hardwoods.

Range: Extreme S. Ontario east to central New York, southwest to Oklahoma, and north to S. Minnesota; naturalized eastward; at 300–2000′ (91–610 m).

The roasted seeds were once used as a coffee substitute; raw seeds, however, are poisonous. The reddish-brown wood makes attractive cabinets, and the fruit pulp has been used in home remedies. Scattered or rare in the wild, this species is planted as an ornamental for the very large leaves and for the stout twigs which are bare except in summer. As the leaves develop late in spring and shed early, the leafless trees often appear to be dead. The generic name, from Greek, means "naked branch."

299, 398, 505 **Honey Mesquite**
Prosopis glandulosa Torr.

Description: Spiny, large, thicket-forming shrub, or small tree with short trunk, open spreading crown of crooked branches, and narrow, beanlike pods.
Height: 20′ (6 m).
Diameter: 1′ (0.3 m).
Leaves: *bipinnately compound;* 3–8″ (7.5–20 cm) long; the short axis bearing *1 pair of side axes or forks,* each fork with *7–17 pairs* of stalkless *leaflets* ⅜–1¼″ (1–3.2 cm) long; narrowly oblong; hairless or nearly so; yellow-green.
Bark: dark brown; rough, thick, becoming sheddy.
Twigs: slightly zigzag; with stout, yellowish, mostly *paired spines* ¼–1″ (0.6–2.5 cm) long at enlarged nodes, which afterwards bear short spurs.
Flowers: ¼″ (6 mm) long; nearly stalkless, light yellow; fragrant; crowded in *narrow clusters* 2–3″ (5–7.5 cm) long; in spring and summer.
Fruit: 3½–8″ (9–20 cm) long, less than ⅜″ (10 mm) wide; *narrow pod* ending in long narrow point, slightly flattened, wavy-margined between seeds; sweetish pulp; maturing in summer, remaining closed; several beanlike seeds within 4-sided cas-

Habitat: Sandy plains and sandhills and along

valleys and washes; in short grass, desert grasslands, and deserts.

Range: E. Texas and SW. Oklahoma west to extreme SW. Utah and S. California; also N. Mexico; naturalized north to Kansas and SE. Colorado; to 4500′ (1372 m).

The seeds are disseminated by livestock that graze on the sweet pods, and the shrubs have invaded grasslands. Cattlemen regard mesquites as range weeds and eradicate them. In sandy soils, dunes often form around shrubby mesquites, burying them except for a rounded mass of branching tips. The deep taproots, often larger than the trunks, are grubbed up for firewood. Southwestern Indians prepared meal and cakes from the pods. As the common name indicates, this species is also a honey plant. The word "mesquite" is a Spanish adaptation of the Aztec name "mizquitl."

304, 442, 507 **Black Locust**
"Yellow Locust" "Locust"
Robinia pseudoacacia L.

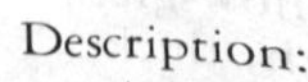

Description: Medium-sized, spiny tree with a forking, often crooked and angled trunk and irregular, open crown of upright branches.

Height: 40–80′ (12–24 m).
Diameter: 1–2′ (0.3–0.6 m).
Leaves: pinnately compound; 6–12″ (15–30 cm) long. *7–19 leaflets* 1–1¾″ (2.5–4.5 cm) long, ½–¾″ (12–19 mm) wide; paired (except at end); *elliptical;* with *tiny bristle tip;* without teeth; hairy when young; drooping and folding at night. Dark blue-green above, pale and usually hairless beneath.
Bark: light gray; thick, deeply furrowed into long rough forking ridges.
Twigs: dark brown, with stout *paired*

spines ¼–½″ (6–12 mm) long at nodes.
Flowers: ¾″ (19 mm) long; *pea-shaped;* with 5 unequal *white petals,* the largest yellow near base; very fragrant; in showy drooping clusters 4–8″ (10–20 cm) long at base of leaves; in late spring.
Fruit: 2–4″ (5–10 cm) long; *narrowly oblong flat pod;* dark brown; maturing in autumn, remaining attached into winter, splitting open; 3–14 dark brown flattened beanlike seeds.

Habitat: Moist to dry sandy and rocky soils, especially in old fields and other open areas, and in woodlands.

Range: Central Pennsylvania and S. Ohio south to NE. Alabama, and from S. Missouri to E. Oklahoma; naturalized from Maine to California and in S. Canada; from 500′ (152 m) to above 5000′ (1524) in southern Appalachians.

Black Locust is widely planted for ornament and shelterbelts, and for erosion control particularly on lands strip-mined for coal. Although it grows rapidly and spreads by sprouts like a weed, it is short-lived. Virginia Indians made bows of the wood and apparently planted the trees eastward. British colonists at Jamestown discovered this species in 1607 and named it for its resemblance to the Carob Tree or Old World Locust (*Ceratonia siliqua* L.). Posts of this durable timber served as cornerposts for the colonists' first homes.

305, 467 **Clammy Locust**
Robinia viscosa Vent.

Description: Small tree with a spreading crown of slender branches, or a shrub; often spiny, with showy pink flowers and *sticky or clammy gland-hairs* on twigs, leafstalks, flower stalks, and pods. Height: 30′ (9 m).

Diameter: 8″ (20 cm).
Leaves: pinnately compound; 5–10″ (13–25 cm) long. *13–21 leaflets ¾–1½″* (2–4 cm) long, ½–¾″ (12–19 mm) wide; paired (except at end); *elliptical;* with *tiny bristle tip;* not toothed; hairy when young. Dark green above, paler and finely hairy beneath.
Bark: dark brown; smooth, becoming furrowed.
Twigs: dark brown, with small *paired spines* at some nodes.
Flowers: ¾″ (19 mm) long; *pea-shaped;* with 5 unequal *pale pink* petals, the largest pale yellow near base; crowded in drooping clusters to 3″ (7.5 cm) long at base of leaves; in late spring.

Fruit: 2–3¼″ (5–8 cm) long; *narrowly oblong flat pod;* brown, sticky with gland-hairs; splitting open; several mottled-brown, flattened, beanlike seeds.

Habitat: Open forests, clearings, and waste places; forming thickets.

Range: W. Virginia southwest to central Alabama; naturalized beyond in E. United States and SE. Canada; at 500–4000′ (152–1219 m).

Discovered in 1776 by the early American botanist, William Bartram, in South Carolina. Easily recognized by the sticky or clammy secretion of the gland-hairs.

302, 510 **Texas Sophora**
"Eves-necklace" "Coralbean"
Sophora affinis Torr. & Gray

Description: Shrub or small tree with rounded crown of stout, spreading branches, pinkish flowers, and pods resembling a string of beads.
Height: 30′ (9 m).
Diameter: 8″ (20 cm).
Leaves: pinnately compound; 6–9″ (15–23 cm) long. *Usually 13–19 leaflets*

¾–1½″ (2–4 cm) long; paired (except at end); *elliptical; short-pointed;* without teeth; covered with gray hairs when young. Yellow-green above, paler and slightly hairy beneath.
Bark: gray-brown; thin, scaly, becoming furrowed.
Twigs: greenish, slightly zigzag, slender, covered with fine hairs.
Flowers: ½″ (12 mm) long; *pea-shaped,* with 5 unequal *white petals tinged with pink* and covered with fine hairs; in lateral clusters 2–4½″ (5–11 cm) long; in early spring.
Fruit: 1–5″ (2.5–13 cm) long and ⅜″ (10 mm) in diameter; *black leathery pod,* slightly hairy, long-pointed at ends, *very narrow between seeds,* thin fleshy sweetish walls; maturing in summer, remaining closed and attached in winter; 4–8 (rarely 1) shiny brown elliptical beanlike seeds.

Habitat: Moist soils along streams and on limestone hills; often in groves.

Range: SW. Arkansas and NW. Louisiana west to S. Oklahoma and central Texas; at 300–1800′ (91–549 m).

The seeds are reported to be poisonous. *Sophora* is from the Arabic name of a tree with pea-shaped flowers.

312 Chinese Scholartree
"Japanese Pagodatree"
Sophora japonica L.

Description: Medium-sized, introduced tree with open, rounded crown of spreading branches, yellowish-white pea-shaped flowers, and pods like a string of beads.

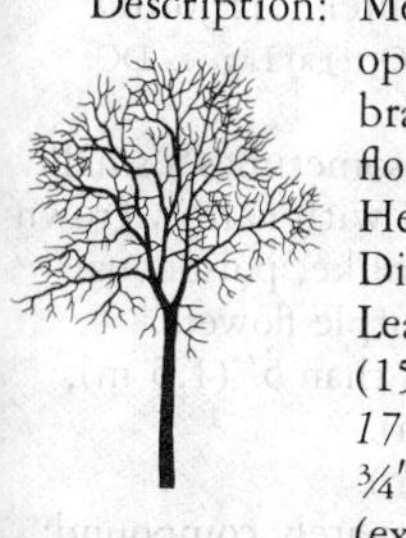

Height: 60′ (18 m).
Diameter: 2′ (0.6 m).
Leaves: pinnately compound; 6–10″ (15–25 cm) long, with slender axis. 7–*17 leaflets* 1–2″ (2.5–5 cm) long, ½–¾″ (12–19 mm) wide; mostly paired (except at end); *ovate or lance-shaped;*

bristle-tipped; without teeth. Shiny dark green above, pale and covered with fine hairs beneath; remaining green into autumn. Leaves have odd odor when crushed.
Bark: gray-brown; furrowed into scaly ridges.
Twigs: dark green; hairless or with fine hairs.
Flowers: ½″ (12 mm) long; *pea-shaped,* with 5 unequal *yellowish-white petals;* in upright (or spreading) loose showy clusters 6–12″ (15–30 cm) long; at twig ends; *in late summer.*
Fruit: 2–3″ (5–7.5 cm) long, ⅜″ (10 mm) in diameter; bean-shaped *yellowish pod;* narrow between seeds like a string of beads; maturing in autumn, often continuing to hang down in winter; 1–6 dark brown beanlike seeds.

Habitat: A shade and street tree in humid temperate regions.

Range: Native of China and Korea. Planted across the United States, especially in the South.

This street tree is unusual in having abundant late summer blossoms. Hardy under city conditions but slow-growing. In the Orient, where it is often grown around temples, a yellow dye is extracted from the flower buds; the bark and other parts reportedly have medicinal properties.

340, 471, 504 **Mescalbean**
"Texas Mountain-laurel" "Frijolillo"
Sophora secundiflora (Ortega) Lag. ex DC.

Description: Evergreen shrub or sometimes a small, much-branched tree with narrow crown and bright red, beanlike, poisonous seeds and showy, purple flowers.

Height: usually less than 5′ (1.5 m), sometimes 20′ (6 m).
Diameter: 6″ (15 cm).
Leaves: *evergreen;* pinnately compound;

3–6″ (7.5–15 cm) long. Usually *5–11 leaflets* ¾–2″ (2–5 cm) long, ⅜–1″ (1–2.5 cm) wide; paired (except at end); *elliptical;* rounded or slightly notched at tip; without teeth; *thick and leathery. Shiny dark green* and becoming hairless above, paler and hairless or nearly so beneath.

Bark: dark gray; fissured into narrow flattened scaly ridges; becoming rough and shaggy.

Twigs: densely covered with white hairs when young, becoming light brown.

Flowers: ¾–1″ (2–2.5 cm) long; pea-shaped, with 5 *unequal bluish-purple or violet petals;* very fragrant; crowded on one side of stalk in clusters 2–4½″ (5–11 cm) long, of many flowers each; at twig ends; in early spring with new leaves, sometimes also in autumn.

Fruit: 1–5″ (2.5–13 cm) long, about ⅝″ (15 mm) in diameter; a *cylindrical pod;* densely covered with brown hairs; pointed at ends; *slightly narrowed between seeds;* hard and thick-walled; maturing in late summer and not opening; usually 3–4 (sometimes 1–8) small *bright red* rounded beanlike *seeds, extremely poisonous.*

Habitat: Moist soils of streams, canyons, and hillsides, mainly on limestone; forming thickets.

Range: Central to SW. and Trans-Pecos Texas and SE. New Mexico; also south to central Mexico; to 6500′ (1981 m).

This species is often cultivated in warm regions for the shiny, evergreen foliage and large showy flowers. However, further growing of this dangerous plant is not recommended. One seed is sufficiently toxic to kill an adult and children have become ill from eating the flowers. Indians made necklaces as well as a narcotic powder from the seeds. The foliage and seeds are also poisonous to livestock.

CALTROP FAMILY
(Zygophyllaceae)

About 200 species of shrubs, herbs, and a few trees in mostly tropical or subtropical regions; 2 native tree species and several species of shrubs and herbs in North America.
Leaves: opposite, pinnately compound with even number of leaflets which are asymmetrical with unequal sides, not toothed, often leathery, with paired stipules.
Twigs: with rings at enlarged nodes.
Flowers: solitary or few; bisexual, regular, generally with 5 sepals and 5 yellow or blue petals, usually with disk, 5–10 (15) stamens often with scales at base, and 1 pistil.
Fruit: usually an angled or winged capsule with few seeds.

298, 470, 553 **Texas Lignumvitae**
"Texas Porliera" "Soapbush"
Guaiacum angustifolium Engelm.

Description: Evergreen shrub or small tree with a compact head of short, stout, crooked branches and blue or purple flowers.

Height: 20′ (6 m).
Diameter: 8″ (20 cm).
Leaves: *evergreen; opposite* or crowded; *pinnately compound;* 1–3″ (2.5–7.5 cm) long; with paired *spiny-pointed* scales or stipules at base, and with hairy axis. *8–16 paired leaflets* ¼–⅝″ (6–15 mm) long, ¹⁄₁₆–⅛″ (1.5–3 mm) wide; narrowly oblong, unequal at base; stalkless; thickened; with network of veins; folding at night and often also at midday. *Shiny dark green* on both surfaces.
Bark: gray or black; becoming rough, fissured, and scaly.
Twigs: gray, short, stout, stiff, with *enlarged ring nodes.*
Flowers: ½–¾″ (12–19 mm) wide;

with 5 *rounded blue or purple petals;* fragrant; on long stalks at leaf bases; in spring and summer.
Fruit: ½" (12 mm) in diameter; a flat *heart-shaped capsule* with 2 (sometimes 3–4) lobes and cells; brown, slightly winged, splitting open; usually *2* shiny yellow-brown *beanlike seeds* with thick reddish cover.

Habitat: Thickets in valleys and canyons.

Range: S. to central and Trans-Pecos Texas; also NE. Mexico; to 3000′ (914 m).

The northernmost of about 5 tropical species of lignumvitae. Another species, Roughbark Lignumvitae (*G. sanctum* L.), occurs rarely on the Florida Keys. The bark has been used as a soap substitute and in folk remedies. The name "lignumvitae" (Latin for "tree of life") refers to the former medicinal use of the wood extract of a related tree.

RUE (CITRUS) FAMILY
(Rutaceae)

Shrubs and trees, rarely herbs, having aromatic leaves, fruit, and bark with pungent citrus odor. About 1000 species in tropical and warm temperate regions; 12 native and 4 naturalized tree and several native shrub species in North America.
Leaves: simple, pinnately or palmately compound; with gland-dots, generally hairless, without stipules.
Flowers: usually regular; mostly white or greenish; small to large and showy; commonly bisexual, sometimes male and female on separate plants; generally with 5 or 4 sepals and 5 or 4 petals, 8–10 (or as few as 3 and sometimes more than 10) stamens, and generally 1 pistil.
Fruit: usually a capsule or berry, sometimes a drupe, follicle, or winged key (samara).

352, 417, 501 **Common Hoptree**
"Wafer-ash"
Ptelea trifoliata L.

Description: Aromatic shrub or small tree with a rounded crown; bark, crushed foliage, and twigs have a slightly lemonlike, *unpleasant odor.*

Height: 20′ (6 m).
Diameter: 6″ (15 cm).
Leaves: palmately compound, 4–7″ (10–18 cm) long, with *3 leaflets* at end of long leafstalk. Leaflets 2–4″ (5–10 cm) long, ¾–2″ (2–5 cm) wide; *ovate* or elliptical; long-pointed at tip; finely wavy-toothed or not toothed; with *tiny gland-dots;* hairy when young. Shiny dark green above, paler and sometimes hairy beneath; turning yellow in autumn.
Bark: brownish-gray; thin; smooth or slightly scaly; bitter.
Twigs: brown, slender, covered with

fine hairs, slightly warty.
Flowers: ⅜″ (10 mm) wide; 4–5 narrow *greenish-white* petals; in terminal branched clusters including male, female, and bisexual flowers; in spring.
Fruit: ⅞″ (22 mm) in diameter; numerous, *disk-shaped,* waferlike with rounded wing, yellow-brown; in drooping clusters; maturing in summer, remaining closed and attached in winter; 2–3 reddish-brown, long-pointed seeds.

Habitat: Dry rocky uplands; also in valleys and canyons.

Range: Extreme S. Ontario east to W. New York and New Jersey, south to Florida, west to Texas, and north to S. Wisconsin; also local west to Arizona and S. Utah and in Mexico; to 8500′ (2591 m) in West.

This widespread species includes many varieties with leaflets of differing sizes and shapes. The common name refers to a reported use in earlier days of the bitter fruit as a substitute for hops in brewing beer. The bitter bark of the root, like other aromatic barks, has been used for home remedies. The northernmost New World representative of the Rue (Citrus) family.

339 Common Prickly-ash

"Toothache-tree"
"Northern Prickly-ash"
Zanthoxylum americanum Mill.

Description: Much-branched shrub, often forming thickets, and rarely a round-crowned tree; *aromatic, spiny,* and with *tiny gland-dots* on foliage, flowers, and fruit.

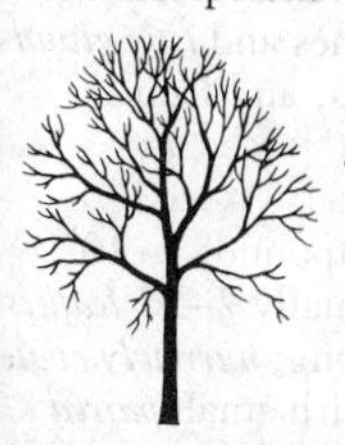

Height: 20′ (6 m) or more.
Diameter: 6″ (15 cm).
Leaves: pinnately compound; 5–10″ (13–25 cm) long; *5–11 paired leaflets,* 1–2″ (2.5–5 cm) long; *elliptical or ovate;*

blunt-pointed at ends; edges straight or slightly wavy; hairy when young; stalkless. Dull green with sunken veins above, paler and hairy on veins beneath.
Bark: gray to brown; smooth.
Twigs: brown or gray; hairy when young; often with *paired short stout spines* less than ⅜" (10 mm) long.
Flowers: less than $\frac{3}{16}$" (5 mm) wide; with 5 spreading fringed yellow-green petals; in short-stalked clusters; male and female on separate plants; in spring before leaves.

Fruit: $\frac{3}{16}$" (5 mm) long; *podlike,* elliptical, brown, slightly fleshy; maturing in late summer and splitting open.

Habitat: Moist soils in valleys and rocky uplands.

Range: S. Ontario east to S. Quebec, south to Pennsylvania, west to central Oklahoma, and north to E. North Dakota; local to Georgia; to 2000′ (610 m).

The northernmost representative of a tropical genus named from Greek words meaning "yellow" and "wood." A drug formerly was obtained from the dried, bitter, aromatic bark. The fresh bark is chewed for relief from toothache, numbing the pain.

---

323 Hercules-club
"Toothache-tree" "Tingle-tongue"
Zanthoxylum clava-herculis L.

Description: Aromatic, spiny, round-crowned tree with spreading branches and *tiny gland-dots* on foliage, flowers, and fruit.

Height: 40′ (12 m).
Diameter: 1′ (0.3 m).
Leaves: pinnately compound; 5–10" (13–25 cm) long. Usually *7–17 leaflets* 1½–3" (4–7.5 cm) long; *narrowly ovate, pointed* at tip, often with small *paired*

spines at base; edges wavy; almost stalkless; remaining attached into winter. Shiny green above, paler and sometimes hairy beneath.
Bark: light gray; thin, with conspicuous corky-based *conical knobs* terminating in a *stout spine* that eventually drops off.
Twigs: brown; hairy when young; with *scattered* stout straight *spines* to ½" (12 mm) long.
Flowers: 3/16" (5 mm) wide; with 5 yellow-green petals; in large much-branched clusters at ends of leafy twigs; male and female usually on separate plants; in spring.

Fruit: ¼" (6 mm) long; *elliptical pods;* brown; *3 or fewer* from a flower; maturing in early summer; 1 shiny black seed hanging from split pod.

Habitat: Moist sandy soils near coast and along streams; often along fences, apparently spread by birds.

Range: Central Virginia to S. Florida, west to E. Texas, and north to SE. Oklahoma; to 500′ (152 m).

The common and Latin species names both refer to the spiny branches. As the alternate names indicate, the bitter aromatic bark is a home remedy for toothache; chewing bark or foliage numbs the pain.

301, 400 **Lime Prickly-ash**
"Wild-lime"
Zanthoxylum fagara (L.) Sarg.

Description: *Evergreen* shrub or *small tree* with spreading rounded crown, often with a leaning trunk; aromatic, spiny, with *tiny gland-dots* on foliage, flowers, and fruit.
Height: 25′ (7.6 m).
Diameter: 8″ (20 cm).
Leaves: *evergreen;* pinnately compound; 3–4″ (7.5–10 cm) long; with *flat or winged axis.* 5–13 leaflets (usually 7–9),

⅜–1″ (1–2.5 cm) long; *elliptical;* wavy-toothed beyond middle; narrowed to base; *thick and leathery;* stalkless. Shiny green above, paler beneath.
Bark: gray; thin, smooth or warty, becoming scaly.
Twigs: dark gray; slender, zigzag, hairless or nearly so; with *paired hooked sharp spines* less than ¼″ (6 mm) long.
Flowers: ⅛″ (3 mm) wide; with 4 yellow-green petals in small clusters to ½″ (12 mm) wide, on old twigs; male and female on separate plants; in early spring.
Fruit: 3⁄16″ (5 mm) long; *podlike,* rounded, brown, warty, clustered along twig, 1-seeded; *1–2 from a flower;* maturing in autumn and splitting open.

Habitat: Moist soil mostly near coast and on plains.
Range: Central and S. Florida, Florida Keys, and S. Texas; also N. Mexico; to 500′ (152 m).

The powdered bark and leaves have a sharp taste and have been used also as a spice. The crushed foliage has an odor of limes, as the common names suggest.

QUASSIA FAMILY
(Simaroubaceae)

Shrubs and trees often with very bitter bark, wood, and other parts. About 150 species in tropical and subtropical regions; 5 native and 1 naturalized tree and 2 native shrub species in North America.

Leaves: alternate; pinnately compound, with leaflets generally not toothed; without stipules.

Flowers: mostly small; often in large branched clusters; generally male and female and usually on separate plants; (sometimes bisexual), regular, with 3- to 8-lobed calyx, 3–8 petals (sometimes none), with same number or twice as many stamens as petals; usually 2–5 simple pistils on disk.

Fruit: generally a capsule, drupe, or winged key (samara).

313, 385, 496 **Ailanthus**
"Tree-of-Heaven"
Ailanthus altissima (Mill.) Swingle

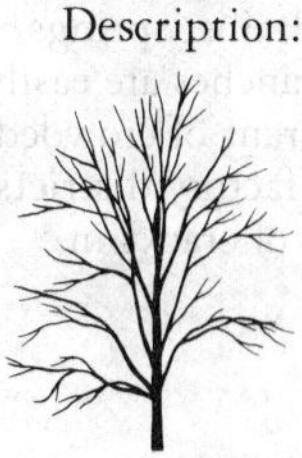

Description: A hardy, introduced tree with a spreading, rounded, open crown of stout branches and coarse foliage; *male flowers* and *crushed foliage* have *disagreeable odor.*

Height: 50–80′ (15–24 m).

Diameter: 1–2′ (0.3–0.6 m).

Leaves: pinnately compound; 12–24″ (30–61 cm) long. *13–25 leaflets* (sometimes more) 3–5″ (7.5–13 cm) long, 1–2″ (2.5–5 cm) wide; paired (except at end); broadly *lance-shaped;* with *2–5 teeth near broad 1-sided base* and gland-dot beneath each tooth; covered with fine hairs when young. Green above, paler beneath.

Bark: light brown; smooth, becoming rough and fissured.

Twigs: light brown, *very stout;* covered

with fine hairs when young; with *brown* pith.

Flowers: ¼″ (6 mm) long; with 5 *yellowish-green petals;* in terminal branched clusters 6–10″ (15–25 cm) long; male and female usually on separate trees; in late spring and early summer.

Fruit: 1½″ (4 cm) long; showy, reddish-green or reddish-brown, narrow, flat, winged, 1-seeded; 1–6 from a flower; maturing in late summer and autumn.

Habitat: Widespread in waste places, spreading rapidly by suckers.

Range: Native of China but widely naturalized across temperate North America; from near sea level to high mountains.

Widely planted as an ornamental and shade tree and in shelterbelts, for the rapid growth and coarse foliage reminiscent of tropical trees. However, it is no longer recommended for good sites where other trees will grow. Male flowers have an objectionable odor, and some people are allergic to their pollen which may produce symptoms of hayfever. The roots, which are classed as poisonous, get into drains, springs, and wells. The weak branches are easily broken by storms. Tolerant of crowded dusty cities and smoky factory districts, often even growing out of cracks in concrete.

MAHOGANY FAMILY
(Meliaceae)

Trees, usually large, and shrubs, with bitter, astringent bark and often aromatic wood. About 1000 species in tropical regions; 1 native and 1 naturalized tree species in North America.
Leaves: alternate, generally pinnately compound; leaflets mostly paired, usually not toothed and often oblique; without stipules. Naked buds with tiny young leaves often in form of hand.
Flowers: small; generally in branched clusters; bisexual—regular, with 4- to 5-lobed calyx, and corolla of 4–5 petals or lobes, 8–10 stamens united in a tube and around a disk, and 1 pistil.
Fruit: generally a capsule or berry with seeds often winged.

341, 469 **Chinaberry**
"China Tree" "Pride of India"
Melia azedarach L.

Description: A naturalized shade tree, deciduous or nearly evergreen in the far South, with dense, spreading crown and clusters of *round, yellow, poisonous fruit; foliage* has *bitter taste and strong odor* when crushed.

Height: 40′ (12 m).
Diameter: 1′ (0.3 m).
Leaves: *bipinnately compound;* 8–18″ (20–46 cm) long; with slender green axis and few paired forks. *Numerous leaflets* 1–2″ (2.5–5 cm) long, ⅜–¾″ (10–19 mm) wide; paired (except at end); *lance-shaped* or ovate; *saw-toothed;* wavy or lobed; hairless or nearly so. Dark green above, paler beneath.
Bark: dark brown or reddish-brown; furrowed, with broad ridges.
Twigs: green, stout, hairless or nearly so.
Flowers: ¾″ (19 mm) wide; with 5 *pale purple petals and* narrow violet tube;

fragrant; on slender stalks in *showy* branched clusters 4–10″ (10–25 cm) long; in spring.

Fruit: ⅝″ (15 mm) in diameter; a *yellow poisonous berry,* becoming slightly wrinkled; thin juicy pulp; maturing in autumn, remaining attached in winter; hard stone contains 3–5 seeds.

Habitat: Dry soils near dwellings, in open areas and clearings; sometimes within forests.

Range: Native of S. Asia. Naturalized from SE. Virginia to Florida, west to Texas, and north to SE. Oklahoma; also in California; usually below 1000′ (305 m).

One of the hardiest members of the tropical Mahogany family, Chinaberry grows rapidly but is short-lived. One cultivated variety known as Umbrella-tree, with a very dense, compact, flattened crown like an umbrella, is a popular ornamental. The fruit stones can be made into beads, but the abundant fruit produces litter and is toxic or narcotic.

SPURGE FAMILY
(Euphorbiaceae)

Shrubs, trees, and a few herbs, often with whitish poisonous sap or latex. About 7000 species nearly worldwide, especially tropical; 6 native and 2 naturalized tree and many native shrub and herb species in North America.
Leaves: generally alternate, simple, and with stipules often in form of glands; sometimes pinnately or palmately compound.
Flowers: male and female, generally on same plant, small or tiny, greenish, in clusters, usually regular, often much reduced, with 5 sepals or with none, and without petals or with as many as 5. Male flowers usually with the same number or twice as many stamens as petals, when present, or reduced to 1, or many, and with disk or glands. Female flowers with superior 3-celled ovary and 3 styles.
Fruit: usually a capsule opening in 3 parts, each 1-seeded; sometimes a drupe, the seeds generally with a fleshy protuberance.

95, 580 **Tallowtree**
"Chinese Tallowtree"
Sapium sebiferum (L.) Roxb.

Description: Small naturalized tree with short trunk, several large weak branches, spreading irregular crown, and *poisonous milky sap.*

Height: 30′ (9 m).
Diameter: 6″ (15 cm).
Leaves: 1½–3″ (4–7.5 cm) long and wide. *Nearly 4-sided* or rounded; long-pointed; *without teeth* or with a few small teeth; hairless; with *long slender leafstalk* to 2″ (5 cm), with *2 glands at upper end.* Shiny green above, paler beneath; turning red and yellow in autumn.
Bark: brownish-gray; fissured.

Twigs: green to gray; slender, hairless, brittle.

Flowers: tiny, *yellow-green;* male flowers numerous in narrow clusters 2–4″ (5–10 cm) long, some also with few female flowers at base; in spring.

Fruit: ½″ (12 mm) in diameter; slightly 3-lobed, 3-celled, dark brown; maturing, splitting, and shedding walls in autumn; 3 elliptical *white waxy seeds* 5⁄16″ (8 mm) long, *remaining attached.*

Habitat: Sandy soils along coast and streams, and near towns.

Range: Native of China. Naturalized from S. North Carolina to N. Florida and west to SE. Texas; usually near sea level.

This ornamental is planted for its poplarlike leaves that turn red and yellow in autumn and for its odd white waxy seeds. A rapidly growing hardy weed tree, it often forms thickets. Both the common name and the Latin species name, meaning "bearing wax," refer to the Chinese custom of making candles by boiling the fruit to remove the wax from the seed coats. An oil is extracted from the seeds.

CASHEW FAMILY
(Anacardiaceae)

Trees, shrubs, and few woody vines with resinous sap often in bark and other parts; in a few species the resin or volatile oil is caustic and poisonous to the skin. About 600 species in tropical and north temperate regions; 15 native and 3 naturalized tree, 6 native shrub, and 1 woody vine species in North America.
Leaves: alternate; pinnately compound, with 3 leaflets, or simple; without stipules.
Flowers: tiny or small; commonly white; many in large branched clusters; bisexual or male and female; mostly regular, with 3–5 sepals united at base and 3–5 petals (or no petals); generally 10 stamens around a disk, and 1 pistil with superior 1-celled (to 5-celled) ovary, 1 style, and 3 stigmas.
Fruit: usually a 1-seeded resinous drupe.

64, 460 **American Smoketree**
"Yellowwood" "Chittamwood"
Cotinus obovatus Raf.

Description: Shrub or small tree with a short trunk, open crown of spreading branches, resinous sap with a strong odor, and deep orange-yellow heartwood.
Height: 30′ (9 m).
Diameter: 1′ (0.3 m).
Leaves: 2–6″ (5–15 cm) long, 1½–3″ (4–7.5 cm) wide. *Obovate* or elliptical; short-stalked; *rounded or blunt at tip;* edges straight or slightly wavy. Dull green above, paler (and covered with silky hairs when young) beneath; turning orange and scarlet in autumn.
Bark: gray to blackish; thin, scaly.
Twigs: slender; with whitish bloom when young, turning brown.
Flowers: ⅛″ (3 mm) wide; with 5

greenish-white petals; in branched clusters 6″ (15 cm) long; male and female usually on separate plants; in spring; flowers mostly sterile; stalks becoming long and covered with *short purplish or brownish hairs.*

Fruit: 3/16″ (5 mm) long; *oblong,* flat, pale brown, dry, 1-seeded; scattered in large branched clusters to 8′ (20 cm) long; maturing in summer.

Habitat: Limestone rocky uplands and ravines; in hardwood forests.

Range: Scattered in SE. Tennessee and N. Alabama, Ozark Plateau of SW. Missouri, Arkansas, and E. Oklahoma, and Edwards Plateau of central Texas; at 700–2000′ (213–610 m).

The wood was once used for making a yellow dye. The masses of smokelike fruit clusters with hairy stalks of sterile flowers give the species its common nam[e]. It is sometimes grown for the handsome autumn foliage and unusual fruit. However, the planted ornamentals are mainly the related Common Smoketree (*Cotinus coggygria* Scop.) of Eurasia, which has elliptical, hairless leaves abruptly narrowed at base.

---

303 **Texas Pistache**
"Wild Pistachio" "American Pistachio"
Pistacia texana Swingle

Description: Aromatic, evergreen shrub or small tree with many branches near the base and a broad, rounded, dense crown.

Height: 20′ (6 m).
Diameter: 8″ (20 cm).
Leaves: *evergreen* or nearly so; pinnately compound; 2–4″ (5–10 cm) long; with flattened axis. *9–19 leaflets* about ½″ (12 mm) long and half as wide; *obovate* or elliptical; *curved* and unequal-sided; without teeth; often slightly hairy. Dark red when young, becoming dark green above, paler beneath.

Bark: gray-brown; smooth, becoming rough.
Twigs: gray-brown; hairy when young, slender.
Flowers: tiny; without sepals and petals; *dark red* or yellowish-red; in branched clusters less than 2½″ (6 cm) long, at leaf base; male and female on separate plants; in spring with new leaves.

Fruit: 3/16″ (5 mm) long; *nutlike; elliptical* and slightly flattened; 1-seeded; *reddish-brown* turning *blackish* when dry; hard.

Habitat: Dry limestone cliffs and along ravines.

Range: Local on Edwards Plateau of S. Texas and NE. Mexico; at 600–1000′ (183–305 m).

This uncommon, handsome evergreen is drought-resistant and grows well in full sunlight and merits planting as an ornamental in warm, dry climates. In spring the new dark red foliage is showy. Goats browse the leaves; however, the small seeds are often empty and unimportant as food for wildlife or people. Pistachio nuts are from the related Common Pistache (*Pistacia vera* L.), native of western Asia.

342, 387, 577 **Shining Sumac**
"Dwarf Sumac" "Winged Sumac"
Rhus copallina L.

Description: Shrub or small tree with a short trunk and open crown of stout, spreading branches.

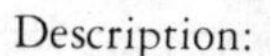

Height: 25′ (7.6 m).
Diameter: 6″ (15 cm).
Leaves: pinnately compound; to 12″ (30 cm) long; with *flat broad-winged axis. 7–17 leaflets* (27 in southeastern variety) 1–3¼″ (2.5–8 cm) long; lance-shaped; usually without teeth; slightly thickened. *Shiny dark green* and nearly hairless above, *paler* and covered with

fine hairs beneath; turning dark reddish-purple in autumn; stalkless.
Bark: light brown or gray; scaly.
Twigs: brown, stout, slightly zigzag, covered with fine hairs; with watery sap.
Flowers: ⅛″ (3 mm) wide; with 5 greenish-white petals; crowded in spreading clusters to 3″ (13 cm) wide, with *hairy branches;* male and female usually on separate plants; in late summer.
Fruit: more than ⅛″ (3 mm) in diameter; 1-seeded; crowded in clusters; rounded and slightly flattened, dark red, *covered with short sticky red hairs;* maturing in autumn, remaining attached in winter.

Habitat: Open uplands, valleys, edges of forests, grasslands, clearings, roadsides, and waste places.

Range: S. Ontario east to SW. Maine, south to Florida, west to central Texas, and north to Wisconsin; to 4500′ (1372 m) in the Southeast.

Shining Sumac is sometimes planted as an ornamental for its shiny leaves and showy fruit. The sour fruit can be nibbled or made into a drink like lemonade. Wildlife eat the fruit, and deer also browse the twigs. It is easily distinguishable from other sumacs by the winged leaf axis and watery sap. Often forms thickets.

315, 386, 566 **Smooth Sumac**
"Scarlet Sumac" "Common Sumac"
Rhus glabra L.

Description: The most common sumac; a large shrub or sometimes a small tree with open, flattened crown of a few stout, spreading branches and with whitish sap.
Height: 20′ (6 m).
Diameter: 4″ (10 cm).

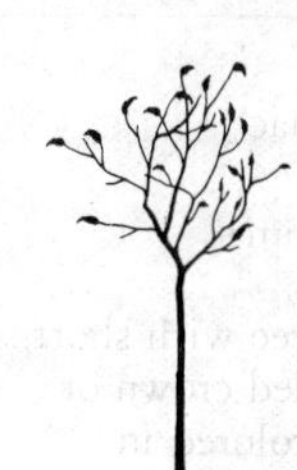

Leaves: pinnately compound; 12″ (30 cm) long; with slender axis. *11–31 leaflets* 2–4″ (5–10 cm) long; *lance-shaped;* saw-toothed; *hairless;* almost stalkless. Shiny green above, *whitish beneath;* turning reddish in autumn.
Bark: brown; smooth or becoming scaly.
Twigs: gray, with whitish bloom; few, very stout, *hairless.*
Flowers: less than ⅛″ (3 mm) wide; with 5 whitish petals; crowded in large upright clusters to 8″ (20 cm) long, with *hairless branches;* male and female usually on separate plants; in early summer.
Fruit: more than ⅛″ (3 mm) in diameter; rounded, 1-seeded, numerous, crowded in upright clusters; dark red, covered with *short sticky red hairs;* maturing in late summer, remaining attached in winter.

Habitat: Open uplands including edges of forests, grasslands, clearings, roadsides, and waste places, especially in sandy soils.

Range: E. Saskatchewan east to S. Ontario and Maine, south to NW. Florida, and west to central Texas; also in mountains from S. British Columbia south to SE. Arizona and in N. Mexico; to 4500′ (1372 m) in the East; to 7000′ (2134 m) in the West.

The only shrub or tree species native to all 48 contiguous states. One cultivated variety has dissected or bipinnate leaves. Raw young sprouts were eaten by the Indians as salad. The sour fruit, mostly seed, can be chewed to quench thirst or prepared as a drink similar to lemonade. It is also consumed by birds of many kinds and small mammals, mainly in winter. Deer browse the twigs and fruit throughout the year.

307 **Prairie Sumac**
"Prairie Flameleaf Sumac"
"Texan Sumac"
Rhus lanceolata (Gray) Britton

Description: Large shrub or small tree with short trunk and open, rounded crown of foliage turning flame-colored in autumn.

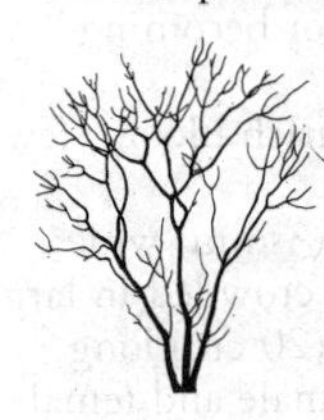

Height: 25′ (7.6 m).
Diameter: 6″ (15 cm).
Leaves: pinnately compound; to 9″ (23 cm) long; with *flat narrowly winged axis.* Usually *13–19 leaflets* 1–2½″ (2.5–6 cm) long, less than ½″ (12 mm) wide; paired (except at end); *narrowly lance-shaped; slightly curved;* long-pointed at tip; blunt and unequal at base; usually without teeth. Shiny dark green above; *paler, covered with fine hairs,* and with prominent veins beneath; turning reddish-purple in autumn.
Bark: gray or brown; smooth or becoming scaly.
Twigs: green or reddish and hairy when young, becoming gray and hairless; stout, ending in whitish hairy bud.
Flowers: 3⁄16″ (5 mm) wide; with 5 greenish-white petals; crowded in upright clusters to 6″ (15 cm) long, composed of many flowers, at ends of twigs; male and female usually on separate plants; in summer.
Fruit: about 3⁄16″ (5 mm) in diameter; slightly flattened, dark red, covered with short sticky red hairs, 1-seeded; numerous, crowded in clusters; maturing in early autumn and falling in early winter.

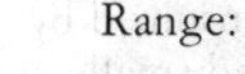

Habitat: Dry rocky slopes and hills, especially limestone; often forming thickets.
Range: Texas and S. New Mexico, local in S. Oklahoma and in NE. Mexico; to 2500′ (762 m); locally to 4000′ (1219 m).

Birds, especially bobwhites, grouse, and pheasants, consume quantities of the fruit in winter, and deer browse the foliage. The leaves contain tannin and have been used in tanning leather.

314, 567, 578 **Staghorn Sumac**
"Velvet Sumac"
Rhus typhina L.

Description: Tall shrub or small tree with irregular, open, flat crown of a few stout, spreading branches; whitish, sticky sap turns black on exposure.
Height: 30′ (9 m).
Diameter: 8″ (20 cm), sometimes larger.
Leaves: pinnately compound; 12–24″ (30–61 cm) long; with stout soft *hairy reddish-tinged axis.* 11–31 leaflets 2–4″ (5–10 cm) long; *lance-shaped;* often slightly curved; saw-toothed; nearly stalkless. Dark green above, whitish (with reddish hairs when young) beneath; turning bright red with purple and orange in autumn.
Bark: dark brown; thin; smooth or becoming scaly.
Twigs: few, very stout, brittle, *dense velvety covering* of long, brown hairs.
Flowers: ⅛–3⁄16″ (3–5 mm) wide; with greenish petals; crowded in upright clusters to 8″ (20 cm) long; *branches densely covered with hairs;* male and female usually on separate plants; in early summer.
Fruit: 3⁄16″ (5 mm) in diameter; rounded, 1-seeded, dark red, covered with *long dark red hairs;* numerous, crowded in upright clusters; maturing in late summer and autumn, remaining attached in winter.

Habitat: Open uplands, edges of forests, roadsides, and old fields.

Range: S. Ontario east to Nova Scotia, south to NW. South Carolina, west to Tennessee, and north to Minnesota; to 5000′ (1524 m) in the Southeast.

Staghorn Sumac reaches tree size more often than related species and commonly forms thickets. In winter, the bare, widely forking, stout, hairy twigs resemble deer antlers "in velvet," hence the alternate common name. Indians made a lemonade-like drink from the crushed fruit of this and related species. Sumac bark and foliage are rich in tannin and were used to tan leather. Grown as an ornamental, especially a variety with dissected leaves, for the autumn foliage and showy fruit.

349 Poison-sumac

"Poison-dogwood" "Poison-elder"
Toxicodendron vernix (L.) Kuntze

Description: *Poisonous* yet attractive narrow-crowned shrub or small tree with waxy whitish berries and dramatic fall foliage.

Height: 25′ (7.6 m).
Diameter: 6″ (15 cm).
Leaves: pinnately compound; 7–12″ (18–30 cm) long; with *reddish* axis. 5–13 leaflets 2½–3½″ (6–9 cm) long; paired except at end; *ovate or elliptical; without teeth;* short-stalked. Shiny dark green above, paler and slightly hairy beneath; turning scarlet or orange in early autumn.
Bark: gray or blackish; thin; smooth or slightly fissured.
Twigs: reddish when young, turning gray with many orange dots; hairless.
Flowers: ⅛″ (3 mm) long; with 5 greenish petals; many, in long, open, branching clusters to 8″ (20 cm) long; male and female on same or separate plants; in early summer.
Fruit: ¼″ (6 mm) in diameter; *rounded* and slightly flat; *whitish,* 1-seeded, *shiny and hairless;* numerous, in drooping branched clusters; maturing in early autumn and often remaining attached until spring.

Habitat: Wet soil of swamps, bogs, seepage

slopes, and frequently flooded areas; in shady hardwood forests.

Range: Extreme S. Quebec and Maine south to central Florida, west to E. Texas, and north to SE. Minnesota; mostly confined to Atlantic and Gulf coastal plains and Great Lakes region; to 1000′ (305 m).

One of the most dangerous North American plants. The clear, very toxic sap turns black on exposure and, for many people, causes a rash upon contact. A black varnish can be made from the sap, as in a related Japanese species. The fruit of Poison-sumac is not toxic to birds or animals and is consumed by many kinds of wildlife, such as bobwhites, pheasants, grouse, and rabbits, especially in winter, when other food is scarce.

CYRILLA FAMILY
(Cyrillaceae)

Shrubs and trees, mostly small, sometimes large. 13 species in tropical America and southeastern United States; 2 native tree species in North America.
Leaves: alternate, simple, mostly leathery, without stipules.
Flowers: tiny; clustered along an axis; bisexual; regular, with 5 sepals united at base and persistent, 5 petals united at base or separate, 5–10 stamens, and 1 pistil composed of superior ovary with 2–4 cells, each with 1–2 ovules, short style, and 2 stigmas.
Fruit: a capsule or berrylike drupe, often angled or winged.

---

45 Buckwheat-tree
"Titi" "Black Titi"
Cliftonia monophylla (Lam.) Britton ex Sarg.

Description: Evergreen, thicket-forming shrub or small tree with short, often crooked trunk, many branches, and a narrow crown.
Height: 20′ (6 m).
Diameter: 6″ (15 cm).
Leaves: *evergreen;* 1–2″ (2.5–5 cm) long, ⅜–¾″ (10–19 mm) wide; *narrowly elliptical; thick and leathery;* with *vein along straight border* and tiny gland-dots; almost stalkless. Shiny dark green above, paler and often whitish beneath.
Bark: dark red-brown or gray; scaly, becoming furrowed, thick, and spongy.
Twigs: numerous; reddish-brown with grayish bloom, slender, stiff.
Flowers: ¼″ (6 mm) wide; with 5–8 *white petals,* sometimes pinkish-tinged; fragrant; in narrow clusters 1–2½″ (2.5–6 cm) long, with narrow reddish-brown bracts; in early spring.

Fruit: ¼″ (6 mm) long; *elliptical, shiny yellow, 2–4-winged;* in showy clusters;

maturing in late summer, turning brown, remaining closed and often attached until spring.

Habitat: Wet, sandy, acid soils of bays and swamps.

Range: SE. Georgia and N. Florida west to SE. Louisiana; to 200′ (61 m).

The persistent winged fruit, similar to the fruit of Buckwheat, makes identification easy. Grown as an ornamental for the fragrant early flowers, shiny evergreen foliage, and showy fruit, it is also a honey-producing plant. The common name "Titi" apparently is of American Indian origin.

67, 404 **Swamp Cyrilla**
"Leatherwood" "Titi"
Cyrilla racemiflora L.

Description: Small tree with short, stout, crooked trunk and spreading crown or a much-branched shrub, with glossy foliage, profuse, tiny, whitish flowers, and clusters of tiny, brown or yellow fruit.
Height: 30′ (9 m).
Diameter: 8″ (20 cm).
Leaves: clustered near end of twig; 1½–3″ (4–7.5 cm) long, ⅜–1″ (1–2.5 cm) wide. *Narrowly oblong,* usually widest beyond middle, blunt or slightly notched at tip; without teeth; slightly thickened; hairless. *Shiny green* above, paler beneath; turning orange and red in autumn. *Deciduous or evergreen in the South* in subtropical and tropical climates.
Bark: gray and smooth, becoming reddish-brown, thin, and scaly; whitish-pink and spongy at base.
Twigs: brown, slender, hairless.
Flowers: ⅛″ (3 mm) wide; with 5 *pointed white petals,* sometimes pinkish-tinged; fragrant; short-stalked; crowded in upright narrow clusters 4–6″ (10–15 cm) long; in early summer.

Fruit: ⅛″ (3 mm) long; beadlike or *egg-shaped,* pointed or rounded, brown or yellow, spongy; in clusters; 2-celled, with 4 or fewer tiny seeds; maturing in late summer, not splitting open.

Habitat: Moist soils of river flood plains and riverbanks, flatwoods, and borders of sandy swamps and ponds.

Range: SE. Virginia south to central Florida and west to SE. Texas; to 500′ (152 m).

This tree also is native in the West Indies and from Central America to Brazil. In the upper mountain forests of Puerto Rico, Swamp Cyrilla is a large dominant tree known as *palo colorado* ("red tree") because of its reddish-brown bark and wood. Bees produce a dark honey from the flowers.

HOLLY FAMILY
(Aquifoliaceae)

Shrubs and trees, small to medium-sized, rarely large; 300–350 species, nearly all in the holly genus (*Ilex*) in tropical and temperate regions, especially tropical America; 14 native tree and 2 native shrub species in North America.

Leaves: alternate, simple, generally leathery and evergreen, sometimes with tiny stipules.

Flowers: small, few clustered along twigs, whitish or greenish, regular, generally male and female on separate plants or bisexual; calyx with 4 (sometimes 5) tiny sepals or teeth, 4 (5) rounded whitish petals sometimes united at base, 4 (5) alternate stamens inserted at base of corolla, without disk, and 1 pistil with superior ovary of 4 (3–5) cells of 1–2 ovules each, usually without style, and 3–5 stalkless stigmas.

Fruit: a round drupe or berry, red, black, or yellow, with stalkless stigmas, bitter pulp, and 3–5 nutlets.

197, 558 **Carolina Holly**
Ilex ambigua (Michx.) Torr.

Description: Deciduous shrub or sometimes a small tree with irregular or rounded crown.
Height: 18′ (5.5 m).
Diameter: 4″ (10 cm).
Leaves: *deciduous;* 1–2″ (2.5–5 cm) long, ⅜–1″ (1–2.5 cm) wide. *Elliptical; finely wavy-toothed;* hairless or slightly hairy. Dull green above, paler beneath.
Bark: dark brown or blackish; smooth or scaly.
Twigs: dark reddish-brown, slender, hairless to densely hairy.
Flowers: ¼″ (6 mm) wide; with 4 (sometimes 5) rounded white petals; on

short twigs; in spring; male and female on separate plants.
Fruit: ¼″ (6 mm) in diameter; *berrylike; elliptical* or sometimes round; *red, translucent*, short-stalked; 4 brown narrow grooved nutlets; maturing in late summer, soon shedding.

Habitat: Moist well-drained soils of upland forests.

Range: North Carolina to central Florida, west to E. Texas and SE. Oklahoma; to 1000′ (305 m).

This holly is recognized by the usually elliptical, rather than round, translucent fruit and by the petals, which sometimes number 5 instead of 4. The Latin species name, meaning "ambiguous" or "doubtful," suggests uncertainty in the classification when the species was first distinguished and named.

---

115 Sarvis Holly
"Serviceberry Holly"
Ilex amelanchier M. A. Curtis

Description: Large shrub, or sometimes a small tree, with foliage recalling serviceberries.
Height: 13′ (4 m).
Diameter: 3″ (7.5 cm).
Leaves: *deciduous* (or nearly evergreen southward); 1½–3½″ (4–9 cm) long, ⅝–1¾″ (1.5–4.5 cm) wide. *Elliptical* or lance-shaped; apparently without teeth or finely saw-toothed; with prominent *network of veins.* Dull green above, slightly hairy beneath.
Bark: brown; smooth.
Twigs: covered with fine hairs.
Flowers: ¼″ (6 mm) wide; with 4 (sometimes to 8) rounded white petals; on long stalks from leaf bases and on older twigs; in spring; male and female on separate plants.
Fruit: ⁵⁄₁₆″ (8 mm) in diameter; *berrylike, dull red,* usually 4 narrow

grooved nutlets; maturing in autumn.

Habitat: Moist soils of stream banks, swamps, and sandhills.

Range: From North Carolina to NW. Florida and SE. Louisiana; to 200′ (61 m).

Rare and local in the Coastal Plain, Sarvis Holly is recognized by the leaves which appear to have no teeth. The common and Latin species names refer to the foliage resembling that of the unrelated serviceberry genus (*Amelanchier*).

213 English Holly
"European Holly"
Ilex aquifolium L.

Description: Cultivated, evergreen tree with dense, conical crown of short, spreading branches and shiny red berries.
Height: 50′ (15 m).
Diameter: 1½′ (0.5 m).
Leaves: *evergreen;* 1¼–2¾″ (3–7 cm) long, ¾–1½″ (2–4 cm) wide. *Elliptical;* spiny-pointed; wavy-edged; with *large spiny teeth;* blunt at base; *stiff* and leathery. *Shiny dark green* above, paler beneath.
Bark: gray; smooth or nearly so.
Twigs: greenish or purplish; angled, hairless or with short hairs.
Flowers: ¼″ (6 mm) wide; with 4 rounded white petals; several on short stalks at base of previous year's leaves; male and female on separate trees; in late spring.
Fruit: ¼–⅜″ (6–10 mm) in diameter; *berrylike; shiny red,* clustered at leaf bases, short-stalked; 4 nutlets; maturing in autumn, remaining attached in winter.

Habitat: Moist soils in humid temperate regions.

Range: Native of S. Europe, N. Africa, and W. Asia. Planted across the United States, mainly in Atlantic, southeastern, and Pacific states.

Numerous horticultural varieties differ in leaf size, shape, spines, color, and in tree habit. Several varieties grown in orchards for Christmas decorations have larger berries and larger leaves than native hollies. To assure fruit, a male plant is needed to pollinate the female. Cultivated since ancient times, it is propagated by cuttings and seeds. The wood is used for veneers and inlays.

52, 556 **Dahoon**
"Dahoon Holly" "Christmas-berry"
Ilex cassine L.

Description: Evergreen shrub or small tree with rounded, dense crown and abundant, bright red berries.
Height: 30′ (9 m).
Diameter: 1′ (0.3 m).
Leaves: *evergreen;* 1½–3½″ (4–9 cm) long, ¼–1¼″ (0.6–3.2 cm) wide. *Oblong or obovate;* slightly thick and leathery; usually *without teeth* or spines; edges often turned under. *Shiny dark green* and becoming hairless above, light green (and densely hairy when young) beneath.
Bark: dark gray; thin, smooth to rough and warty.
Twigs: slender, densely covered with silky hairs, becoming brown.
Flowers: 3⁄16″ (5 mm) wide; with 4 rounded white petals; on short stalks mostly at base of new leaves in spring; male and female on separate plants.
Fruit: ¼″ (6 mm) in diameter; *berrylike, round, shiny red* (sometimes yellow or orange), short-stalked; mealy bitter pulp; 4 narrow grooved brown nutlets; maturing in autumn, remaining attached in winter.

Habitat: Wet soils along streams and swamps, sometimes sandy banks or brackish soils.

Range: SE. North Carolina south to S. Florida, and west to S. Louisiana; to 200′

(61 m); also Bahamas, Cuba, Puerto Rico and 1 variety in Mexico.

Planted as an ornamental for the evergreen foliage and profuse red fruit used in Christmas decorations. The common name apparently is of American Indian origin.

207, 560 **Possumhaw**
"Winterberry" "Swamp Holly"
Ilex decidua Walt.

Description: Deciduous shrub or small tree with spreading crown and bright red berries.

Height: 20′ (6 m).
Diameter: 6″ (15 cm).
Leaves: *deciduous;* 1–3″ (2.5–7.5 cm) long, ⅜–1¼″ (1–3 cm) wide. *Spoon-shaped* or narrowly obovate; mostly *clustered on short spur twigs;* alternate on vigorous twigs; *finely wavy-toothed. Dull green* above, paler and hairy on veins beneath.
Bark: light brown to gray; thin; smooth or warty.
Twigs: light gray, slender, hairless.
Flowers: ¼″ (6 mm) wide; with 4 rounded white petals; on slender stalks at end of spur twigs; in spring; male and female on separate plants.
Fruit: ¼″ (6 mm) in diameter; *berrylike; red;* bitter pulp; 4 narrow grooved nutlets; short-stalked; in clusters; maturing in autumn, remaining attached in winter.

Habitat: Moist soils along streams and in swamps.

Range: Maryland south to central Florida, west to Texas, and north to SE. Kansas; to about 1200′ (366 m).

Possumhaw is conspicuous in winter, with its many, small, red berries along leafless, slender, gray twigs. Opossums, raccoons, other mammals, songbirds, and gamebirds eat the fruit of this and related species.

143, 557 **Mountain Winterberry**
"Mountain Holly"
Ilex montana Torr. & Gray

Description: Deciduous, spreading shrub or small tree with narrow crown and relatively large orange-red berries.
Height: 30′ (9 m).
Diameter: 8″ (20 cm).
Leaves: 2½–5″ (6–13 cm) long; ¾–2¼″ (2–6 cm) wide. *Elliptical* or ovate; abruptly pointed at tip; *finely saw-toothed. Dull green* above, paler and hairy on veins beneath; turning yellow in autumn.
Bark: light brown to gray; thin; smooth or warty.
Twigs: brown to gray; slightly zigzag.
Flowers: nearly ¼″ (6 mm) wide; with 4 (sometimes 5) rounded white petals; on slender stalks; in late spring; male and female on separate plants.
Fruit: ⅜–½″ (10–12 mm) in diameter; *berrylike;* bright *orange-red* (sometimes red or yellow); 4 narrow grooved nutlets; short-stalked; maturing in autumn, often remaining attached into winter.

Habitat: Moist soils in mixed hardwood forests.

Range: W. Massachusetts and New York south to N. Georgia; local southwest to NW. Florida and Louisiana; from 200′ (61 m) to 6000′ (1829 m) in southern Appalachians.

The showy berries persist into winter as the common name implies. Closely related to Carolina Holly, it is sometimes classed as a variety of that species, having larger leaves and fruit.

54

Myrtle Dahoon
"Myrtle-leaf Holly"
Ilex myrtifolia Walt.

Description: Evergreen shrub or small tree with broad, dense crown of many crooked branches and *small, narrow leaves.*
Height: 18′ (5.5 m).
Diameter: 6″ (15 cm).
Leaves: *evergreen;* ½–1¼″ (1.2–3.2 cm) long, ⅛–⅜″ (3–10 mm) wide. *Linear,* bristle-tipped, short-pointed at base; *thick and stiff; edges turned under* and usually without teeth; with wide midvein and *obscure side veins; crowded,* very short-stalked; becoming nearly hairless. Dark green above, paler beneath.
Bark: whitish-gray; rough and warty.
Twigs: brown, slender, stiff, hairy when young.
Flowers: 3⁄16″ (5 mm) wide; with 4 rounded white petals; on short stalks at leaf bases; in spring; male and female on separate plants.
Fruit: ¼″ (6 mm) in diameter; *berrylike; red* (rarely orange or yellow); thin bitter pulp; 4 narrow grooved nutlets; short-stalked; maturing in autumn, remaining attached in winter.

Habitat: Wet, mostly poor or acid sandy soils, bordering ponds and swamps; in pine or baldcypress forests.

Range: North Carolina to central Florida, west to SE. Louisiana; to 200′ (61 m).

Myrtle Dahoon is closely related to Dahoon and has been considered a variety of that species. However, the latter has larger, broader leaves and grows in richer, wet soil. The names both refer to the resemblance of the leaves to those of the unrelated true Myrtle (*Myrtus communis* L.), native in the Mediterranean region.

212, 559 **American Holly**
"Holly" "White Holly"
Ilex opaca Ait.

Description: Evergreen tree with narrow, rounded, dense crown of spiny leaves, small white flowers, and bright red berries.

Height: 40–70′ (12–21 m).
Diameter: 1–2′ (0.3–0.6 m).
Leaves: *evergreen;* spreading in 2 rows; 2–4″ (5–10 cm) long, ¾–1½″ (2–4 cm) wide. *Elliptical;* spiny-pointed and *coarsely spiny-toothed;* thick, *stiff and leathery. Dull green* above, yellow-green beneath.
Bark: light gray; thin; smooth or rough and warty.
Twigs: brown or gray; stout, covered with fine hairs when young.
Flowers: ¼″ (6 mm) wide; with 4 rounded white petals; in short clusters at base of new leaves and along twigs; in spring; male and female on separate trees.
Fruit: ¼–⅜″ (6–10 mm) in diameter; *berrylike; bright red* (rarely orange or yellow); bitter pulp; 4 brown nutlets; scattered; short-stalked; maturing in autumn, remaining attached in winter.

Habitat: Moist or wet well-drained soils, especially flood plains; in mixed hardwood forests.

Range: E. Massachusetts south to central Florida, west to S. central Texas, and north to SE. Missouri; to 4000′ (1219 m); higher in southern Appalachians.

The evergreen fruiting branches from wild and planted trees are popular Christmas decorations. Many improved varieties are grown for ornament, shade, and hedges. The whitish, fine-textured wood is especially suited for inlays in cabinetwork, handles, carvings, and rulers, and can be dyed various shades, even black. Many kind of songbirds, gamebirds, and mammal eat the bitter berries of this and other hollies.

208, 561 Yaupon
"Cassena" "Christmas-berry"
Ilex vomitoria Ait.

Description: Evergreen, much-branched, thicket-forming shrub or small tree with rounded, open crown, small shiny leaves, and abundant, round, shiny red berries.
Height: 20′ (6 m).
Diameter: 6″ (15 cm).
Leaves: *evergreen;* usually ¾–1¼″ (2–3 cm) long, ¼–½″ (6–12 mm) wide. *Elliptical;* blunt at tip; rounded at base; *finely wavy-toothed; thick* and stiff; short-stalked. *Shiny green* above, paler beneath.
Bark: red-brown; thin, finely scaly.
Twigs: gray; branching at right angles; slightly angled and hairy when young, becoming rough.
Flowers: 3/16″ (5 mm) wide; with 4 spreading rounded white petals; on short stalks at base of old leaves; male and female on separate plants.
Fruit: ¼″ (6 mm) in diameter; *berrylike; shiny red,* clustered along twigs, short-stalked; bitter pulp; 4 narrow grooved nutlets; maturing in autumn, often remaining attached in winter.

Habitat: Moist soils, especially along coasts and in valleys, sometimes in sandhills.

Range: SE. Virginia south to central Florida, west to Texas, and north to SE. Oklahoma; to 500′ (152 m).

The ornamental twigs with shiny evergreen leaves and numerous red berries are favorite Christmas decorations. Yaupon is sometimes grown for ornament and trimmed into hedges. The leaves contain caffeine, and American Indians used them to prepare a tea to induce vomiting and as a laxative. Tribes from the interior traveled to the coast in large numbers each spring to partake of this tonic.

BITTERSWEET FAMILY (Celastraceae)

Shrubs, woody vines, and mostly small trees. Widespread, about 700 species; 7 native tree species and several shrub species in North America.
Leaves: alternate or opposite, sometimes whorled, simple, with tiny stipules or none.
Flowers: tiny; usually in clusters with stalks mostly jointed; greenish; bisexual or male and female; regular, with 4–5 sepals united at base and persisting at base of fruit, and 4–5 petals, 4–5 alternate stamens inserted on or below the large disk, and 1 pistil with superior ovary of 2–5 cells each with 2 ovules, short style, and stigma often with 2–5 lobes.
Fruit: a capsule, berry, or drupe; the seed generally with colored covering.

172, 555, 581 Eastern Burningbush
"Eastern Wahoo" "Euonymus"
Euonymus atropurpureus Jacq.

Description: Shrub or rarely a small tree with spreading, irregular crown and red or purple capsules suggesting a burning bush.
Height: 20′ (6 m).
Diameter: 4″ (10 cm).
Leaves: opposite; 2–4½″ (5–11 cm) long, 1–2″ (2.5–5 cm) wide. *Elliptical;* abruptly long-pointed at tip; finely saw-toothed. Green above, paler and often with fine hairs beneath; turning light yellow in autumn.
Bark: gray; smooth, becoming slightly fissured.
Twigs: dark purplish-brown, slender, sometimes 4-angled or slightly winged.
Flowers: ⅜″ (10 mm) wide; with *4 dark red* or *purple petals;* 7–15 flowers clustered on slender, widely forking stalks; in late spring and early summer.

Fruit: ⅝″ (15 mm) wide; *red or purple capsules deeply 4-lobed* and 4-celled, each lobe splitting open; smooth; several hanging on slender stalk; maturing in autumn, remaining attached into winter; in each cell, 1–2 rounded light brown seeds with *red covering.*

Habitat: Moist soils, especially in thickets, valleys, and forest edges.

Range: Extreme S. Ontario to central New York, south to N. Georgia, west to central Texas, and north to SE. North Dakota; to 2000′ (610 m).

The powdered bark was used by American Indians and pioneers as a purgative. "Wahoo" was the native term. The Latin species name, meaning "dark purple," refers to the color of the fruit.

BLADDERNUT FAMILY
(Staphyleaceae)

About 50 species of trees and shrubs in north temperate regions and tropical mountains; 2 native tree species in North America.
Leaves: generally opposite; pinnately compound with odd number of leaflets or only 3; finely toothed; with paired stipules and rings at nodes.
Flowers: small; in clusters at twig ends; white or greenish; generally bisexual; regular, with 5 sepals often persisting at base of fruit, 5 petals, 5 alternate stamens inserted around large cuplike disk, and 1 pistil with superior 2- to 3-celled ovary, many ovules, and 2–3 persistent styles.
Fruit: a berry or inflated capsule with few seeds.

353, 425 **American Bladdernut**
Staphylea trifolia L.

Description: Shrub or small tree with paired leaves of 3 leaflets, striped twigs, and drooping, bladderlike seed capsules.

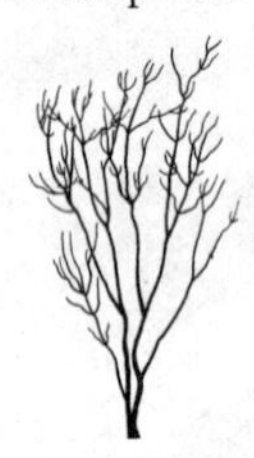

Height: 20′ (6 m).
Diameter: 4″ (10 cm).
Leaves: opposite; palmately compound; 6–9″ (15–23 cm) long; with *long slender stalk* and *3 leaflets,* 2 paired and nearly stalkless and 1 long-stalked. Leaflets 1½–3¼″ (4–8 cm) long; *elliptical; long-pointed at tip; finely saw-toothed.* Green (and hairy on veins) above, paler and hairy beneath.
Bark: gray; smooth, becoming slightly fissured.
Twigs: green to brown or gray; often *striped,* with *rings at nodes.*
Flowers: ½″ (12 mm) long; *bell-shaped;* with *5 white petals;* long-stalked; in terminal drooping clusters to 4″ (10 cm) long; in spring.

Fruit: 1¼–2″ (3–5 cm) long; drooping

elliptical capsule, swollen, slightly 3-lobed and 3-celled, green to brown, opening at long-pointed tip; few rounded shiny yellow-brown seeds; maturing in late summer.

Habitat: Moist soils in understory of hardwood forests.

Range: S. Ontario to extreme S. Quebec and New Hampshire, south to NW. Florida, west to E. Oklahoma, and north to SE. Minnesota; to 2000′ (610 m).

The generic name, *Staphylea,* is from Greek and means "cluster of grapes," referring to the flowers. The Latin species name, *trifolia,* meaning "three-leaf," refers to the leaflets.

MAPLE FAMILY
(Aceraceae)

Trees and shrubs; about 125 species, nearly all in the maple genus (*Acer*), in north temperate regions south into tropical mountains; 13 native tree species in North America.
Leaves: deciduous; opposite; long-stalked; mostly simple, broad, and palmately lobed and veined; toothed; sometimes pinnately compound; without stipules; the sap sometimes sweetish or milky.
Flowers: commonly male and female on same or separate trees or bisexual, in often-branched clusters; small; with 5 or 4 colored sepals separate or sometimes united at base, corolla of 5 or 4 overlapping petals or corolla absent, 4–10 stamens (usually 8) from edge of large disk, and 1 pistil with 2-celled 2-lobed ovary with 2 ovules in each cell, and 2-forked style.
Fruit: paired, flat, long-winged, 1-seeded keys (samaras).

---

262 **Florida Maple**
"Southern Sugar Maple" "Hammock Maple"
Acer barbatum Michx.

Description: Medium-sized tree with spreading, rounded crown and paired, palmately lobed leaves.

Height: 60′ (18 m).
Diameter: 2′ (0.6 m).
Leaves: opposite; 1½–3″ (4–7.5 cm) long and wide. *3 or 5 shallow, short-pointed or blunt lobes; wavy-edged* with 5 main veins from base; long-stalked. Dark green above, *paler, whitish, and hairy beneath;* turning yellow and red in autumn.
Bark: light gray; smooth, becoming rough and furrowed.
Twigs: reddish-brown, slender, usually

hairy when young.
Flowers: ⅛″ (3 mm) long; with bell-shaped 5-lobed yellow calyx; in drooping clusters on slender hairy stalks; usually male and female on separate trees; in early spring.
Fruit: ¾–1″ (2–2.5 cm) long including long wing; *paired forking keys,* brown, 1-seeded; maturing in autumn.

Habitat: Moist soils of valleys and upland slopes.

Range: SE. Virginia south to central Florida, west to E. Texas, and north to E. Oklahoma; to 2000′ (610 m).

Florida Maple is the southeastern relative of Sugar Maple (which has larger leaves) and is often classed as a variety of that species. The two intergrade where their ranges meet. Unlike Sugar Maples, Florida Maples are not tapped commercially for sugar.

263 Chalk Maple
"White-bark Maple"
Acer leucoderme Small

Description: Small tree with rounded crown of slender branches and with distinctly *whitish bark.*

Height: 30′ (9 m).
Diameter: 1′ (0.3 m).
Leaves: opposite; 2–3½″ (5–9 cm) long and wide. Palmately divided into 5 *(sometimes 3) blunt or short- or long-pointed lobes; edges wavy* or coarsely toothed; 5 main veins from base; long-stalked. Becoming dark green and hairless above, yellow-green and with *soft hairs beneath;* turning red in autumn.
Bark: *whitish*-gray or chalky and *smooth;* becoming dark brown and furrowed at base.
Twigs: green to reddish; slender, hairless, shiny.
Flowers: ⅛″ (3 mm) long; with bell-shaped 5-lobed yellow calyx; in drooping clusters on long slender

stalks; usually male and female on same or separate trees; in early spring.
Fruit: ¾″ (19 mm) long including long wing; *paired widely forking keys;* becoming hairless; reddish or brown; 1-seeded; maturing in summer or autumn.

Habitat: Moist soils of streams and rocky riverbanks.

Range: North Carolina south to NW. Florida and west to E. Texas and SE. Oklahoma; at 200–1000′ (61–305 m).

This rare tree is sometimes planted as a shade tree in the Southeast. Both the common name and the scientific name, meaning "white skin," describe the smooth bark.

335, 494 **Boxelder**
"Ashleaf Maple" "Manitoba Maple"
Acer negundo L.

Description: Small to medium-sized tree with a short trunk and a broad, rounded crown of light green foliage.

Height: 30–60′ (9–18 m).
Diameter: 2½′ (0.8 m).
Leaves: opposite; *pinnately compound;* 6″ (15 cm) long; with slender axis. 3–7 *leaflets* sometimes slightly lobed, 2–4″ (5–10 cm) long, 1–1½″ (2.5–4 cm) wide; paired and short-stalked (except at end); *ovate or elliptical, long-pointed* at tip, short-pointed at base; *coarsely saw-toothed,* sometimes lobed. *Light green* and mostly hairless above, paler and varying in hairiness beneath; turning yellow (or sometimes red) in autumn.
Bark: light gray-brown; with many narrow ridges and fissures, becoming deeply furrowed.
Twigs: *green,* often whitish or purplish; slender, ringed at nodes, mostly hairless.
Flowers: $^{3}/_{16}$″ (5 mm) long; with very small *yellow-green* calyx of 5 lobes or

sepals; several clustered on slender drooping stalks; male and female on separate trees; before leaves in spring.
Fruit: 1–1½″ (2.5–4 cm) long; *paired, slightly forking keys* with flat narrow body and *long curved wing; pale yellow,* 1-seeded; maturing in summer and remaining attached in winter.

Habitat: Wet or moist soils along stream banks and in valleys with various hardwoods; also naturalized in waste places and roadsides.

Range: S. Alberta east to extreme S. Ontario and New York, south to central Florida, and west to S. Texas; also scattered from New Mexico to California and naturalized in New England; to 8000′ (2438 m) in the Southwest.

Boxelder is classed with maples having similar paired key fruits, but is easily distinguishable by the pinnately compound leaves. Hardy and fast-growing, it is planted for shade and shelterbelts but is short-lived and easily broken in storms. Common and widely distributed, it is spreading in the East as a weed tree. Plains Indians made sugar from the sap. The common name indicates the resemblance of the foliage to that of elders (*Sambucus*) and the whitish wood to that of Box (*Buxus sempervirens* L.).

260, 622 **Black Maple**
"Hard Maple" "Rock Maple"
Acer nigrum Michx. f.

Description: Large tree with rounded, dense crown and paired, palmately 3-lobed leaves.
Height: 80′ (24 m).
Diameter: 2–3′ (0.6–0.9 m).
Leaves: opposite; 4–5½″ (10–14 cm) long and wide. With 3 (sometimes 5) *broad long-pointed lobes; edges wavy* or with few blunt teeth; 3 or 5 main veins

from base; *sides often drooping;* leafstalks long and hairy. Dull green above, *yellow-green and with soft hairs beneath;* turning yellow in autumn.
Bark: dark gray or blackish; becoming deeply furrowed.
Twigs: brown, slender, hairy when young.
Flowers: 3/16″ (5 mm) long; with bell-shaped 5-lobed yellow calyx; in drooping clusters on long slender hairy stalks; male and female on same or separate trees; with leaves in early spring.

Fruit: 1–1¼″ (2.5–3.2 cm) long including long wing; *paired forking keys;* brown, 1-seeded; maturing in autumn.

Habitat: Moist soils of valleys and uplands; in mixed hardwood forests.

Range: S. Ontario east to S. Quebec and Vermont, southwest to Tennessee and Missouri and north to SE. Minnesota; local in adjacent states; to 2500′ (762 m), slightly higher in south.

Closely related to Sugar Maple, Black Maple is often treated as a variety of that species. The ranges are similar, though Black Maple is more common throughout Iowa, while Sugar Maple extends farther into Canada in the Northeast. The sweet sap is similarly tapped for maple syrup.

255, 623 **Striped Maple**
"Moosewood"
Acer pensylvanicum L.

Description: Small tree with short trunk and open crown of striped, upright branches and coarse foliage; often a shrub.
Height: 30′ (9 m).
Diameter: 8″ (20 cm).
Leaves: opposite; 5–7″ (13–18 cm) long and nearly as wide. With *3 short broad* long-pointed *lobes at tip; doubly saw-toothed;* with *3 main veins* from base;

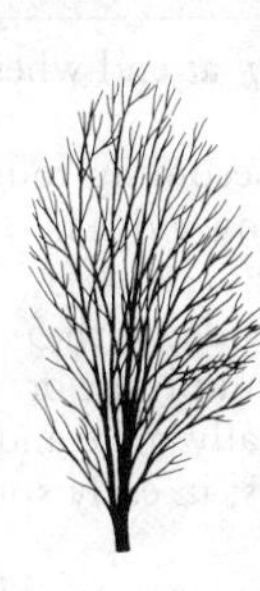

with rust-colored hairs when young and in vein angles; stout leafstalk. Light green above, paler beneath; turning yellow in autumn.
Bark: bright green with white stripes, becoming reddish-brown with long pale vertical lines; thin; smooth or warty.
Twigs: *green, becoming striped* with *whitish lines.*
Flowers: ⅜" (10 mm) wide; bell-shaped; with 5 *bright yellow petals;* slender-stalked; male and female usually in separate clusters to 6" (15 cm) long; *drooping* at end of leafy twigs; in late spring.

Fruit: 1¼" (3 cm) long; many *paired, widely forking keys;* long-winged, light brown, 1-seeded; maturing in autumn.

Habitat: Moist upland soils in understory of hardwood forests.

Range: S. Ontario east to Nova Scotia, south to N. Georgia, and west to S. Minnesota; to 5500′ (1676 m).

Striped Maple is easily recognized, even in winter, by the striped twigs and bark, which make it a popular ornamental. Rabbits, beavers, deer, and moose eat the bark, especially in winter.

259, 388 **Norway Maple**
Acer platanoides L.

Description: Introduced shade tree with rounded crown of dense foliage and with milky sap in leafstalks.

Height: 60′ (18 m).
Diameter: 2′ (0.6 m).
Leaves: opposite; 4–7" (10–18 cm) long and wide. *Palmately 5-lobed;* the shallow lobes and edges with *scattered long teeth;* 5 or 7 main veins from notched base. Dull green with sunken veins above, paler and hairless (except in vein angles) beneath; turning bright yellow in autumn. Long slender

leafstalk, with *milky sap* at end when broken off.

Bark: gray or brown; becoming rough and furrowed into narrow ridges.

Twigs: brown, hairless.

Flowers: 5⁄16″ (8 mm) wide; with 5 *greenish-yellow* petals; in upright or spreading clusters; usually male and female on separate trees; in early spring before leaves.

Fruit: 1½–2″ (4–5 cm) long; *paired keys* with long wing and *flattened body; spreading widely,* light brown, hanging on long stalk; maturing in summer.

Habitat: A street tree, escaping along roadsides; in humid temperate regions.

Range: Native across Europe from Norway to Caucasus and N. Turkey. Widely planted across the United States.

Norway Maple is fast-growing and tolerant of city smoke and dust. Varieties have columnar and low, rounded habits and reddish and variegated foliage. The species name, meaning "like *Platanus,*" indicates the similarity of the leaves to those of Sycamore and Planetree, to which it is not related.

---

253 Planetree Maple
"Sycamore Maple"
Acer pseudoplatanus L.

Description: Large, introduced shade tree with widely spreading, rounded crown and large, paired, *palmately 5-lobed leaves.*

Height: 70′ (21 m).

Diameter: 2′ (0.6 m).

Leaves: opposite; 3½–6″ (9–15 cm) long and wide. The 5 shallow lobes short-pointed and *wavy saw-toothed;* with 5 *main veins* from notched base; long slender leafstalk. Dull dark green with sunken veins above, pale with raised, sometimes hairy veins beneath; turning brown in autumn.

Bark: gray; smooth or with broad flaky scales.
Twigs: gray, hairless.
Flowers: 3/16″ (5 mm) wide; with 5 *greenish-yellow* petals; male and bisexual; in narrow branched *drooping* clusters 5″ (13 cm) long; in early spring.
Fruit: 1¼–2″ (3–5 cm) long; *paired keys* with *elliptical body* and long wing; light brown; maturing in summer.

Habitat: Hardy in exposed places and adapted to seashore gardens, tolerant of salt spray; sometimes escaping along roadsides.

Range: Native of Europe and W. Asia; planted across the United States.

Although fast-growing, this species is not as hardy northward as Norway Maple. An important timber and shade tree in Europe, where it is called Sycamore. The species name, meaning "false *Platanus*," refers to the resemblance of the foliage to that of the sycamore or planetree genus.

261, 366, 369, 495, 593

Red Maple
"Scarlet Maple" "Swamp Maple"
Acer rubrum L.

Description: Large tree with narrow or rounded, compact crown and *red flowers, fruit, leafstalks,* and *autumn foliage.*

Height: 60–90′ (18–27 m).
Diameter: 2½′ (0.8 m).
Leaves: opposite; 2½–4″ (6–10 cm) long and nearly as wide. Broadly ovate, with *3 shallow short-pointed lobes* (sometimes with 2 smaller lobes near base); irregularly and wavy *saw-toothed,* with 5 main veins from base; long red or green leafstalk. Dull green above, *whitish* and hairy *beneath;* turning red, orange, and yellow in autumn.
Bark: gray; thin, smooth, becoming fissured into long thin scaly ridges.
Twigs: reddish, slender, hairless.
Flowers: ⅛″ (3 mm) long; *reddish;*

crowded in nearly stalkless clusters along twigs; male and female in separate clusters; in *late winter* or very early spring before leaves.
Fruit: ¾–1″ (2–2.5 cm) long including long wing; *paired forking keys; red turning reddish-brown;* 1-seeded; maturing in spring.

Habitat: Wet or moist soils of stream banks, valleys, swamps, and uplands and sometimes on dry ridges; in mixed hardwood forests.

Range: Extreme SE. Manitoba east to E. Newfoundland, south to S. Florida, west to E. Texas; to 6000′ (1829 m).

Red Maple is a handsome shade tree, displaying red in different seasons. Pioneers made ink and cinnamon-brown and black dyes from a bark extract. It has the greatest north-south distribution of all tree species along the East Coast.

264, 493, 600 **Silver Maple**
"Soft Maple" "White Maple"
Acer saccharinum L.

Description: Large tree with short, stout trunk, few large forks, spreading, open, irregular crown of long, curving branches, and graceful cut-leaves.

Height: 50–80′ (15–24 m).
Diameter: 3′ (0.9 m).
Leaves: opposite; 4–6″ (10–15 cm) long and nearly as wide. Broadly ovate, *deeply 5-lobed* and *long-pointed* (middle lobe often 3-lobed); *doubly saw-toothed,* with 5 main veins from base; becoming hairless; slender drooping reddish leafstalk. Dull green above, *silvery-white beneath;* turning pale yellow in autumn.
Bark: gray; becoming furrowed into long scaly shaggy ridges.
Twigs: light green to brown; long, spreading and often slightly drooping,

hairless; with slightly unpleasant odor when crushed.
Flowers: ¼″ (6 mm) long; *reddish buds turning greenish-yellow;* crowded in nearly stalkless clusters; male and female in separate clusters; in *late winter* or very early spring before leaves.
Fruit: 1½–2½″ (4–6 cm) long including long broad wing; *paired, widely forking* keys; *light brown,* 1-seeded; maturing in spring.

Habitat: Wet soils of stream banks, flood plains, and swamps; with other hardwoods.

Range: S. Ontario east to New Brunswick, south to NW. Florida, west to E. Oklahoma, north to N. Minnesota; to 2000′ (610 m), higher in mountains.

Its rapid growth makes Silver Maple a popular shade tree; however, its form is not generally pleasing, its brittle branches are easily broken in windstorms, and the abundant fruit produces litter. Sugar can be obtained from the sweetish sap, but yield is low.

258, 374, 592 **Sugar Maple**
"Hard Maple" "Rock Maple"
Acer saccharum Marsh.

Description: Large tree with rounded, dense crown and striking, multicolored foliage in autumn.
Height: 70–100′ (21–30 m).
Diameter: 2–3′ (0.6–0.9 m).
Leaves: opposite; 3½–5½″ (9–14 cm) long and wide; palmately lobed with 5 *deep long-pointed lobes; few narrow long-pointed teeth;* 5 main veins from base; leafstalks long and often hairy. Dull dark green above, paler and often hairy on veins beneath; turning deep red, orange, and yellow in autumn.
Bark: light gray; becoming rough and deeply furrowed into narrow scaly ridges.

Twigs: greenish to brown or gray; slender.
Flowers: 3/16″ (5 mm) long; with bell-shaped 5-lobed *yellowish-green* calyx; male and female in drooping clusters on long slender hairy stalks; with new leaves in early spring.
Fruit: 1–1¼″ (2.5–3 cm) long including long wing; *paired forking keys;* brown, 1-seeded; maturing in autumn.

Habitat: Moist soils of uplands and valleys, sometimes in pure stands.

Range: Extreme SE. Manitoba east to Nova Scotia, south to North Carolina, and west to E. Kansas; local in NW. South Carolina and N. Georgia; to 2500′ (762 m) in north and 3000–5500′ (914–1676 m) in southern Appalachians.

Maples, particularly Sugar Maple, are among the leading furniture woods. This species is used also for flooring, boxes and crates, and veneer. Some trees develop special grain patterns, including birdseye maple with dots suggesting the eyes of birds, and curly and fiddleback maple, with wavy annual rings. Such variations in grain are in great demand. The boiled concentrated sap is the commercial source of maple sugar and syrup, a use colonists learned from the Indians. Each tree yields between 5 and 60 gallons of sap per year; about 32 gallons of sap make 1 gallon of syrup or 4½ pounds of sugar.

---

254 Mountain Maple
"Moose Maple"
Acer spicatum Lam.

Description: Shrub or small tree with short trunk and slender, upright branches.
Height: 25′ (7.6 m).
Diameter: 6″ (15 cm).
Leaves: opposite; 2½–4½″ (6–11 cm)

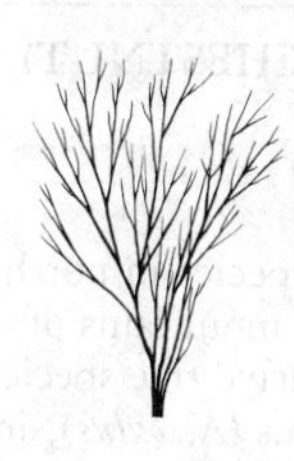

long and wide; with 3 (sometimes 5) *short broad lobes;* short-pointed; *coarsely saw-toothed;* 3 or 5 main veins from base; long leafstalks often turn red. Light green and becoming hairless above, hairy beneath; turning bright red and orange in autumn.

Bark: brown; thin; scaly or slightly furrowed.

Twigs: light gray, slender; hairy when young.

Flowers: ¼″ (6 mm) wide; *greenish-yellow;* short-stalked; in *narrow upright hairy clusters* to 5″ (13 cm); at end of leafy twig; male and female in separate clusters; in early summer.

Fruit: ¾–1″ (2–2.5 cm) long; *paired forking long-winged keys;* clustered along slender stalk; 1-seeded; *red or yellow* when immature, turning brown in autumn.

Habitat: Moist rocky uplands, especially mountains; in understory of hardwood forests.

Range: E. Saskatchewan east to Newfoundland, south to Pennsylvania, and west to NE. Iowa; also in southern Appalachians to N. Georgia; to 6000′ (1829 m).

Mountain Maple is hardy and adapted to partial shade. The Latin species name, meaning "spiked," refers to the long spikelike flower clusters. Rabbits, beavers, deer, and moose browse the bark, and ruffed grouse eat the buds.

BUCKEYE (HORSECHESTNUT) FAMILY
(Hippocastanaceae)

Trees and shrubs; 15 species in north temperate regions and mountains of tropical America. 6 native tree species, all in the buckeye genus (*Aesculus*), in North America.
Leaves: deciduous (evergreen in tropics); opposite; long-stalked; palmately compound, with 5–9 (sometimes 3) elliptical or lance-shaped leaflets, mostly saw-toothed; without stipules.
Flowers: many in large upright branched clusters; showy, often large; irregular; slightly bell-shaped; both bisexual and male; with tubular 5-lobed calyx and corolla of 4–5 generally unequal rounded white, pink, red, or yellow petals, 6–8 long curved stamens, and 1 pistil composed of 3-celled ovary with 2 ovules in each cell and long curved style.
Fruit: a large rounded brown capsule, often spiny, with hard thick wall splitting into 3 parts. 1–3 large, rounded or slightly angled, shiny brown seeds with gray scar at base, poisonous or inedible.

---

355 Texas Buckeye
"White Buckeye"
Aesculus arguta Buckl.

Description: Shrub or small tree with rounded to oblong crown of stout branches.
Height: 20′ (6 m).
Diameter: 6″ (15 cm), sometimes larger.
Leaves: opposite; palmately compound; with slender leafstalks 2–5″ (5–13 cm) long. *7–11 leaflets* 2½–5″ (6–13 cm) long, ⅜–1¼″ (1–3 cm) wide; *lance-shaped;* long-pointed at both ends; *unevenly saw-toothed;* nearly stalkless.

Shiny yellow-green above, paler and hairy beneath.
Bark: light gray; smooth, becoming scaly or furrowed.
Twigs: reddish-brown; stout, becoming hairless.
Flowers: ½–¾″ (12–19 mm) long; narrowly *bell-shaped,* with *4 unequal pale yellow petals* and 7–8 *longer stamens;* in upright branched clusters 4–6″ (10–15 cm) long; in spring.
Fruit: ¾–1¾″ (2–4 cm) in diameter; light brown spiny capsule, splitting on 2–3 lines; 1 or 2 large shiny brown *poisonous* seeds; maturing in summer.

Habitat: Sandy soils of slopes, hills, and stream bluffs; forming thickets and in forest understory.
Range: Extreme SE. Nebraska south to W. Missouri, S. and SW. Oklahoma, and central Texas; at 500–2000′ (152–610 m).

Texas Buckeye is distinguishable by the large number of narrow leaflets and usually shrubby size. The Latin species name, meaning "sharp," refers to the sharply toothed leaflets. The common name of the genus is suggested by the large, rounded, dark brown seed with a white spot or scar at the base. It is sometimes classed as a western shrubby variety of Ohio Buckeye.

359, 382, 530 **Ohio Buckeye**
"Fetid Buckeye"
"American Horse-chestnut"
Aesculus glabra Willd.

Description: Tree with upright flowers at ends of twigs and an irregular, rounded crown; sometimes shrubby. Twigs and leaves often have a slightly *unpleasant odor* when crushed.

Height: 30–70′ (9–21 m).
Diameter: 1–2′ (0.3–0.6 m).
Leaves: opposite; *palmately compound;*

with slender leafstalks 2–6″ (5–15 cm) long. 5–7 *leaflets* 2½–6″ (6–15 cm) long, ¾–2¼″ (2–6 cm) wide; *elliptical; unevenly saw-toothed;* nearly stalkless. Yellow-green above, paler and often hairy beneath; turning orange or yellow in autumn.

Bark: ashy-gray; scaly, becoming rough and furrowed into thick scaly plates; with an unpleasant odor.

Twigs: reddish-brown; stout, becoming hairless.

Flowers: ¾–1″ (2–2.5 cm) long; narrowly *bell-shaped;* with *4 nearly equal pale yellow or greenish-yellow petals* and 7 *longer stamens;* in upright branched clusters 4–6″ (10–15 cm) long; giving off an *unpleasant odor;* in spring.

Fruit: 1–2″ (2.5–5 cm) in diameter; a pale brown *spiny capsule,* splitting on 2–3 lines; 1–3 large dark brown *poisonous* seeds; maturing in summer or autumn.

Habitat: Rich moist soils of valleys and mountain slopes, also on drier sites; sometimes a thicket-forming shrub on stream banks; in mixed hardwood forests.

Range: W. Pennsylvania south to central Alabama, west to SE. Oklahoma, and north to central Iowa; at 500–2000′ (152–610 m).

The state tree of Ohio, the Buckeye State. Pioneers carried a buckeye seed in their pockets to ward off rheumatism. The seeds and young foliage are poisonous, and the toxic bark was formerly used medicinally. Sometimes planted as an ornamental for the showy autumn foliage. The wood is used for furniture, boxes, flooring, and musical instruments.

360, 407, 625 **Horsechestnut**
Aesculus hippocastanum L.

Description: Introduced shade and ornamental tree with spreading, elliptical to rounded crown of stout branches and coarse foliage.
Height: 70′ (21 m).
Diameter: 2′ (0.6 m).
Leaves: opposite; *palmately compound;* with leafstalks 3–7″ (7.5–18 cm) long. 7 *leaflets* (sometimes 5) spreading fingerlike, 4–10″ (10–25 cm) long, 1–3½″ (2.5–9 cm) wide; *obovate* or elliptical; broadest toward abrupt point; tapering to stalkless base; *saw-toothed.* Dull dark green above, paler beneath.
Bark: gray or brown; thin, smooth, becoming fissured and scaly.
Twigs: light brown; stout, hairless; ending in large blackish sticky bud.
Flowers: 1″ (2.5 cm) long; narrowly *bell-shaped;* with 4–5 spreading *narrow white petals, red- and yellow-spotted* at base; many flowers in *upright* branched clusters 10″ (25 cm) long; in late spring.
Fruit: 2–2½″ (5–6 cm) in diameter; a brown *spiny* or warty *capsule,* splitting into 2–3 parts; 1–2 large rounded shiny brown *poisonous seeds;* maturing in late summer.

Habitat: A shade and street tree in rich moist soils.

Range: Native of SE. Europe; widely planted across the United States and escaped in the Northeast.

Horsechestnut is showy when bearing masses of whitish flowers for a few weeks in spring. Easily propagated from seed and tolerant of city conditions, although the stout branches are broken by winds. Turks reportedly used the seeds to concoct a remedy given to horses suffering from cough, hence the common and Latin species names.

358, 384, 531 Yellow Buckeye
"Sweet Buckeye" "Big Buckeye"
Aesculus octandra Marsh.

Description: Tree with rounded crown and upright clusters of showy yellow flowers.
Height: 70–90′ (21–27 m).
Diameter: 2–3′ (0.6–0.9 m).
Leaves: opposite; *palmately compound;* with slender leafstalks 3½–7″ (9–18 cm) long. 5–7 *leaflets* 4–8″ (10–20 cm) long, 1½–3″ (4–7.5 cm) wide; *elliptical to obovate; evenly saw-toothed;* short-stalked. Dark green and usually hairless above, yellow-green and often hairy beneath.
Bark: brown to gray; thin, fissured into large scaly plates.
Twigs: light brown; stout, often hairy.
Flowers: 1¼″ (3 cm) long; with *4 very unequal yellow petals* and 7–8 *shorter stamens;* in upright branched terminal clusters 4–6″ (10–15 cm) long; in spring.
Fruit: 2–3″ (5–7.5 cm) in diameter; a pale brown, *smooth* or slightly pitted *capsule,* splitting on 2–3 lines; 1–3 large shiny brown *poisonous seeds;* maturing in early autumn.

Habitat: Rich, moist, deep soils from river bottoms to deep mountain valleys or slopes; in mixed forests.

Range: SW. Pennsylvania south to N. Alabam and N. Georgia and north to extreme S. Illinois; at 500–6300′ (152–1920 m).

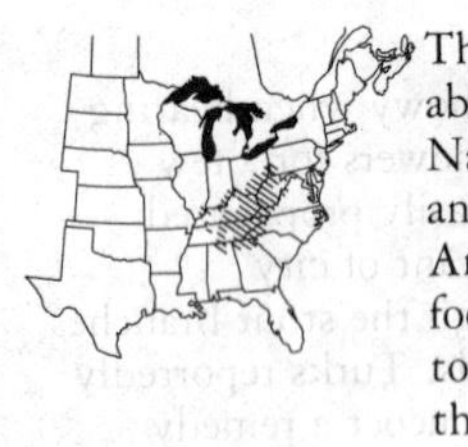

The largest of the buckeyes, it is abundant in Great Smoky Mountains National Park. The seeds are poisonous and young shoots are toxic to livestock American Indians made a nutritious food from the seeds, after removing th toxic element by roasting and soaking them.

357, 367 Red Buckeye
"Scarlet Buckeye" "Firecracker-plant"
Aesculus pavia L.

Description: Shrub or small tree with bright red flowers and irregular crown of short, crooked branches.
Height: 25′ (7.6 m).
Diameter: 8″ (20 cm).
Leaves: opposite; *palmately compound;* with slender leafstalks 3–6″ (7.5–15 cm) long. *5 (sometimes 7) leaflets* 2½–6″ (6–15 cm) long, 1¼–2½″ (3–6 cm) wide; narrowly *elliptical; irregularly saw-toothed;* short-stalked. Dark green with sunken veins and nearly hairless above, dull green and sometimes densely covered with whitish hairs beneath.
Bark: brown-gray to light gray; smooth.
Twigs: reddish-brown; stout.
Flowers: 1¼″ (3 cm) long; with 4 *unequal bright red* (sometimes yellow-red or yellow) *petals* and 6–8 stamens about as long as petals; many, in narrow upright branched clusters 4–8″ (10–20 cm) long; in late spring.
Fruit: 1½–2″ (4–5 cm) in diameter; a *smooth* light brown *capsule,* splitting on 2–3 lines; 1–3 large rounded shiny brown *poisonous seeds;* maturing in early autumn.

Habitat: Moist soils, especially along river bluffs, borders of streams, swamps, and in flood plains; in understory of mixed forests.

Range: SE. North Carolina southwest to N. Florida, west to central Texas, and north to S. Illinois; to 1500′ (457 m).

Planted as a handsome ornamental for the showy red flowers, suggesting firecrackers. American Indians threw powdered seeds and crushed branches of this and other buckeyes into pools of water to stupify fish. The fish then rose to the surface and were easily caught. Pioneers used the gummy roots as a soap substitute and made home remedies from the bitter bark. The scientific name, *pavia,* is an old word for "buckeye."

356, 383 **Painted Buckeye**
"Dwarf Buckeye" "Georgia Buckeye"
Aesculus sylvatica Bartr.

Description: A rounded shrub or sometimes a small tree, with spreading crown and showy yellow or reddish flowers.
Height: 25′ (7.6 m).
Diameter: 8″ (20 cm).
Leaves: opposite; *palmately compound;* with slender leafstalk 4–6″ (10–15 cm) long. 5 *(sometimes 7) leaflets* 4–6″ (10–15 cm) long, 1½–2½″ (4–6 cm) wide; *narrowly elliptical or lance-shaped; finely saw-toothed;* short-stalked. Yellow-green above, green and often hairy beneath.
Bark: gray or brown; thin, scaly.
Twigs: light reddish-brown; stout, hairless.
Flowers: 1–1¼″ (2.5–3 cm) long; with *4 unequal bright yellow or sometimes reddish petals* and 6–7 *shorter stamens;* in upright branched clusters 4–6″ (10–15 cm) long; in spring.
Fruit: 1–1½″ (2.5–4 cm) in diameter; a light brown *smooth capsule,* usually splitting along 3 lines; 1–3 large dark brown *poisonous seeds;* maturing in autumn.

Habitat: Moist well-drained slopes in understory of mixed forests.

Range: SE. Virginia southeast to N. Alabama; at 100–1200′ (30–366 m).

This handsome ornamental was discovered more than two centuries ago by William Bartram, an early American botanist. The Latin species name means "of the woods." The common name refers to the large showy flowers, which are variable in color, suggesting an artist's paintbrush.

SOAPBERRY FAMILY
(Sapindaceae)

About 1500 species of trees, shrubs, and woody vines with tendrils; rarely herbs; in tropical and subtropical regions. 7 native tree species, 3 of woody vines, and 3 of herbaceous vines in North America.
Leaves: alternate; generally pinnately compound, sometimes with 3 leaflets; without stipules (except in vines); the leaflets commonly alternate.
Flowers: numerous; tiny; in branched clusters; usually male and female or bisexual; regular or irregular, with 5 sepals, usually 5 petals often with scale or gland at base within, generally 8 or 10 stamens inserted in a disk, and 1 pistil with superior ovary usually 3-celled with 1–2 ovules in each cell, and style.
Fruit: often large, 3-celled capsule, berry, drupe, or winged key, the seed often with covering. Fruit and seeds of a few species are edible; those of some are poisonous.

308, 563 **Western Soapberry**
"Wild Chinatree" "Jaboncillo"
Sapindus drummondii Hook. & Arn.

Description: *Poisonous* tree with rounded crown of upright branches, or large, spreading shrub.

Height: 20–40′ (6–12 m).
Diameter: to 1′ (0.3 m).
Leaves: pinnately compound; 5–8″ (13–20 cm) long; with long slender axis. *11–19 leaflets* 1½–3″ (4–7.5 cm) long, ⅜–¾″ (10–19 mm) wide; paired (except at end); *lance-shaped, curved and slightly one-sided, long-pointed* at tip, blunt and unequal at base; *without teeth;* slightly thickened; short-stalked. *Dull yellow-green* above, slightly hairy with prominent veins beneath.

Bark: light gray; becoming rough and furrowed.
Twigs: *yellow-green;* covered with fine hairs.
Flowers: ⅛″ (3 mm) wide; usually with 5 rounded *yellowish-white* petals; almost stalkless; generally male and female; in upright clusters 6–9″ (15–23 cm) long; in late spring or summer.
Fruit: ⅜–½″ (10–12 mm) in diameter; *berrylike;* sometimes paired; *yellow* or orange, *nearly transparent;* 1-seeded; leathery; maturing in autumn, turning black and remaining attached in winter; *poisonous.* Single round dark brown seed.

Habitat: Moist soils along streams and on limestone uplands, in and bordering hardwood forests; westward in plains and mountains, grassland, upper desert, and oak woodland zones.

Range: SW. Missouri south to Louisiana, west to S. Arizona, and northeast to SE. Colorado; also N. Mexico; to 6000′ (1829 m).

The poisonous fruit, containing the alkaloid saponin, has been used as a soap substitute for washing clothes. The foliage is unpalatable and may be poisonous to livestock, and the fruit causes a rash on some people's skin. Necklaces and buttons are made from the round dark brown seeds, and baskets are made from the wood, which splits easily.

318 Mexican-buckeye
"Texas-buckeye" "Spanish-buckeye"
Ungnadia speciosa Endl.

Description: Shrub or small tree with irregular crown of upright branches and showy pink flowers.
Height: 25′ (7.6 m).
Diameter: 8″ (20 cm).
Leaves: pinnately compound; 5–12″

(13–30 cm) long; with slender axis. 5–9 *leaflets* 3–5″ (7.5–13 cm) long, 1–1½″ (2.5–4 cm) wide; paired (except at end); *ovate; long-pointed* at tip; rounded and often unequal at base; *wavy saw-toothed;* hairy when young; *leathery;* stalkless or nearly so. *Shiny dark green* above, light green and nearly hairless beneath.

Bark: light gray; thin, fissured.

Twigs: light brown, slender, slightly zigzag, covered with fine hairs.

Flowers: 1″ (2.5 cm) wide; with *4–5 unequal pink or purplish-pink petals;* male, female, and bisexual; slender-stalked; in clusters crowded along twigs; before or with leaves in spring.

Fruit: 1½–2″ (4–5 cm) wide; a long-stalked drooping *capsule;* broadly pear-shaped, *3-lobed, reddish-brown,* rough and leathery, 3-celled; 3 *round poisonous* seeds; maturing and splitting open in autumn.

Habitat: Moist soils of rocky canyons, slopes, and ridges.

Range: Central Texas west to Trans-Pecos Texas and S. New Mexico; also in NE. Mexico; to 5000′ (1524 m).

From a distance the plants in full flower resemble redbuds or peaches. The sweetish but poisonous seeds are sometimes used by children as marbles. Livestock seldom browse the toxic foliage, but bees produce fragrant honey from the flowers. Although not a true buckeye, it is so called because of the similar large capsules and seeds. This distinct plant, alone in its genus, commemorates Baron Ferdinand von Ungnad, Austrian ambassador at Constantinople, who introduced Horsechestnut into western Europe in 1576.

BUCKTHORN FAMILY
(Rhamnaceae)

Shrubs, woody vines, and small to large trees, rarely herbs, often spiny. About 700 species worldwide; 15 native and 3 naturalized tree, about 50 shrub, and 1 woody vine species in North America.
Leaves: mostly alternate, also opposite; simple; often with 3 or more veins from base; usually with tiny stipules.
Flowers: small; greenish or yellowish; mostly in clusters along twigs; usually bisexual; regular, with concave cuplike base with 5 or 4 sepals touching by edges (not overlapping) in bud; with 5 or 4 small petals (sometimes not present) concave and very narrow with narrow base, 5 or 4 opposite stamens enclosed by petals, and 1 pistil with superior 2- to 4-celled ovary within the disk, style, and 1–5 stigmas.
Fruit: a berry, drupe, or capsule, often opening in 3 parts.

---

41 **Bluewood**
"Capul Negro" "Brasil"
Condalia hookeri M. C. Johnst.

Description: Spiny, much-branched, thicket-forming shrub or small tree with thin crown of widely spreading, stiff branches and very small leaves.

Height: 20′ (6 m).
Diameter: 6″ (15 cm).
Leaves: alternate or 2–4 on short twigs, ⅝–1″ (1.5–2.5 cm) long, ⅜–½″ (10–12 mm) wide. *Spoon-shaped; not toothed* or with 2–4 small teeth toward *rounded tip with tiny point;* hairy when young. Shiny *yellow-green* above, paler with raised veins beneath; falling in winter.
Bark: gray to brown; smooth, becoming furrowed into narrow flat scaly ridges.
Twigs: green to gray, slightly zigzag,

slender, *stiff,* often finely hairy, ending in narrow *spine.*
Flowers: about ⅛″ (3 mm) wide; *cup-shaped;* with 5 pointed *green* sepals (petals absent); almost stalkless; 1–3 flowers at leaf base; in late spring.
Fruit: 3⁄16″ (5 mm) in diameter; *shiny black;* with thick purplish *sweet juicy* pulp; 1-seeded; maturing in summer.

Habitat: Dry soils of plains.

Range: Central and S. Texas and NE. Mexico; to 2000′ (610 m).

Jelly can be made from the edible fruit, but the impenetrable thickets with spiny twigs make them difficult to harvest. Birds also consume the fruit. The common name refers to a blue dye obtained from the wood, which is also prized for fuel.

---

145 Carolina Buckthorn
"Indian-cherry" "Yellowwood"
Rhamnus caroliniana Walt.

Description: Shrub or small tree with spreading crown of many slender branches.
Height: 30′ (9 m).
Diameter: 6″ (15 cm).
Leaves: 2–5″ (5–13 cm) long, ¾–1½″ (2–4 cm) wide. *Elliptical; finely wavy-toothed;* with many nearly straight side veins; covered with rust-colored hairs when young. Becoming dark green above, paler and often hairy beneath; turning yellow in autumn.
Bark: gray, often with blackish patches; thin, slightly fissured.
Twigs: green or reddish; slender, becoming hairless, foul-smelling when crushed; ending in hairy naked buds of tiny brown leaves.
Flowers: 3⁄16″ (5 mm) wide; bell-shaped; with 5 pointed greenish-yellow sepals; in clusters of several flowers at leaf bases; in late spring and early summer.
Fruit: ⅜″ (10 mm) in diameter;

berrylike; ripening from *red to shiny black;* with thin juicy sweetish pulp; usually 3 seeds; maturing in late summer and autumn, often remaining attached.

Habitat: Moist soils of stream valleys and upland limestone ridges.

Range: Extreme S. Ohio and W. Virginia south to central Florida, west to central Texas, and north to central Missouri; to 2000′ (610 m); also NE. Mexico.

Songbirds and other wildlife consume the berries, which apparently have medicinal properties. Although included in the buckthorn genus, this species has no spines. It was discovered in South Carolina, hence the common and Latin species names. Another name, "Polecat-tree," describes the odor of crushed foliage and flowers.

---

201 European Buckthorn

"European Waythorn" "Rhineberry"
Rhamnus cathartica L.

Description: Introduced, ornamental, spiny shrub or sometimes small tree.

Height: 20′ (6 m).
Diameter: 4″ (10 cm).
Leaves: *mostly opposite* or clustered on short spurs; ¾–2½″ (2–6 cm) long, ¾–2″ (2–5 cm) wide. *Broadly elliptical; finely wavy-toothed;* with *few long curved* side veins; hairless or nearly so; slender-stalked. Green above, paler beneath; falling in late autumn.
Bark: brown; smooth, peeling off in thin curly strips to reveal reddish inner bark.
Twigs: gray; *often paired* and with *short spurs;* slender, hairless or nearly so; ending in narrow pointed *scaly buds* or sharp *spines;* sometimes thornless.
Flowers: 3⁄16″ (5 mm) wide; *bell-shaped* with *4* spreading pointed *greenish-yellow sepals;* clustered on short stalks at leaf

bases; male and female generally on separate plants; in late spring.

Fruit: 5⁄16″ (8 mm) in diameter; *berrylike; black;* with bitter pulp; usually 4 seeds; maturing in late summer and autumn.

Habitat: Dry soils in open woods, clearings, and along fences and roadsides; spread by birds.

Range: Native of Europe and Asia; naturalized locally from S. Ontario east to Nova Scotia, south to North Carolina, west to NE. Kansas, and north to North Dakota.

The hardy, thorny shrubs can be sheared into good hedges. Widely planted in Europe for centuries. The fruit has been used medicinally as a cathartic, hence the Latin species name.

75, 545, 610 **Glossy Buckthorn**
"Alder Buckthorn"
Rhamnus frangula L.

Description: Introduced, ornamental shrub or small tree with glossy foliage and red-to-black berries.
Height: 20′ (6 m).
Diameter: 4″ (10 cm).
Leaves: 1½–2¾″ (4–7 cm) long, ¾–2″ (2–5 cm) wide. *Elliptical;* usually *widest above middle; not toothed; with several almost straight parallel side veins;* nearly hairless. *Shiny dark green* above, paler beneath; turning clear yellow in autumn.
Bark: grayish; thin, slightly fissured and warty.
Twigs: slender; covered with fine hairs; *thornless,* ending in naked bud of tiny hairy leaves.
Flowers: ⅛″ (3 mm) wide; *bell-shaped;* with 5 pointed *greenish-yellow* sepals; in short clusters at leaf bases; in late spring and early summer.
Fruit: 5⁄16″ (8 mm) in diameter;

berrylike; turning from *red to black;* clustered; 2–3 seeds; maturing in late summer and autumn.

Habitat: Hardy in various soils, escaping especially in wet areas along fences and in bogs.

Range: Native to Europe, W. Asia, and N. Africa. Naturalized from S. Manitoba east to Nova Scotia, south to New Jersey and west to Illinois.

Glossy Buckthorn is a handsome ornamental with shiny, alderlike leaves, turning yellow in autumn. It has long been planted, both as a tree and as a tall hedge, especially in Europe; however, it spreads rapidly and may become a pest. The bark and berries were once used medicinally as a laxative, and the bark yields a yellow dye. The Latin species name *frangula,* meaning "to break," refers to the brittle wood and is still the official drug name.

BASSWOOD (LINDEN) FAMILY
(Tiliaceae)

Trees, or in tropical regions also shrubs and herbs, deciduous except in tropics, commonly with fibrous bark; more than 400 species; represented by the basswood or linden genus (*Tilia*) with 3 native species and by a few tropical herbs in North America.
Leaves: alternate in 2 rows, simple, often oblique or unequal and with 3 or more main veins from base, often toothed, commonly with star-shaped hairs, and with paired stipules.
Flowers: in branched clusters (hanging from a strap-shaped stalk in basswood), bisexual, regular, with calyx of 5 sepals usually separate, 5 petals (sometimes none), many stamens usually united at base in groups of 5–10, and 1 pistil with 2- to 10-celled ovary, 1 style, and as many stigmas as cells.
Fruit: nutlike (rounded and 1- to 3-seeded in basswood), capsule, or drupelike.

158, 406, 612 **American Basswood**
"American Linden" "Bee-tree"
Tilia americana L.

Description: Large tree with long trunk and a dense crown of many small, often drooping branches and large leaves; frequently has two or more trunks, and sprouts in a circle from a stump.

Height: 60–100′ (18–30 m).
Diameter: 2–3′ (0.6–0.9 m).
Leaves: in 2 rows; 3–6″ (7.5–15 cm) long and almost as wide. *Broadly ovate* or rounded; long-pointed at tip; *notched* at base; *coarsely saw-toothed; palmately veined;* long slender leafstalks. Shiny dark green above, light green and *nearly hairless* with tufts of hairs in vein angles beneath; turning pale yellow or brown in autumn.

Bark: dark gray; smooth, becoming furrowed into narrow scaly ridges.
Twigs: reddish or green, slender, slightly zigzag, hairless.
Flowers: ½–⅝" (12–15 mm) wide; with 5 *yellowish-white* petals; fragrant; in long-stalked clusters hanging from middle of leafy greenish bract; in early summer.

Fruit: ⅜" (10 mm) in diameter; nutlike, *elliptical* or rounded, gray, covered with fine hairs; hard; 1–2 seeds; maturing in late summer and autumn, often persisting into winter.

Habitat: Moist soils of valleys and uplands; in hardwood forests.

Range: SE. Manitoba east to SW. New Brunswick and Maine, south to W. North Carolina, and west to NE. Oklahoma; to 3200′ (975 m).

American Basswood, the northernmost basswood species, is a handsome shade and street tree. When flowering, the trees are full of bees, hence the name "Bee-tree"; this species is favored by bees over others and produces a strongly flavored honey. The soft, light wood is especially useful for making food boxes, yardsticks, furniture, and pulpwood. Indians made ropes and woven mats from the tough fibrous inner bark.

---

159 Carolina Basswood

"Linn" "Bee-tree" "Linden"
Tilia caroliniana Mill.

Description: Tree with narrow, dense crown, large leaves, mostly hairy beneath, and pale yellow flowers.
Height: 30–60′ (9–18 m).
Diameter: 1–2′ (0.3–0.6 m).
Leaves: in 2 rows; 2½–6" (6–15 cm) long and almost as wide. *Broadly ovate;* abruptly long-pointed at tip; *unequal and nearly straight* or notched *at base;*

coarsely saw-toothed; palmately veined; long slender leafstalks. Dark green becoming hairless and sometimes shiny above, pale green with easily rubbed-off, soft *coat of rust-colored hairs* (or sometimes nearly hairless) beneath.
Bark: gray; furrowed into scaly ridges.
Twigs: reddish-brown, slender, slightly zigzag, usually *hairy.*
Flowers: ½″ (12 mm) wide; with 5 *pale yellow* hairy petals; fragrant; in long-stalked clusters hanging from middle of leafy greenish bract; in late spring and early summer.
Fruit: ¼″ (6 mm) in diameter; nutlike, gray, densely covered with hairs; usually 1-seeded; maturing in late summer and early autumn.

Habitat: Moist soils of valleys and rocky uplands; in hardwood forests.

Range: North Carolina south to central Florida, west to central Texas, and north to SE. Oklahoma; to 2000′ (610 m).

Carolina Basswood, found at low altitudes, is the southernmost basswood species. Uncommon and usually of small size, the wood is of limited value, but like all basswoods, it is a useful honey plant.

---

156 European Linden
"Common Linden"
Tilia ×*europaea* L.

Description: Large, cultivated, shade tree with dense pyramid-shaped crown.
Height: 70′ (21 m).
Diameter: 2′ (0.6 m).
Leaves: in 2 rows; 2–4″ (5–10 cm) long and wide. *Heart-shaped or rounded,* abruptly long-pointed at tip, unequal and slightly notched at base; *sharply saw-toothed;* palmately veined; long-stalked. Dull dark green above, pale green and *nearly hairless* with tufts of

hairs in vein angles beneath.
Bark: gray; smooth, becoming fissured into scaly ridges.
Twigs: brown; slender, hairless.
Flowers: ½″ (12 mm) wide; with 5 *pale yellow petals;* fragrant; in long-stalked clusters hanging from middle of leafy greenish bract; in early summer. 
Fruit: ¼″ (6 mm) in diameter; nutlike, *round,* short-pointed, gray, covered with fine hairs; hard; maturing in late summer.

Habitat: Moist soils in humid temperate regions.

Range: Cultivated across the United States, especially in the Northeast and the Pacific Northwest.

A hybrid of two European species, Littleleaf Linden (*Tilia cordata* Mill.) and Bigleaf Linden (*Tilia platyphyllos* Scop.), European Linden is one of several introduced shade trees known as lindens and is related to the native basswoods. The flowers are a source of honey.

---

155 White Basswood

"Linden" "Bee-tree"
Tilia heterophylla Vent.

Description: Large tree with a dense crown of large leaves with whitish lower surfaces. 
Height: 60–80′ (18–24 m).
Diameter: 2′ (0.6 m).
Leaves: in 2 rows; 3–7″ (7.5–18 cm) long and almost as wide. *Broadly ovate,* long-pointed at tip, *unequal and nearly straight* at base; *coarsely saw-toothed, palmately veined;* with long, slender leafstalks. Shiny dark green above, with *thick coat of white hairs* beneath.
Bark: gray; becoming furrowed into scaly ridges.
Twigs: gray or brown, slender, slightly zigzag, hairless.
Flowers: ½–⅝″ (12–15 mm) wide; with 5 *yellowish-white* petals; fragrant;

in long-stalked clusters hanging from middle of leafy greenish bract; in early summer.
Fruit: 3/8" (10 mm) in diameter; nutlike, *round,* gray, covered with fine hairs, hard; 1–2 seeds; maturing in late summer.

Habitat: Moist soils of valleys and uplands in hardwood forests.

Range: SW. Pennsylvania south to NW. Florida, and west to N. Arkansas and Missouri; also local north to W. New York; at 200–5000′ (61–1524 m).

From some distance the trees can be distinguished by the whitish lower leaf surfaces that are upturned by breezes. Sprouts arise from the base of a tree, sometimes forming a ring around the trunk. Like American Basswood, this species is useful for making boxes, furniture, and pulpwood and yields a fragrant honey.

STERCULIA FAMILY
(Sterculiaceae)

Shrubs, vines, and trees; about 700 species in tropical and subtropical regions; 2 native and 1 naturalized tree and a few native shrub species in North America.
Leaves: alternate; simple, often palmate-veined and palmately lobed, sometimes palmately compound; with star-shaped hairs; the leafstalk often with enlargement at tip, with stipules.
Flowers: usually in branched clusters along or at end of twigs, or sometimes along trunks; generally bisexual and regular with parts in groups of 5; calyx of 3–5 lobes, and 5 petals (sometimes none present), 5 stamens united in a tube or separate, sometimes with 5 sterile stamens, and 1 pistil composed of superior ovary generally with 5 (sometimes 1–4) cells with 2 or more ovules, and 1–5 styles often lobed; sometimes the stamens and pistils on a long stalk.
Fruit: a capsule or berry or 5 follicles.

---

265 Chinese Parasoltree
"Bottletree" "Japanese Varnish-tree"
Firmiana simplex (L.) W. F. Wight

Description: Ornamental and naturalized tree, with crown rounded like an umbrella, and with large, lobed leaves.

Height: 30′ (9 m).
Diameter: 6″ (15 cm).
Leaves: 6–12″ (15–30 cm) long and wide. *Heart-shaped* or rounded; with *3 or 5 long-pointed lobes;* 5 main veins from base; without teeth; very long-stalked. Dull green above, often hairy beneath.
Bark: *gray-green; smooth.*
Twigs: gray-green; stout.
Flowers: ½″ (12 mm) long; numerous; with 5 *yellow sepals, turning red* (petals absent); male and female in separate

upright clusters 8–16" (20–41 cm) long, with scurfy, hairy branched stems· in late spring and early summer.
Fruit: 2–4" (5–10 cm) long; 5 produced from 1 flower; podlike, long-pointed, greenish turning light brown, stalked; splitting open *like a leaf, exposing on edges* several *pealike seeds;* maturing in late summer.

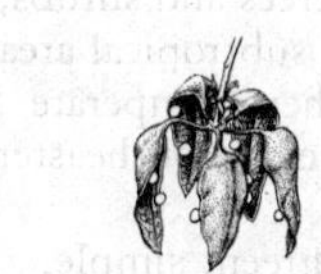

Habitat: Escaping along roadsides and in mixed hardwood forests.

Range: Native of China; cultivated and naturalized locally across S. United States from North Carolina south to N. Florida and west to California.

The opening fruit releases a brownish-black liquid. It is reported that a tea can be prepared from the roasted seeds. This fast-growing tree of a tropical family is planted as an ornamental for its large leaves as far north as Washington, D.C.

TEA FAMILY
(Theaceae)

About 500 species of trees and shrubs, mostly in tropical and subtropical areas but also in warm northern temperate regions; 4 native species in southeastern United States.
Leaves: alternate, evergreen, simple, usually leathery, sometimes with lines parallel to midvein, without stipules.
Flowers: often large, showy and aromatic; generally solitary or a few clustered along twigs; bisexual; regular; often with 2 scales at base, with calyx of 5–7 sepals usually separate, overlapping, and persisting at base of fruit; corolla of 5 commonly white or pink petals, separate or united at base, overlapping; many stamens often united to corolla in 5 opposite groups; and 1 pistil composed of 2- to 5-celled ovary generally superior, with 2 or more ovules in each cell, and 2–5 persistent styles, often united at base.
Fruit: usually a hard capsule with central persistent column or a berry or a drupe.

214, 437 **Franklinia**
"Franklin-tree"
Franklinia alatamaha Bartr. ex Marsh.

Description: Shrub or small tree with beautiful, large, white flowers and an open crown of upright branches.

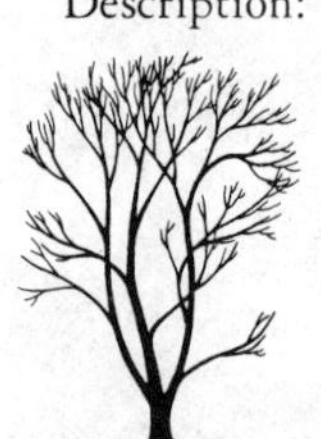

Height: 20′ (6 m).
Diameter: 4″ (10 cm).
Leaves: 5–7″ (13–18 cm) long, 1½–2¾″ (4–7 cm) wide. *Oblong* or obovate; short-pointed; *widest beyond middle; saw-toothed;* short-stalked. *Shiny green* above, paler (with gray hairs when young) and becoming nearly hairless beneath; turning brilliant orange or red in autumn.
Bark: dark brown; smooth, thin,

becoming gray and furrowed.
Twigs: dark reddish-brown, stout, slightly angled, hairless.
Flowers: 3″ (7.5 cm) wide; cup-shaped; with 5 *rounded* wavy-bordered *white petals,* hairy on outer surface, and with many yellow stamens; borne *singly* and nearly stalkless at leaf bases; in autumn.
Fruit: ¾″ (19 mm) in diameter; *hard capsule;* 5-celled, and *splitting oddly* both from tip to middle in 5 *parts* and from base to middle; brown or gray; several flat seeds; maturing in autumn.

Habitat: Formerly on a sandhill beside a river.
Range: Extinct in its native range and known only in cultivation. Originally found near Fort Barrington, close to the coast of SE. Georgia.

Franklinia is grown for its handsome flowers, borne in September, and its showy autumnal foliage as well as for its historical interest. It was discovered in 1765 by John Bartram and his son William, botanists from Philadelphia; it apparently even then occupied only two or three acres. In 1773 and 1778, William returned to obtain seeds and plants, but it has not been seen wild since 1790. It may have been exterminated by wholesale collecting for shipment to London nurseries or destroyed by natural causes such as flooding. Named by William Bartram for Benjamin Franklin and for the Altamaha River (then spelled Alatamaha), where it was discovered.

119, 438 **Loblolly-bay**
"Gordonia" "Bay"
Gordonia lasianthus (L.) Ellis

Description: Evergreen tree or shrub with showy, large, white, fragrant flowers; narrow, compact crown of upright branches; and shiny, leathery foliage.
Height: 60′ (18 m).

Diameter: 1½′ (0.5 m).
Leaves: *evergreen;* 4–6″ (10–15 cm) long, 1–2″ (2.5–5 cm) wide. *Narrowly elliptical* or lance-shaped; short-pointed; *finely saw-toothed; thick and leathery;* short-stalked. *Shiny dark green,* turning red before falling irregularly throughout the year.
Bark: dark red-brown or gray; thick, deeply furrowed into narrow flat ridges.
Twigs: dark brown, stout, hairless.
Flowers: 2½″ (6 cm) wide; cup-shaped; with 5 *large rounded white petals,* waxy and *silky* on outer surface, and with many yellow stamens; fragrant; borne *singly* on *long reddish stalks* at leaf bases; in summer.
Fruit: ½″ (12 mm) in diameter; an *egg-shaped capsule,* pointed, gray, hairy, hard; 5-celled, and *splitting along 5 lines* to below middle; several long-winged brown seeds; maturing in autumn.

Habitat: Wet soil of bays and edges of swamps; also in sandhills; with various hardwoods and conifers.

Range: E. North Carolina south to central Florida and west to S. Mississippi; to 500′ (152 m).

The bark was once used locally for tanning leather. The Latin species name means "hairy-flowered." This genus, honoring James Gordon (1728–91), British nurseryman, includes about 30 species; all the others are in southeastern Asia and Indomalaysia.

193, 436 **Virginia Stewartia**
"Silky-camellia" "Round-fruit Stewartia"
Stewartia malacodendron L.

Description: Shrub or small tree with beautiful, large, purple-centered, white or yellow flowers.
Height: 20′ (6 m).
Diameter: 4″ (10 cm).

Leaves: 2½–4″ (6–10 cm) long, 1–2″ (2.5–5 cm) wide. *Broadly elliptical;* short-pointed; finely *saw-toothed* and hairy on edges; short-stalked. Dark green above, light green and covered with soft hairs beneath.
Bark: dark brown; smooth.
Twigs: gray; slender, hairy when young.
Flowers: 4″ (10 cm) wide; cup-shaped; with 5 *rounded white or yellow petals* with wavy edges, waxy and *silky* on outer surface, and with *many purple stamens and 1 style;* borne *singly* on short stalks from leaf bases near tip of twig; in late spring and summer.
Fruit: ⅝″ (15 mm) in diameter; a *rounded capsule,* very short-pointed, brown, hairy, hard; 5-celled and splitting into 5 parts; several shiny seeds, not angular or winged; maturing in autumn.

Habitat: Moist soil, especially near streams and along banks; in understory of hardwood forests in Coastal Plain.

Range: E. Virginia southwest to NW. Florida and SE. Texas; local in Arkansas; to 500′ (152 m); sometimes slightly higher.

This genus, related to the evergreen camellias, honors John Stuart (1713–92), the Earl of Bute and a patron of botany. The species name is Greek for "soft tree," referring to the silky hairs covering the lower leaf surface.

---

194 Mountain Stewartia

"Mountain-camellia" "Angle-fruit Stewartia"

Stewartia ovata (Cav.) Weatherby

Description: Shrub or small tree with gray, hairless twigs, bright green foliage, and large white flowers.
Height: 20′ (6 m).
Diameter: 3″ (7.5 cm).

Leaves: 2½–4½" (6–11 cm) long, 1½–2¼" (4–6 cm) wide. *Ovate* to elliptical; long-pointed; edges finely *saw-toothed* and hairy; short-stalked. Dark green and hairless above, light gray-green and slightly hairy beneath; turning orange and red in autumn.
Bark: gray-brown; furrowed.
Twigs: gray, slender, hairless.
Flowers: 4" (10 cm) wide; cup-shaped; with 5 *unequal rounded waxy white petals* with wavy edges, *many whitish* or yellowish *stamens* (rarely purple), and 5 *styles;* borne *singly* on short stalks at leaf bases; in early summer.
Fruit: ¾" (19 mm) in diameter; an *egg-shaped pointed capsule,* brown, hairy, hard; *deeply 5-angled,* 5-celled, and splitting into 5 parts; several angled or winged seeds with dull surfaces; maturing in autumn.

Habitat: Moist soils of stream borders and bluffs; in understory of hardwood forests, chiefly in mountains.
Range: E. and S. Virginia and E. Kentucky, south to extreme N. Georgia and NE. Mississippi; usually at 1000–2500' (305–762 m); sometimes lower.

Though found in 8 states, this uncommon species is mostly confined to the southern Appalachians. The scientific name refers to the egg-shaped leaves.

ELAEAGNUS FAMILY
(Elaeagnaceae)

Much-branched shrubs, sometimes small trees, often with spiny twigs, and with dense covering of silvery, brownish, or golden tiny scales or star-shaped hairs on twigs, lower leaf surfaces, and fruits. About 50 species in north temperate regions, mostly in plains and along coasts; 4 native shrub species including 1 which is sometimes a small tree (omitted from this guide) and 1 naturalized species reaching tree size in North America.
Leaves: deciduous; alternate, opposite, or whorled; short-stalked; simple; without teeth; often slightly thickened; without stipules.
Flowers: small, single or in clusters at leaf bases, bisexual (or male and female on same or separate plants), with tubular 4-lobed calyx and no corolla, 4–8 stamens inserted in tube, and 1 pistil with 1-celled ovary, 1 ovule, and 1 style.
Fruit: drupelike; dry, with fleshy covering, not opening, 1-seeded.

100, 392 **Russian-olive**
"Oleaster"
Elaeagnus angustifolia L.

Description: Introduced shrub or small tree often with crooked or leaning trunk, dense crown of low branches, *silvery foliage,* and sometimes *spiny twigs.*

Height: 20′ (6 m).
Diameter: 4″ (10 cm).
Leaves: 1½–3¼″ (4–8 cm) long, ⅜–¾″ (10–19 mm) wide. *Lance-shaped* or oblong; without teeth; short-stalked. Dull gray-green with obscure veins above, *silvery,* scaly, and *brown-dotted* beneath.
Bark: gray-brown; thin, fissured and *shedding in long strips.*

Twigs: *silvery,* scaly when young, becoming reddish-brown; long and slender; often ending in short *spine.*
Flowers: 3/8″ (10 mm) long; *bell-shaped;* with 4 calyx lobes, yellow inside, *silvery* outside (petals absent); fragrant; short-stalked; scattered along twigs at leaf bases; in late spring or early summer.
Fruit: 3/8–1/2″ (10–12 mm) long; berrylike, *elliptical, yellow to brown with silvery scales,* becoming shiny; thin, yellow, mealy, *sweet* edible pulp; large brown stone; scattered along twig; maturing in late summer and autumn.

Habitat: Moist soils, from salty to alkaline; spreading in valleys.
Range: Native of S. Europe and Asia; planted and naturalized from British Columbia east to Ontario and from New England west to California; to 5000′ (1524 m) or above.

The fruit is consumed by songbirds, such as cedar waxwings, robins, and grosbeaks, and by pheasants and quail. Tolerant of cold, drought, and city smoke, Russian-olive is a popular ornamental. The plants sprout and spread from roots, sometimes becoming pests. Not related to the Olive (*Olea europaea* L.), which also has narrow gray leaves.

GINSENG FAMILY
(Araliaceae)

About 700 species of shrubs and trees, sometimes large or unbranched; also woody vines and few herbs. Mostly tropical, especially in Indomalaysia and America, also in temperate regions; 1 native tree, 2 native shrub, and 5 native herb species in North America.
Leaves: alternate; large; crowded at tip of few stout twigs; palmately compound or simple and mostly not toothed or palmately lobed, rarely bi- or tripinnately compound; usually thickened; often with star-shaped hairs; mostly with long petioles; stipules often forming a sheath.
Flowers: tiny; greenish, yellowish, or whitish; in heads or umbels; generally compound; bisexual, or male and female on separate plants; regular, with 5-toothed calyx, and 5 (sometimes 10) petals from a disk, 5 (sometimes 10) stamens alternate with petals, and 1 pistil with inferior ovary of 2–5 (sometimes 1–15) cells each with 1 ovule, and 2–5 styles sometimes united.
Fruit: a berry.

322, 538 **Devils-walkingstick**
"Hercules-club" "Prickly-ash"
Aralia spinosa L.

Description: Spiny, aromatic, thicket-forming shrub or small tree with 1 (sometimes several) stout and usually unbranched trunk, very large compound leaves, and big clusters of tiny flowers; sometimes with a few spreading branches and a thin crown.

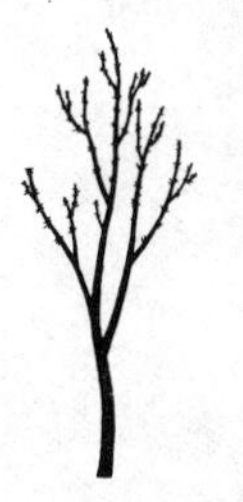

Height: 30′ (9 m).
Diameter: 8″ (20 cm).
Leaves: *clustered* at ends of twigs; *bipinnately compound;* 15–30″ (38–76 cm) long and nearly as wide; with

prickly branched axis. Numerous leaflets 2–3½" (5–9 cm) long; *ovate* or broadly elliptical; *finely saw-toothed;* nearly hairless. Dark green above, paler and *often with prickles* on midvein beneath; turning light yellow in autumn.

Bark: dark brown; thin, fissured, often with scattered stout spines.

Twigs: light brown, green inside; *very stout;* with many straight slender sharp *prickles* and large pith.

Flowers: less than ⅛" (3 mm) long and wide; with 5 *tiny white petals;* in upright clusters 8–16" (20–41 cm) long; with many hairy branches; in late summer.

Fruit: ¼" (6 mm) in diameter; *berrylike; black* skin; thin purplish juicy pulp; 3–5 seeds; maturing in autumn.

Habitat: Moist soils mostly near streams in understory of hardwood forests; often forms dense groves from root sprouts.

Range: New Jersey and New York south to central Florida, west to E. Texas, and north to SE. Missouri; naturalized north to New England, S. Ontario, and Wisconsin; to 3500′ (1067 m); sometimes to 5000′ (1524 m) in southern Appalachians.

Occasionally planted in the Victorian era as a grotesque ornamental. The aromatic spicy roots and fruit were used by early settlers in home remedies, including a cure for toothaches.

DOGWOOD FAMILY
(Cornaceae)

About 120 species of shrubs and trees in north and south temperate zones and tropical mountains; rarely herbs; 15 species of native trees and shrubs, sometimes becoming trees, including dogwood (*Cornus*) and tupelo (*Nyssa*); about 10 native shrubs and 2 native herbs in North America.
Leaves: deciduous or evergreen, opposite or alternate, generally not toothed, without stipules.
Flowers: bisexual, or male and female usually on separate plants; tiny or small; with calyx of 4–5 sepals (none in female flowers of silktassel genus, *Garrya*), and corolla of usually 4–5 petals (none in *Garrya*), 4–10 stamens, and 1 pistil with inferior ovary usually 1- or 2-celled with 1 style (2 in *Garrya*).
Fruit: mostly a drupe (sometimes a berry), sour or bitter, 1- or 2-seeded.

77, 409 **Alternate-leaf Dogwood**
"Pagoda Dogwood" "Blue Dogwood"
Cornus alternifolia L. f.

Description: Shrub or small tree with short trunk and *flat-topped, spreading crown* of long, horizontal branches.

Height: 25′ (7.6 m).
Diameter: 6″ (15 cm).
Leaves: *alternate; clustered* at ends of twigs; 2½–4½″ (6–11 cm) long, 1–2″ (2.5–5 cm) wide. *Elliptical;* appearing not toothed, but with tiny teeth on edges; *5–6 long curved veins* on each side of midvein; slender-stalked. Green and nearly hairless above, paler or whitish with pressed hairs beneath; turning yellow or red in autumn.
Bark: gray or brown; smooth or fissured into narrow ridges.

Twigs: greenish; slender, *branching singly*.
Flowers: ¼" (6 mm) wide; with 4 *spreading white petals;* in upright branched and flat clusters about 2" (5 cm) wide at end of leafy twigs; in late spring.
Fruit: about ¼" (6 mm) in diameter; *berrylike, blue-black;* numerous, on red stalks; thin bitter pulp; stone containing 1–2 seeds; maturing in late summer.

Habitat: Moist soils in understory of hardwood and coniferous forests.

Range: S. Manitoba east to Newfoundland, south to NW. Florida and west to N. Arkansas; to 6500′ (1981 m) in southern Appalachians.

Unlike all other native dogwoods, this species has alternate rather than opposite leaves. The name "Pagoda Dogwood" alludes to the flat-topped crown, with horizontal layers of branches. The bitter berrylike fruits of this and other dogwoods are consumed in quantities in fall and winter by wildlife.

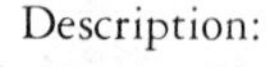

79 Roughleaf Dogwood

Cornus drummondii C. A. Meyer

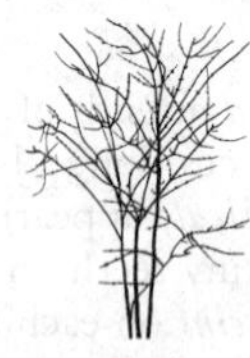

Description: Thicket-forming shrub or sometimes a small tree with short trunk and open, spreading crown.
Height: 20′ (6 m).
Diameter: 6" (15 cm).
Leaves: opposite; 1½–3½" (4–9 cm) long, 1¼–2" (3–5 cm) wide. *Elliptical;* without teeth; *3–5 long curved veins* on each side of midvein; densely covered with white hairs when young. *Green and rough* with tiny stiff hairs above, pale and covered with *soft hairs* beneath.
Bark: gray-brown or reddish brown; thin, finely fissured.

Twigs: light green and covered with fine hairs when young, turning gray or reddish-brown; with brownish pith; slender.
Flowers: ¼″ (6 mm) wide; with 4 *spreading white petals;* in upright, branched, somewhat flat clusters 2–3″ (5–7.5 cm) wide; at ends of leafy twigs; in late spring.
Fruit: ¼″ (6 mm) in diameter; berrylike; *white;* with thin bitter pulp; in loose clusters on red stalks; stone containing 1–2 seeds; maturing in late summer and autumn.

Habitat: Along streams and in dry uplands, forming thickets at forest borders in prairie grasslands and in understory of hardwood forests.

Range: Extreme S. Ontario and S. Michigan south to Louisiana, west to central Texas, and north to SE. South Dakota; to 2000′ (610 m).

This dogwood is easily recognized by the rough, upper leaf surfaces and white fruit. It spreads from root sprouts and provides cover for wildlife; various small birds, such as Bell's vireo, nest in the thickets.

76, 454, 554, 579 Flowering Dogwood
"Eastern Flowering Dogwood"
"Dogwood"
Cornus florida L.

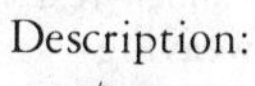

Description: A lovely, small, flowering tree with short trunk and crown of spreading or nearly horizontal branches.

Height: 30′ (9 m).
Diameter: 8″ (20 cm).
Leaves: opposite; 2½–5″ (6–13 cm) long, 1½–2½″ (4–6 cm) wide. *Elliptical;* edges slightly wavy, appearing not toothed but with tiny teeth visible under a lens; 6–7 *long curved veins* on each side of midvein; short-stalked. Green and nearly hairless

above, paler and covered with fine hairs beneath; turning bright red above in autumn.
Bark: dark reddish-brown; rough, broken into *small square plates.*
Twigs: green or reddish, slender, becoming hairless.
Flowers: 3⁄16" (5 mm) wide; with 4 yellowish-green petals; many of these tiny flowers tightly crowded in a *head* ¾" (19 mm) wide, bordered by 4 large broadly elliptical *white petal-like bracts* (*pink* in some cultivated varieties) 1½–2" (4–5 cm) long; in early spring before leaves. The flower heads (with bracts) 3–4" (7.5–10 cm) across are commonly called flowers.
Fruit: ⅜–⅝" (10–15 mm) long; berrylike, *elliptical,* shiny *red;* several at end of long stalk; thin mealy bitter pulp; stone containing 1–2 seeds; maturing in autumn.

Habitat: Both moist and dry soils of valleys and uplands in understory of hardwood forests; also in old fields and along roadsides.

Range: S. Ontario east to SW. Maine, south to N. Florida, west to central Texas, and north to central Michigan; to 4000′ (1219 m), almost 5000′ (1524 m) in southern Appalachians.

Flowering Dogwood is one of the most beautiful eastern North American trees with showy early spring flowers, red fruit, and scarlet autumn foliage. The hard wood is extremely shock-resistant and useful for making weaving-shuttles. It is also made into spools, small pulleys, mallet heads, and jeweler's blocks. Indians used the aromatic bark and roots as a remedy for malaria and extracted a red dye from the roots.

78 Red-osier Dogwood
"Kinnikinnik" "Red Dogwood"
Cornus stolonifera Michx.

Description: Large, spreading, thicket-forming shrub, with several stems, clusters of small white flowers, and small whitish fruit; rarely a small tree.
Height: commonly 3–10′ (1–3 m), rarely to 15′ (4.6 m).
Diameter: 3″ (7.5 cm).
Leaves: *opposite;* 1½–3½″ (4–9 cm) long, ⅝–2″ (1.5–5 cm) wide. *Elliptical* or ovate; short- or long-pointed; without teeth; 5–7 *long curved sunken veins* on each side of midvein. *Dull green* above, whitish green and covered with fine hairs beneath; turning reddish in autumn.
Bark: gray or brown; smooth or slightly furrowed into flat plates.
Twigs: *purplish-red,* slender, hairy when young, with rings at nodes.
Flowers: ¼″ (6 mm) wide; with 4 *spreading white petals;* many, crowded in upright *flattish clusters* 1¼–2″ (3–5 cm) wide; in late spring and early summer.
Fruit: ¼–⅜″ (6–10 mm) in diameter; *whitish,* juicy; stone with 2 seeds; maturing in late summer.

Habitat: Moist soils, especially along streams; forming thickets and in understory of forests.

Range: Central Alaska east to Labrador and Newfoundland, south to N. Virginia, and west to California; also N. Mexico; to 5000′ (1524 m); to 9000′ (2743 m) in the Southwest.

Red-osier Dogwood is useful for erosion control on stream banks. The common name recalls the resemblance of the reddish twigs to those of some willows called osiers, used in basketry. The Latin species name, meaning "bearing stolons," refers to the rooting of branch tips touching the ground and forming new shoots.

198 Water Tupelo
"Tupelo-gum" "Cotton-gum"
Nyssa aquatica L.

Description: Large *aquatic* tree with *swollen base,* long, straight trunk, narrow, open crown of spreading branches, and large, shiny leaves.
Height: 100′ (30 m).
Diameter: 3′ (0.9 m).
Leaves: 5–8″ (13–20 cm) long, 2–4″ (5–10 cm) wide, sometimes larger. *Ovate;* often with a *few large teeth;* slightly thickened; with long hairy leafstalks. Shiny dark green above, paler and hairy beneath.
Bark: dark brown or gray; furrowed into scaly ridges.
Twigs: reddish-brown; stout, hairy when young.
Flowers: *greenish;* on long stalks back of new leaves in early spring; many *male* flowers ¼″ (6 mm) long in heads ⅝″ (15 mm) wide; *solitary female* flowers ⅜″ (10 mm) long; male and female usually on separate trees.
Fruit: 1″ (2.5 cm) long; *oblong, berrylike, dark purple;* with thin sour pulp; stone with 10 winglike ridges; maturing in early autumn.

Habitat: Swamps and flood plains of streams, close to the water, where submerged a few months each winter and spring; often in pure stands.

Range: SE. Virginia south to N. Florida, west to SE. Texas, and north to S. Illinois; to 500′ (152 m).

This aquatic tree was named *Nyssa* after one of the ancient Greek water nymphs or goddesses of lakes and rivers. The name Tupelo is from Creek Indian words meaning "swamp tree." The spongy wood of the roots has served locally as a substitute for cork in floats of fish nets.

66 Ogeechee Tupelo
"Ogeechee-lime" "Sour Tupelo"
Nyssa ogeche Bartr. ex Marsh.

Description: Small tree with 1 or more crooked, leaning trunks and narrow, rounded crown; sometimes a much-branched shrub.
Height: 30–40′ (9–12 m).
Diameter: 1′ (0.3 m).
Leaves: 4–5½″ (10–14 cm) long, 2–2½″ (5–6 cm) wide. *Oblong to elliptical;* blunt-pointed; *not toothed* (rarely with a few teeth); *slightly thickened.* Shiny dark green and slightly hairy above, paler beneath and with *velvety hairs* on veins.
Bark: dark brown; thin; irregularly fissured into scaly plates.
Twigs: light reddish-brown; covered with rust-colored hairs; slender.
Flowers: *greenish;* hairy; on short stalks in back of new leaves in early spring; *many tiny male flowers in heads* ½″ (12 mm) wide; *solitary female* flowers 3⁄16″ (5 mm) long; male and female usually on separate trees.
Fruit: 1–1½″ (2.5–4 cm) long; berrylike, *oblong,* shiny or dull *red;* with thick *juicy sour pulp;* stone with 10–12 papery wings; maturing in late summer and remaining attached in autumn.

Habitat: Permanently wet soils bordering streams, swamps, and lakes, and in swamp forests; usually within 2′ (0.6 m) of water level and in areas flooded for long periods.

Range: Extreme S. South Carolina, S. Georgia, and N. Florida; to 250′ (76 m).

Ogeechee-lime preserve is made from the edible but sour fruit, and the juice is used like that of limes. Natural propagation is by sprouts from roots and stumps as well as from seeds. The common and Latin species names refer to the Ogeechee River in southeast Georgia, where the tree was discovered.

72, 540, 583 **Black Tupelo**
"Blackgum" "Pepperidge"
Nyssa sylvatica Marsh.

Description: Tree with a dense, conical or sometimes flat-topped crown, many slender, nearly *horizontal branches,* and glossy foliage turning scarlet in autumn.
Height: 50–100′ (15–30 m).
Diameter: 2–3′ (0.6–0.9 m).
Leaves: 2–5″ (5–13 cm) long, 1–3″ (2.5–7.5 cm) wide. *Elliptical* or oblong; *not toothed* (rarely with a few teeth); slightly thickened; often crowded on short twigs. *Shiny green* above, pale and often hairy beneath; turning *bright red* in early autumn.
Bark: gray or dark brown; thick, rough, *deeply furrowed into rectangular* or irregular *ridges.*
Twigs: light brown; slender, often hairy, with some short spurs.
Flowers: greenish; at end of long stalks at base of new leaves in early spring; *many tiny male* flowers in heads ½″ (12 mm) wide; *2–6 female* flowers 3⁄16″ (5 mm) long; male and female usually on separate trees.
Fruit: ⅜–½″ (10–12 mm) long; berrylike, *elliptical, blue-black;* with thin bitter or sour pulp; stone slightly 10- to 12-ridged; maturing in autumn.

Habitat: Moist soils of valleys and uplands in hardwood and pine forests.

Range: Extreme S. Ontario east to SW. Maine, south to S. Florida, west to E. Texas, and north to central Michigan; local in Mexico; to 4000′ (1219 m), sometimes higher in southern Appalachians.

A handsome ornamental and shade tree, Black Tupelo is also a honey plant. The juicy fruit is consumed by many birds and mammals. Swamp Tupelo (var. *biflora* (Walt.) Sarg.), a variety with narrower oblong leaves, occurs in swamps in the Coastal Plain from Delaware to eastern Texas.

HEATH FAMILY
(Ericaceae)

Widespread, especially on acid soils, about 1500 species, mostly shrubs, sometimes trees; 15 native tree and many native shrub species in North America.
Leaves: usually alternate, simple, elliptical, and not toothed; often thick and evergreen, without stipules.
Flowers: small or large and showy, bisexual, regular or slightly irregular, with 4- to 7-lobed calyx, generally persisting on fruit; corolla of 4–7 lobes or petals, often bell- or funnel-shaped, 8–10 stamens from a disk, and 1 pistil with superior or inferior ovary of usually 5 cells and many ovules and 1 style.
Fruit: a capsule, berry, or drupe.

---

120 Texas Madrone
"Texas Madroño"
Arbutus texana Buckl.

Description: Small evergreen tree with short, often crooked trunk and rounded crown of stout, crooked, spreading branches, or a shrub.

Height: 20′ (6 m).
Diameter: 8″ (20 cm).
Leaves: *evergreen;* 1–3½″ (2.5–9 cm) long, ⅝–1½″ (1.5–4 cm) wide. *Elliptical* or ovate; without teeth, or sometimes wavy- or saw-toothed; *thick and stiff;* with slender, hairy leafstalks. *Shiny green* above, paler and slightly hairy beneath.
Bark: on *branches, pinkish to reddish-brown, smooth,* thin, peeling off in thin papery scales; on trunks, dark brown and divided into square plates.
Twigs: *red,* densely covered with hairs when young; becoming dark red-brown and scaly.
Flowers: ¼″ (6 mm) long; *jug-shaped* or

urn-shaped, *white or pink-tinged* corolla; short-stalked; in upright branched clusters about 2½" (6 cm) long and wide; in early spring.
Fruit: ⅜" (10 mm) in diameter; *berrylike,* dark *red* to yellowish-red, and finely warty; mealy sweetish pulp; large stone containing many flat seeds; maturing in autumn.

Habitat: Canyons and rocky slopes of mountains, in oak woodlands, and on rocky plains.

Range: Central Texas (Edwards Plateau) to Trans-Pecos Texas, SE. New Mexico (Guadalupe Mountains), and NE. Mexico; at 2000–6000′ (610–1829 m).

It is reported that the fruit of this uncommon species is edible and that those of related European species have narcotic properties. The wood has been used locally for tool handles. The local names, "Naked Indian" and "Lady's Leg," refer to the smooth flesh-colored bark.

65, 443 **Elliottia**
"Southern-plume"
Elliottia racemosa Muhl. ex Ell.

Description: Shrub or small tree with conical crown and showy, white clusters of flowers.
Height: 20′ (6 m).
Diameter: 4" (10 cm).
Leaves: 2¾–4" (7–10 cm) long, 1–1½" (2.5–4 cm) wide. *Oblong* or elliptical; without teeth; short-stalked. Dull green above, paler and slightly hairy beneath.
Bark: light gray; smooth, thin.
Twigs: light brown, upright, slender, hairy when young.
Flowers: ¾" (19 mm) wide; with 4–5 *narrow* curved *white petals;* fragrant; on long slender stalks; in upright narrow clusters, 4–10" (10–25 cm) long; in early summer.

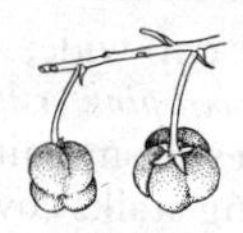

Fruit: ⅜–½″ (10–12 mm) in diameter; slightly flat, *rounded capsule,* splitting open on 4–5 lines; with many winged seeds.

Habitat: Moist, sandy, acid soils near streams and sometimes on dry ridges, in woods.

Range: Local in 8 counties of E. and SE. Georgia; at 100–400′ (30–122 m).

One of the rarest native trees, Elliottia is alone in a distinct genus. It has been proposed for addition to the list of endangered species. It is apparently now extinct at one locality in South Carolina where it had been collected in 1853. However, fairly extensive local stands have been found in Georgia. *Elliottia* honors Stephen Elliott (1771–1830), botanist of South Carolina and probably the discoverer of this plant.

57, 440, 462 **Mountain-laurel**
"Calico-bush" "Ivybush"
Kalmia latifolia L.

Description: Evergreen, many-stemmed, thicket-forming shrub or sometimes a small tree with short, crooked trunk; stout, spreading branches; a compact, rounded crown; and beautiful, large, *pink flower clusters.*

Height: 20′ (6 m).

Diameter: 6″ (15 cm).

Leaves: *evergreen;* alternate or *sometimes opposite or in 3's;* 2½–4″ (6–10 cm) long, 1–1½″ (2.5–4 cm) wide. *Narrowly elliptical* or lance-shaped; *hard whitish point* at tip; without teeth; *thick and stiff.* Dull dark green above, yellow-green beneath.

Bark: dark reddish-brown; thin, fissured into long narrow ridges and shredding.

Twigs: reddish-green with sticky hairs when young; later turning reddish-brown, peeling, and exposing darker layer beneath.

Flowers: ¾–1″ (2–2.5 cm) wide; *saucer-shaped,* with *5-lobed pink or white corolla* with purple lines, from pointed deep pink buds; on long stalks covered with sticky hairs; in upright branched flat clusters 4–5″ (10–13 cm) wide; in spring.

Fruit: ¼″ (6 mm) wide; a rounded dark brown capsule; with *long threadlike style* at tip; covered with sticky hairs; 5-celled, splitting open along 5 lines; many tiny seeds; maturing in autumn and remaining attached.

Habitat: Dry or moist acid soils; in understory of mixed forests on upland mountain slopes and in valleys; also in shrub thickets called "heath balds" or "laurel slicks."

Range: SE. Maine south to N. Florida, west to Louisiana, and north to Indiana; to 4000′ (1219 m), higher in southern Appalachians.

Mountain-laurel is one of the most beautiful native flowering shrubs and is well displayed as an ornamental in many parks. The stamens of the flowers have an odd, springlike mechanism which spreads pollen when tripped by a bee. The leaves, which are poisonous to livestock, are seldom browsed. Honey from the flowers is believed to be poisonous. The wood has been used for tool handles and turnery, and the burls, or hard knotlike growths, for briar tobacco pipes. Linnaeus named this genus for his student Peter Kalm (1716–79), a Swedish botanist who traveled in Canada and the eastern United States.

102 Tree Lyonia

"Staggerbush" "Titi"

Lyonia ferruginea (Walt.) Nutt.

Description: Evergreen shrub or small tree with crooked trunk, irregular crown, and

tiny, nearly round flowers.
Height: 20′ (6 m).
Diameter: 6″ (15 cm).
Leaves: *evergreen;* 1–3½″ (2.5–9 cm) long, ½–1″ (1.2–2.5 cm) wide. *Oblong* or narrowly elliptical; *hard point at tip;* edges wavy and turned under; *thick and stiff.* Shiny green above, gray or *rusty scurfy* with tiny scales beneath.
Bark: reddish-brown; furrowed into long, narrow, scaly ridges.
Twigs: gray or *rusty scurfy,* becoming dark red and smooth; slender.
Flowers: ⅛″ (3 mm) wide; with *rounded,* hairy, *pinkish or white corolla* with *5 teeth* around small opening; several clustered on slender curved stalks at base of leaves; in early spring.
Fruit: ¼″ (6 mm) long; brown *egg-shaped* capsule, long-stalked, *splitting along lines into 5 parts;* many-seeded; maturing in early autumn and remaining attached.

Habitat: Sandy soils of mixed woods and on dunes.

Range: Extreme S. South Carolina, SE. Georgia, and Florida; to 300′ (91 m).

This tree is often called "Staggerbush" because of its crooked trunks. The Latin species name, meaning "rust-colored," refers to the leaf lower surfaces.

137, 405, 511, 585 **Sourwood**
"Sorrel-tree" "Lily-of-the-valley-tree"
Oxydendrum arboreum (L.) DC.

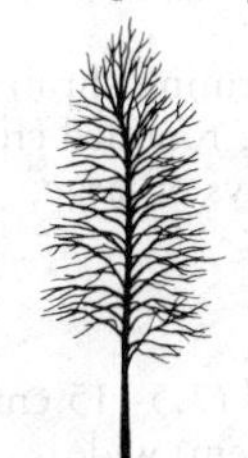

Description: Tree with conical or rounded crown of spreading branches, clusters of flowers recalling Lily-of-the-valley, and glossy foliage that turns red in autumn.
Height: 50′ (15 m).
Diameter: 1′ (0.3 m).
Leaves: 4–7″ (10–18 cm) long, 1½–2½″ (4–6 cm) wide. *Elliptical* or lance-shaped; *finely saw-toothed;* with *sour taste. Shiny yellow-green* above, paler and

slightly hairy on veins beneath; turning red in autumn.
Bark: brown or gray; thick; fissured into narrow, scaly ridges.
Twigs: light yellow-green, slender, hairless.
Flowers: buds and young flowers hanging down short-stalked on 1 side of slender axes, with urn-shaped white corolla, ¼″ (6 mm) long, slightly 5-lobed; in terminal drooping clusters 4–10″ (10–25 cm) long; in midsummer.
Fruit: ⅜″ (10 mm) long; a *narrowly egg-shaped capsule;* gray and covered with fine hairs; *upright on curved stalks* along drooping axes; 5-celled, splitting along 5 lines; many-seeded; maturing in autumn, remaining attached into winter.

Habitat: Moist soils in valleys and uplands with oaks and pines.

Range: SW. Pennsylvania and SE. Maryland, south to NW. Florida, west to Louisiana, north to S. Indiana; 5000′ (1524 m) or slightly higher in southern Appalachians.

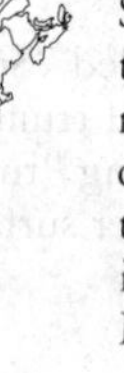

Sourwood is an attractive ornamental throughout the year. Both the genus name, meaning "sour tree," and the common name refer to the acid taste of the foliage, although Sourwood honey is esteemed. Abundant in Great Smoky Mountains National Park.

62, 468 **Catawba Rhododendron**
"Mountain-rosebay" "Purple-laurel"
Rhododendron catawbiense Michx.

Description: Evergreen, thicket-forming shrub or small tree with broad, rounded crown and spectacular displays of large, purplish blossoms.
Height: 20′ (6 m).
Diameter: 4″ (10 cm).
Leaves: *evergreen;* 3–6″ (7.5–15 cm) long, 1½–2½″ (4–6 cm) wide.

Elliptical, blunt at tip, rounded at base; thick and leathery with edges often rolled under; hairless; long stout leafstalks. *Shiny dark green* above, whitish beneath.
Bark: brown or gray; fissured into narrow scaly ridges.
Twigs: greenish, stout.
Flowers: 2¼″ (6 cm) wide; with *bell-shaped corolla* of 5 *rounded lobes;* waxy *lilac-purple* (sometimes pink); slender-stalked; in upright branched rounded clusters; in late spring and early summer.
Fruit: ½–⅝″ (12–15 mm) long; a long-stalked *narrowly egg-shaped capsule,* densely covered with reddish-brown hairs; 5-celled, splitting along 5 lines; many-seeded; maturing in late summer and autumn.

Habitat: Rocky slopes and ridges in understory of mountain forests, and in shrub thickets called "heath balds" or "laurel slicks."

Range: W. Virginia south to N. Georgia and NE. Alabama, north to E. Kentucky; usually at 1500–6500′ (457–1981 m); locally down to 200′ (61 m).

Catawba or Purple Rhododendron is plentiful in Great Smoky Mountains National Park, attracting thousands of visitors each year. On high mountain ridges, plants usually flower in June; at lower altitudes, flowers appear earlier, although the times may vary slightly. The common and scientific names, of American Indian origin, are from the Catawba River in North Carolina.

63, 439, 465 **Rosebay Rhododendron**
"Rosebay" "Great-laurel"
Rhododendron maximum L.

Description: Evergreen, thicket-forming shrub or tree with short, crooked trunk, broad, rounded crown of many stout, crooked branches, and large white blossoms.

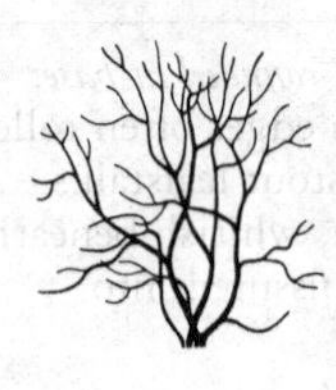

Height: 20′ (6 m).
Diameter: 6″ (15 cm).
Leaves: *evergreen;* 4–10″ (10–25 cm) long, 1–3″ (2.5–7.5 cm) wide. *Oblong* or narrowly elliptical, *short-pointed* at both ends; *thick and leathery* with edges rolled under; short stout leafstalks. *Shiny dark green* above, whitish and covered with fine hairs beneath.
Bark: red-brown; scaly, thin.
Twigs: green with reddish gland-hairs, becoming reddish-brown and scaly; stout.
Flowers: 1½″ (4 cm) wide; *bell-shaped corolla* of 5 *rounded lobes;* waxy *white* or sometimes light pink (rarely reddish); the largest or upper lobe with many green spots; in upright, branched, rounded clusters; in summer.
Fruit: ½″ (12 mm) long; long-stalked, *narrowly egg-shaped capsule;* dark reddish-brown, with gland-hairs; 5-celled and splitting open along 5 lines; many seeds; maturing in autumn and remaining attached.

Habitat: Moist soils, especially along streams in understory of mountain forests, forming dense thickets.

Range: Maine southwest to W. New York and south, mostly in mountains, to N. Georgia; to 6000′ (1829 m) in southern Appalachians.

Rosebay Rhododendron is abundant in the Great Smoky Mountains National Park. Often grown as an ornamental, it is one of the hardiest and largest evergreen rhododendrons. The wood is occasionally used for tool handles, and a home remedy has been prepared from the leaves. Honey from rhododendrons is poisonous.

206, 424 Tree Sparkleberry

"Farkleberry" "Tree-huckleberry"
Vaccinium arboreum Marsh.

Description: A shrub or tree with short trunk, irregular crown of crooked branches, small, glossy, elliptical leaves, and shiny black berries.
Height: 25′ (7.6 m).
Diameter: 6″ (15 cm).
Leaves: deciduous and, in southern part of range, evergreen or nearly so; ½–1¾″ (1.2–4.5 cm) long, ¼–1″ (0.6–2.5 cm) wide. *Elliptical* or obovate, *blunt or rounded* at tip; with tiny teeth or not toothed; prominent *network of fine veins;* slightly thickened. Shiny dark green above, paler and sometimes slightly hairy on veins beneath.
Bark: light reddish-brown; thin, covered with fine scales.
Twigs: reddish, slender, stiff, crooked; hairy when young.
Flowers: ¼″ (6 mm) long and wide; with *bell-shaped,* slightly *5-lobed white* corolla; on drooping stalks in unbranched clusters 2–3″ (5–7.5 cm) long; in spring.
Fruit: ¼″ (6 mm) in diameter; *shiny black berries;* thin, dry, mealy pulp; 8–10 seeds; maturing in autumn, remaining attached into winter.

Habitat: Sandy and rocky dry uplands, in forest understory and clearings.

Range: Virginia south to central Florida, west to SE. Texas, and north to SE. Kansas; to 2500′ (762 m).

This is the tallest of the genus of blueberries, often called huckleberries. The fruit has thin, slightly sweet pulp and large seeds. Although not palatable to humans, the berries are consumed by wildlife.

SAPODILLA (OR SAPOTE) FAMILY (Sapotaceae)

Trees and few shrubs with white latex or milky sap. About 700 species, mostly in tropical and warm temperate regions; 8 native and 1 naturalized tree and 2 native shrub species in North America.

Leaves: alternate, simple, generally not toothed, thick, usually without stipules.

Flowers: small; generally white, green, or light brown; crowded or solitary at base of leaves or below at nodes; bisexual; regular, with hairy calyx of 4–8 overlapping lobes, corolla with short tube and 4–8 short lobes, stamens very short, generally 4–8 (to many) inserted on corolla opposite the lobes, often with sterile stamens, alternate and 1 pistil with superior ovary containing generally 4–5 (sometimes 1–14) cells with 1 ovule and 1 short style.

Fruit: a berry with 1 (sometimes few) large elliptical shiny seed with large scar and milky pulp, sometimes edible.

---

44 Gum Bumelia

"Woolly Buckthorn" "Chittamwood"

Bumelia lanuginosa (Michx.) Pers.

Description: Tree or thicket-forming shrub with straight trunk and narrow crown of short, stiff branches, often with stout *spines*.

Height: 50′ (15 m).

Diameter: 1′ (0.3 m).

Leaves: alternate or *clustered on short side twigs;* 1–3″ (2.5–7.5 cm) long, ⅜–1″ (1–2.5 cm) wide. *Elliptical* or obovate, rounded at tip, *widest beyond middle,* tapering toward long-pointed base; without teeth; slightly thickened. Shiny dark green above, *densely covered*

with gray or rust-colored hairs beneath; falling irregularly in late autumn.
Bark: dark gray; furrowed into narrow scaly ridges.
Twigs: slender, often zigzag, covered with gray or rust-colored hairs when young; often ending in straight *spines,* with single spines to ¾″ (19 mm) long at base of some leaves; gummy, milky sap.
Flowers: ⅛″ (3 mm) wide; with *bell-shaped 5-lobed white corolla;* clustered on slender stalks at leaf bases; in summer.
Fruit: ⅜–½″ (10–12 mm) long; *elliptical purplish-black berry;* sweetish pulp; 1 seed; maturing in autumn.

Habitat: Valleys and rocky slopes of uplands in hardwood forests; forming shrubby thickets in the Southwest.

Range: E. and S. Kansas and central Missouri, southeast to central Florida and west to S. and W. Texas; local in SW. New Mexico and SE. Arizona; also N. Mexico; to 2500′ (762 m); a southwestern variety to 5000′ (1524 m).

Gum from cuts in the trunk is sometimes chewed by children. The fruit is edible but if eaten in quantity is said to cause dizziness and stomach disturbances. *Bumelia* is an ancient Greek name for European Ash (*Fraxinus excelsior* L.); the specific name, meaning "woolly," describes the young leaves. The wood has been used locally for making tool handles and cabinets.

---

43 Buckthorn Bumelia
"Buckthorn" "Smooth Bumelia"
Bumelia lycioides (L.) Pers.

Description: Often spiny shrub or small tree with a spreading, open crown; evergreen southward.
Height: 20′ (6 m).
Diameter: 6″ (15 cm).

Leaves: alternate or *clustered on short side twigs;* usually 2–4″ (5–10 cm) long, ½–1¼″ (1.2–3 cm) wide. *Elliptical* or reverse lanceolate, *widest beyond middle,* tapering to long-pointed base; without teeth; thin; with prominent *network of veins.* Shiny green above, paler and hairless or nearly so beneath.
Bark: reddish-gray; thin; smooth or scaly.
Twigs: red-brown, slender, often with short straight or curved *spines;* sap slightly milky.
Flowers: ⅛″ (3 mm) wide; with calyx covered with rust-colored hairs, and *bell-shaped, 5-lobed white corolla;* clustered on short stalks at leaf base; in summer.
Fruit: ⅝″ (15 mm) long; *elliptical black berry;* thin, bittersweet pulp; 1 large seed; maturing in autumn.

Habitat: Borders of streams, swamps, and lakes in forests; also rocky bluffs and dunes.

Range: SE. Virginia south to N. Florida, west to SE. Texas, and north to S. Indiana; to 1000′ (305 m).

The thin fruit pulp is edible but is too bitter to be palatable. The Latin species name means "like *Lycium*" (wolfberry), an unrelated genus of spiny shrubs of similar appearance. Birds consume the berries, and livestock browse the foliage.

42 Tough Bumelia

"Narrowleaf Bumelia"
"Tough Buckthorn"
Bumelia tenax (L.) Willd.

Description: Shrub or small tree with spreading crown of straight, *flexible, tough branches,* often with long, *straight, stout spines.*
Height: 20′ (6 m).
Diameter: 4″ (10 cm).
Leaves: *evergreen* except northward; alternate or *clustered on short side twigs;*

1–2½″ (2.5–6 cm) long, ⅜–1¼″ (1–3 cm) wide. Reverse lance-shaped, rounded at tip, *widest beyond middle,* tapering to long-pointed base; without teeth. Shiny dark green with *prominent network of veins* above, *yellowish* or whitish and covered with *silky hairs* beneath; turning yellow and falling in winter.
Bark: reddish-brown; thick, irregularly furrowed into narrow, flat, scaly ridges.
Twigs: slender, covered with *silky hairs* when young; becoming reddish-brown and hairless; often with spines to 1″ (2.5 cm) long; sap slightly milky.
Flowers: ⅛″ (3 mm) wide; with calyx covered with *silky hairs,* and *bell-shaped, 5-lobed greenish corolla;* many on slender stalks at leaf bases; in late spring and early summer.
Fruit: ⅜–½″ (10–12 mm) long; *round or elliptical berry; black;* with thin sweet pulp; 1 seed; maturing in autumn.

Habitat: Dry sandy soil, pine forests, usually near coast.

Range: South Carolina south to S. Florida; usually below 100′ (30 m).

This very local species was discovered in South Carolina by Alexander Garden (1728–91); he sent a specimen to Linnaeus, who named it in 1767. The common and scientific names both refer to the tough branches.

EBONY FAMILY
(Ebenaceae)

Trees, sometimes shrubs, nearly all in the persimmon genus (*Diospyros*). Widespread, about 400 species in tropical and warm temperate regions; 2 native tree species in North America.
Leaves: alternate, simple, not toothed, thick, without stipules.
Flowers: small, male and female generally on different plants; solitary or few along twigs, regular, calyx persisting at base of fruit and 3- to 7-lobed; urn- or bell-shaped tubular corolla with 3–7 lobes, stamens generally double or triple the lobes and inserted in tube, and 1 pistil with superior ovary of 2–16 cells each, with 2 ovules and 2–8 styles and stigmas.
Fruit: a berry, sometimes edible, with few large seeds.

---

40 **Texas Persimmon**
"Black Persimmon" "Mexican Persimmon"
Diospyros texana Scheele

Description: Much-branched shrub or tree with short trunk and narrow crown.

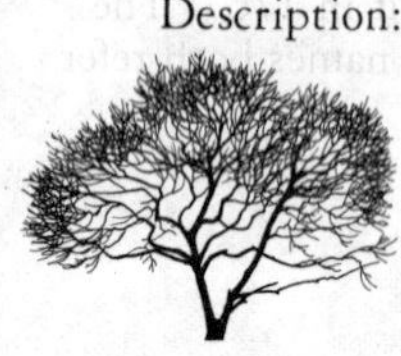

Height: 40′ (12 m).
Diameter: 1′ (0.3 m).
Leaves: ¾–1½″ (2–4 cm) long, ⅜–¾″ (10–19 mm) wide. *Obovate* or oblong, rounded or slightly notched at tip, *widest above middle,* tapering to long-pointed base; *thick, with edges rolled under.* Shiny dark green and hairless (or nearly so) above, paler and hairy beneath; falling in winter.
Bark: light *reddish-gray;* thin, *smooth, mottled* and peeling, exposing gray inner layers.
Twigs: gray, slender, slightly zigzag, covered with fine hairs when young.
Flowers: ¼″ (6 mm) long and wide; *bell-shaped,* with *5-lobed white,* finely

hairy *corolla;* short-stalked on previous year's twigs back of new leaves; in early to late spring. Male and female on separate trees: 1–3 male flowers together; female flowers solitary.

Fruit: ¾" (19 mm) in diameter; *black berry;* black juicy sweet pulp; 3–8 seeds; maturing in late summer.

Habitat: Dry rocky uplands and in canyons, plains, and open woodlands.

Range: SE. to central and Trans-Pecos Texas, south to NE. Mexico; to 4000′ (1219 m).

The edible black persimmons stain teeth, lips, and hands, and have been used as a dye. When immature, they are strongly astringent. The heartwood, found only in very large trunks, is black, like that of related ebony (*Diospyros*), while the sapwood is clear yellow.

83, 393, 571 **Common Persimmon**
"Simmon" "Possumwood"
Diospyros virginiana L.

Description: Tree with a dense cylindrical or rounded crown, or sometimes a shrub, best-known by its sweet, orange fruit in autumn.

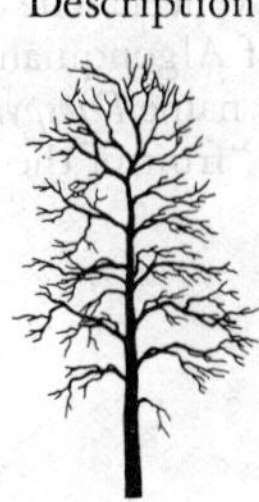

Height: 20–70′ (6–21 m).
Diameter: 1–2′ (0.3–0.6 m).
Leaves: 2½–6" (6–15 cm) long, 1½–3" (4–7.5 cm) wide. *Ovate to elliptical;* long-pointed; without teeth; slightly thickened. Shiny dark green above, whitish-green and hairless to densely hairy beneath; turning yellow in autumn.
Bark: brown or blackish; thick, deeply furrowed into *small square scaly plates.*
Twigs: brown to gray, slightly zigzag, often hairy.
Flowers: with *bell-shaped, 4-lobed white corolla;* fragrant; scattered and almost stalkless at leaf bases. Male and female

on separate trees in spring. Male, 2–3 together, ⅜″ (10 mm) long. Female, solitary, ⅝″ (15 mm) long.

Fruit: ¾–1½″ (2–4 cm) in diameter; a rounded or *slightly flat, orange* to purplish-brown *berry;* 4–8 large flat seeds; maturing in autumn before frost and often remaining attached into winter; orange pulp becoming soft and juicy at maturity.

Habitat: Moist alluvial soils of valleys and in dry uplands; also at roadsides and in old fields, clearings, and mixed forests.

Range: S. Connecticut south to S. Florida, west to central Texas, and north to extreme SE. Iowa; to 3500′ (1067 m).

When ripe, the sweet fruit of Persimmon somewhat recalls the flavor of dates. Immature fruit contains tannin and is strongly astringent. Persimmons are consumed fresh and are used to make puddings, cakes, and beverages. American Indians made persimmon bread and stored the dried fruit like prunes. Opossums, raccoons, skunks, deer, and birds also feed upon the fruit. Principal uses of the wood are for golf-club heads, shuttles for textile weaving, and furniture veneer. The word "persimmon" is of Algonquian origin, while the genus name *Diospyros,* from the Greek, means "fruit of the god Zeus."

SNOWBELL (STORAX) FAMILY
(Styracaceae)

Mostly small, sometimes large, trees and shrubs; about 150 species in tropical and warm temperate regions; 6 native tree species and 3 native shrub species in North America.
Leaves: alternate, simple, not toothed or with teeth; thin or thick, with star-shaped or scaly hairs; without stipules.
Flowers: sometimes showy, in racemes, bisexual, regular, with 4- to 5-toothed calyx persisting after fruit is formed, and generally white corolla with short tube and 4–6 almost separate lobes, 8–12 stamens (double the number of lobes) united toward base on corolla; 1 pistil with superior or partially inferior ovary, incompletely 3- to 5-celled below and 1-celled above, with 2 to many ovules, 1 style, and 1–5 stigmas.
Fruit: a 1-seeded drupe or capsule.

192, 427 **Carolina Silverbell**
"Snowdrop-tree" "Opossum-wood"
Halesia carolina L.

Description: Shrub or small tree with irregular, spreading, open crown and drooping, bell-shaped white flowers; in southern Appalachians becoming a large tree with straight axis and rounded crown.

Height: 30′ (9 m); in southern Appalachians 80′ (24 m).
Diameter: 1′ (0.3 m); and 2′ (0.6 m) or more in south.
Leaves: 3–6″ (7.5–15 cm) long, 1½–2½″ (4–6 cm) wide. *Elliptical,* abruptly long-pointed; *finely saw-toothed.* Dull dark green and becoming hairless above, covered with white hairs when young and often with tiny star-shaped hairs on veins beneath; turning yellow in autumn.
Bark: reddish-brown; furrowed into loose, broad, scaly ridges.

Twigs: brown, slender, with star-shaped hairs when young.

Flowers: ½–1″ (1.2–2.5 cm) long; with *bell-shaped, 4-lobed white corolla* (rarely pink), opening before reaching full size; in drooping clusters of 2–5 flowers on long stalks on previous year's twig back of new leaves; in early to midspring.

Fruit: 1¼–2″ (3–5 cm) long; *oblong,* podlike, with *4 long broad wings, long-pointed,* dark brown, dry; stone containing 1–3 seeds; maturing in late summer and autumn, remaining closed and attached into winter.

Habitat: Moist soils along streams in understory of hardwood forests.

Range: S. West Virginia south to N. Florida, northwest to S. Illinois; local in Arkansas and SE. Oklahoma; to 5500′ (1676 m) in southern Appalachians.

Common and of largest size in the southern Appalachians, where it is known as "Mountain Silverbell." It can be easily seen in the Great Smoky Mountains National Park. The wood serves for lumber, though of limited supply.

195, 426 **Two-wing Silverbell**
"Snowdrop-tree"
Halesia diptera Ellis

Description: Shrub or small tree with spreading crown and white, bell-shaped, drooping flowers.

Height: 30′ (9 m).

Diameter: 8″ (20 cm).

Leaves: 2½–4½″ (6–11 cm) long, 1½–2½″ (4–6 cm) wide. *Broadly elliptical,* abruptly long-pointed; *finely saw-toothed;* short-stalked. Light green and becoming nearly hairless above, paler with tiny star-shaped hairs beneath.

Bark: reddish-brown; furrowed into

irregular ridges; shedding in narrow plates.
Twigs: light green; slender, hairy when young.
Flowers: ⅝–1¼″ (1.5–3.2 cm) long; with *bell-shaped, deeply 4-lobed, white corolla;* covered with fine hairs; in drooping clusters of 3–6 on long stalks on previous year's twigs back of new leaves; in early to midspring.
Fruit: 1½–2″ (4–5 cm) long; *oblong, podlike,* with *2 long broad wings,* dark brown, dry; large stone; maturing in late summer and autumn, and remaining closed.

Habitat: Wet soils bordering streams and swamps in understory of hardwood and pine forests.

Range: Extreme S. South Carolina south to NW. Florida, west to SE. Texas; to 500′ (152 m).

The common and scientific names both refer to the two-winged fruit. This immature sour green fruit is consumed by wildlife, including squirrels.

215, 489 **Little Silverbell**
"Florida Silverbell"
Halesia parviflora Michx.

Description: Shrub or small tree with small, bell-shaped, white flowers.
Height: 25′ (7.6 m).
Diameter: 6″ (15 cm).
Leaves: 2½–4″ (6–10 cm) long, 1–2″ (2.5–5 cm) wide. *Elliptical,* abruptly long-pointed; *finely saw-toothed* or wavy; densely covered with star-shaped gray hairs when young. Becoming dull green and hairless above, paler or whitish beneath.
Bark: dark brown; narrowly furrowed.
Twigs: slender, covered with gray hairs when young.
Flowers; ⅜–½″ (10–12 mm) long; with bell-shaped 4-lobed white corolla,

covered with fine hairs; in drooping clusters on long stalks on previous year's twigs back of new leaves; in early spring.

Fruit: 1–1½" (2.5–4 cm) long; *club-shaped, podlike,* short-pointed, widest beyond middle, with *4 narrow wings,* dry; large stone; maturing in autumn, remaining closed.

Habitat: Moist sandy valleys and uplands in understory of forests.

Range: Local from South Carolina to N. Florida and Mississippi; to 500′ (152 m).

As the common and Latin species names indicate, this uncommon small tree is distinguished by its relatively small flowers, which are not as ornamental as the large-flowered species of silverbell.

---

196 Bigleaf Snowbell

"Snowbell" "Storax"

Styrax grandifolius Ait.

Description: Shrub or sometimes a small tree with rounded crown, large, nearly round leaves, and showy, bell-shaped, white flowers.

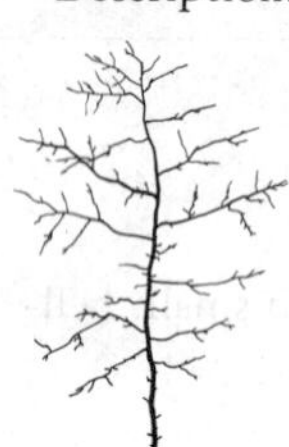

Height: 20′ (6 m).

Diameter: 4" (10 cm).

Leaves: 2½–5½" (6–14 cm) long, 1½–4" (4–10 cm) wide. *Broadly elliptical* to *nearly round,* abruptly short-pointed at tip; *not toothed or with few teeth.* Dark green and hairless or nearly so above, paler and usually *densely coated with gray, tiny, star-shaped hairs* beneath.

Bark: dark gray; smooth.

Twigs: gray; slender, slightly zigzag; scurfy with star-shaped hairs.

Flowers: ¾–1" (2–2.5 cm) wide; with *5 long,* narrow, *white corolla lobes;* short-stalked; fragrant; many, slightly *drooping along unbranched axis* to 5" (13 cm) long; at end of short side twigs; in early spring.

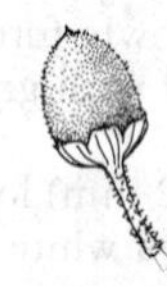

Fruit: ¼″ (6 mm) in diameter; *rounded, hairy,* brown, hard and dry; 1 large seed; maturing in late summer and autumn and *remaining closed.*

Habitat: Moist or wet soils of valleys and uplands; in understory of hardwood forests.

Range: Virginia to N. Florida, and west to E. Texas; local to S. Ohio; to 1000′ (305 m).

The genus name, from the Greek, is that of the European relative, Storax (*Styrax officinalis* L.). That species yields the resin known as storax, used as incense.

SWEETLEAF FAMILY
(Symplocaceae)

Generally small, but sometimes large, trees and shrubs; about 350 species; nearly all in the sweetleaf genus (*Symplocos*), in tropical and subtropical Asia and America; 1 native tree species in southeastern United States.
Leaves: alternate, simple, thick, without teeth or toothed, generally hairless and shiny and often yellow-green, without stipules, with very short leafstalk.
Flowers: small, crowded in clusters, bisexual, regular, with 5-toothed calyx persisting at tip of fruit, white corolla with 5–10 lobes divided almost to base, usually many stamens inserted in tube and united in groups, and 1 pistil with inferior ovary containing 2–5 cells of 2 ovules each, 1 style, and 1 stigma often 2- to 5-lobed.
Fruit: an elliptical drupe or berry with ring and calyx at tip.

55, 401 **Sweetleaf**
"Horse-sugar" "Yellowwood"
Symplocos tinctoria (L.) L'Hér.

Description: Shrub or small tree with short trunk, open crown of spreading branches, and *foliage with sweetish taste.*
Height: 20′ (6 m).
Diameter: 4″ (10 cm).
Leaves: deciduous, or *evergreen southward;* 3–5″ (7.5–13 cm) long, 1–2″ (2.5–5 cm) wide. *Elliptical* or lance-shaped; slightly saw-toothed or without teeth; *slightly thickened. Shiny dark green* above, paler and covered with fine hairs beneath.
Bark: gray; smooth or slightly fissured.
Twigs: light green, slender, often hairy when young.
Flowers: ⅜″ (10 mm) wide; *bell-shaped,* with deeply *5-lobed white or light yellow*

corolla; fragrant; crowded and nearly *stalkless* in clusters on twigs behind new leaves; in early spring.
Fruit: ¼" (6 mm) long; *elliptical,* with *5 teeth at tip;* berrylike; green, turning *brown;* thin pulp; large 1-seeded stone; several in stalkless clusters along twig; maturing in summer and autumn.

Habitat: Moist valley soils in understory of hardwood forests.

Range: S. Delaware to central Florida and west to E. Texas; generally below 1000′ (305 m); to 3500′ (1067 m) in W. North Carolina.

The common names Sweetleaf and "Horse-sugar" refer to the tasty foliage, which livestock eat greedily. The name "Yellowwood" and the Latin species name allude to a yellow dye once obtained from the bark and leaves. The bark, like others with bitter aromatic properties, was used by early settlers as a tonic.

OLIVE FAMILY
(Oleaceae)

Widespread, especially in Asia; about 500 species of trees and shrubs, sometimes woody vines; 22 native and 3 naturalized tree and about 10 native shrub species in North America.
Leaves: opposite, simple (pinnately compound in ash, *Fraxinus*), generally without teeth and thick, without stipules.
Flowers: usually small, sometimes showy, commonly in clusters, generally bisexual or male and female on separate plants (in ash, *Fraxinus*), regular with 4-lobed calyx, and tubular corolla generally 4-lobed, 2 stamens inserted on corolla, and 1 pistil with superior ovary of 2 cells each usually with 2 ovules, 1 style, and 1–2 stigmas.
Fruit: a berry, drupe, capsule, or winged key (samara).

84, 419 **Fringetree**
"Old-man's-beard"
Chionanthus virginicus L.

Description: Shrub or small tree with short trunk, narrow, oblong crown, and showy masses of fragrant, lacy, white flowers.

Height: 30′ (9 m).
Diameter: 6″ (15 cm).
Leaves: *opposite;* 4–8″ (10–20 cm) long, 1–3″ (2.5–7.5 cm) wide. Narrowly elliptical; *without teeth; slightly thickened;* short, stout *purplish* leafstalks. *Shiny* dark green above, paler and slightly hairy beneath; turning yellow in autumn.
Bark: brown, with reddish-tinged scales.
Twigs: light green, *purplish at nodes,* stout, often slightly angled; hairy when young; slightly *rough and warty.*
Flowers: 1″ (2.5 cm) long; with delicate corolla of *4 very narrow whitish lobes* with

purplish dots inside at base; fragrant; *many hanging loosely* on slender stalks in 3's; in branched drooping clusters 4–8″ (10–20 cm) long; on previous year's twigs back of new leaves in late spring. Male and female flowers generally on separate plants.
Fruit: ¾″ (19 mm) long; *elliptical, dark blue* or blackish, often with whitish bloom; thin juicy pulp; large stone; in drooping *grapelike clusters* on long stalks; maturing in autumn.

Habitat: Moist soils of valleys and bluffs; in understory of hardwood forests.

Range: S. New Jersey south to central Florida, west to E. Texas, and north to S. Missouri; to 4500′ (1372 m) in southern Appalachians.

One of the last trees to bear new leaves in spring, it appears dead until the leaves and flowers appear. The genus name *Chionanthus,* meaning "snow" and "flower," describes the blossoms.

112 Swamp-privet
"Common Adelia" "Texas Forestiera"
Forestiera acuminata (Michx.) Poir.

Description: Shrub or small tree with slender, often leaning trunks, forming thickets at water edges, and with paired, *diamond-shaped leaves.*

Height: 25′ (7.6 m).
Diameter: 4″ (10 cm).
Leaves: opposite; 1½–4″ (4–10 cm) long, ½–1″ (1.2–2.5 cm) wide. *Diamond-shaped* or ovate, long-pointed at both ends; with small teeth beyond middle; almost hairless; long-stalked. *Yellow-green* above, paler beneath.
Bark: dark brown; thin, smooth.
Twigs: light brown, slender.
Flowers: *tiny, greenish-yellow,* without petals; in small clusters along twig in early spring before leaves. Male and female flowers on separate plants.

Fruit: ⅜–⅝″ (10–15 mm) long; *narrowly oblong* or slightly curved, *dark purple or blackish;* thin pulp; large stone; in short clusters of several flowers each; maturing in summer.

Habitat: Wet soils bordering streams, swamps, and lakes; at edge of swamp forests.

Range: S. South Carolina to N. Florida, west to E. Texas, and north to central Illinois and SW. Indiana; to 500′ (152 m).

This relative of privet (*Ligustrum*) grows in swamps. The scientific name describes the long-pointed or "acuminate" leaves. The genus name *Forestiera* honors Charles de Forestier, a French physician and naturalist who died about 1820. The plants withstand flooding and are useful for erosion control. Wild ducks consume the fruit.

---

104 Florida-privet

"Florida Forestiera" "Wild-olive"

Forestiera segregata (Jacq.) Krug & Urban

Description: Evergreen shrub or small tree with 1 or more short, much-branched trunks, forming dense, irregular crowns and thickets.

Height: 20′ (6 m).

Diameter: 4″ (10 cm).

Leaves: *evergreen;* opposite; ¾–2¼″ (2–6 cm) long, ½–¾″ (12–19 mm) wide. *Narrowly elliptical* or diamond-shaped; *blunt-pointed;* without teeth; with few side veins and slightly curved up on sides; *somewhat thickened;* very short-stalked. Shiny green or gray-green above, dull light green with tiny dots beneath.

Bark: whitish-gray; smooth.

Twigs: greenish and covered with fine hair when young, becoming light gray; slender, stiff.

Flowers: ⅛″ (3 mm) long; *greenish-yellow,* without calyx or corolla, short-

stalked; in small clusters at leaf bases or on previous year's twigs back of leaves in spring. Male and female flowers generally on separate plants.
Fruit: ¼–⅜″ (6–10 mm) long; *elliptical, purplish or blackish;* thin bitter pulp; large stone; maturing in summer.

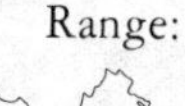

Habitat: Moist soil, including sand dunes, hammocks, and understory of pine forests.

Range: Along and near coasts of SE. Georgia south to Florida Keys and north to W. Florida; near sea level; also Bermuda and West Indies.

This species extends throughout the Caribbean to Puerto Rico and the Virgin Islands, where it is called "Ink-bush."

329, 492, 602 White Ash
Fraxinus americana L.

Description: Large tree with straight trunk and dense, conical or rounded crown of foliage with whitish lower surfaces.
Height: 80′ (24 m).
Diameter: 2′ (0.6 m).
Leaves: opposite; pinnately compound; 8–12″ (20–30 cm) long. *Usually* 7 (5–9) *leaflets* 2½–5″ (6–13 cm) long, 1¼–2½″ (3–6 cm) wide; paired (except at end); *ovate* or elliptical; finely saw-toothed or *almost without teeth.* Dark green above, *whitish* and sometimes hairy beneath; turning purple or yellow in autumn.
Bark: dark gray; thick, with deep diamond-shaped furrows and forking ridges.
Twigs: gray or brown, stout, mostly hairless.
Flowers: ¼″ (6 mm) long; purplish, without corolla; many in small clusters before leaves in early spring; male and female on separate trees.
Fruit: 1–2″ (2.5–5 cm) long; brownish

key with *narrow wing not extending down cylindrical body;* hanging in clusters; maturing in late summer and autumn.

Habitat: Moist soils of valleys and slopes, especially deep well-drained loams; in forests with many other hardwoods.

Range: S. Ontario east to Cape Breton Island, south to N. Florida, west to E. Texas, and north to E. Minnesota; to 2000′ (610 m) in the north; to 5000′ (1524 m) in the south.

The wood of White Ash is particularly suited for making baseball bats, tennis racquets, hockey sticks, polo mallets, oars, and playground equipment. A variation with hairs covering twigs, leafstalks, and underleaf surfaces has been called Biltmore Ash.

---

331 **Berlandier Ash**
"Mexican Ash"
Fraxinus berlandierana A. DC.

Description: Small tree with short trunk and rounded crown of spreading branches.
Height: 30′ (9 m).
Diameter: 1′ (0.3 m).
Leaves: opposite; pinnately compound; 3–7″ (7.5–18 cm) long. *3 or 5 leaflets* 1½–4″ (4–10 cm) long, ⅝–1½″ (1.5–4 cm) wide; paired (except at end); *lance-shaped,* sometimes elliptical; *few teeth* or coarsely saw-toothed. Shiny green above, pale and sometimes hairy in vein angles beneath.
Bark: gray; thick, furrowed into narrow ridges.
Twigs: light green, slender, hairless.
Flowers: ⅛″ (3 mm) long; without corolla; in small clusters of many flowers each; before leaves in early spring. Male flowers yellowish and female flowers greenish, on separate trees.
Fruit: 1–1¼″ (2.5–3 cm) long; light brown *key* with *narrow wing extending*

nearly to base of narrow body, sometimes 3-winged; hanging in clusters; maturing in late spring or summer.

Habitat: Moist soils along streams and in canyons.

Range: Central and S. Texas and NE. Mexico; to 1000′ (305 m).

Berlandier Ash, a southwestern relative of Green Ash, has fewer and smaller leaflets and smaller fruit and is adapted to a warmer, less humid climate.

325 Carolina Ash
"Water Ash" "Pop Ash"
Fraxinus caroliniana Mill.

Description: Tree with 1, or sometimes more than 1, trunk, often enlarged at base and leaning, and a rounded or narrow crown.
Height: 30–50′ (9–15 m).
Diameter: 1′ (0.3 m).
Leaves: opposite; pinnately compound; 6–12″ (15–30 cm) long; *5 or 7* (sometimes 9) *leaflets* 2–4½″ (5–11 cm) long, 1–2″ (2.5–5 cm) wide; paired (except at end); *elliptical* or ovate; *long-pointed; coarsely saw-toothed;* slightly thickened; slender-stalked. Green above, paler or whitish and often slightly hairy beneath.
Bark: light gray; thin and scaly, becoming rough and furrowed.
Twigs: light green to brown, usually hairless, slender.
Flowers: ⅛″ (3 mm) long; without corolla; in small clusters of many flowers each; before leaves in early spring. Male flowers yellowish, female flowers greenish, on separate trees.
Fruit: 1¼–2″ (3–5 cm) long; yellow-brown *key* with *broad elliptical wing extending to base of flat body,* sometimes 3-winged; hanging in clusters, maturing in summer and autumn.

Habitat: Wet soils of swamps and riverbanks flooded part of year; in swamp forests.

Range: NE. Virginia to S. Florida and west to SE. Texas; to 500′ (152 m); also a variety in Cuba.

Carolina Ash is the ash that ranges farthest to the southeast. Its large, broadly winged, flat keys are distinctive. The soft lightweight wood is not used commercially.

---

334 Black Ash
"Basket Ash" "Hoop Ash"
Fraxinus nigra Marsh.

Description: Tree with narrow, rounded crown of upright branches.

Height: 30–50′ (9–15 m).
Diameter: 1′ (0.3 m).
Leaves: opposite; pinnately compound; 12–16″ (30–41 cm) long. *7–11 leaflets* 3–5″ (7.5–13 cm) long, 1–1½″ (2.5–4 cm) wide; paired (except at end); *broadly lance-shaped; finely saw-toothed; stalkless.* Dark green above, paler beneath with tufts of rust-colored hairs along midvein; turning brown in autumn.
Bark: gray; corky, fissured into soft scaly plates that rub off easily.
Twigs: gray, stout, becoming hairless.
Flowers: ⅛″ (3 mm) long; *purplish,* without calyx or corolla; in small clusters of many flowers each; before leaves in early spring. Male and female flowers on separate trees.

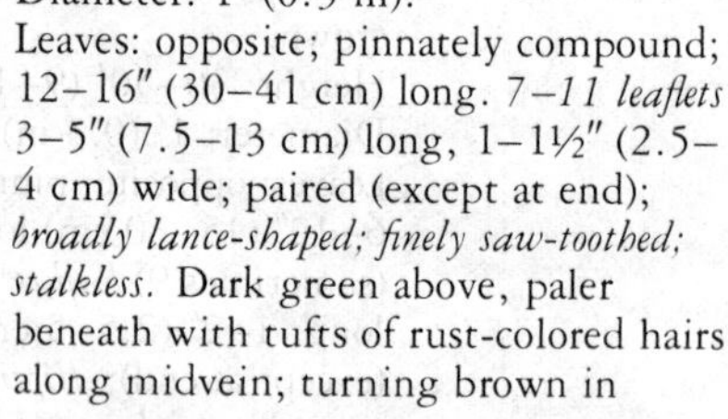

Fruit: 1–1½″ (2.5–4 cm) long; key with *broad oblong wing extending to base of flat body;* hanging in clusters; maturing in late summer.

Habitat: Wet soils of swamps, peat bogs, and streams, especially cold swamps where drainage is poor; in coniferous and hardwood forests.

Range: SE. Manitoba east to Newfoundland, south to West Virginia, and west to

Iowa; local in NE. North Dakota and N. Virginia; to 3500′ (1067 m).

The northernmost native ash, Black Ash takes its name from the dark brown heartwood. Baskets, barrel hoops, and woven chair bottoms are made from thin tough strips of split wood, giving rise to the other names.

328 Green Ash

"Swamp Ash" "Water Ash"
Fraxinus pennsylvanica Marsh.

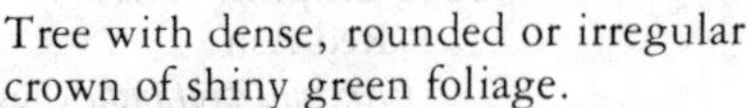

Description: Tree with dense, rounded or irregular crown of shiny green foliage.

Height: 60′ (18 m).
Diameter: 1½′ (0.5 m).
Leaves: opposite; pinnately compound; 6–10″ (15–25 cm) long; 5–9 (usually 7) *leaflets* 2–5″ (5–13 cm) long, 1–1½″ (2.5–4 cm) wide; paired (except at end); *lance-shaped* or ovate; coarsely *saw-toothed* or almost without teeth; mostly hairless. *Shiny green above,* green or paler and slightly hairy beneath; turning yellow in autumn.
Bark: gray; furrowed into scaly ridges, with reddish inner layer.
Twigs: green, becoming gray and hairless; slender.
Flowers: ⅛″ (3 mm) long; greenish; without corolla; in small clusters of many flowers each; before leaves in early spring. Male and female flowers on separate trees.
Fruit: 1¼–2¼″ (3–6 cm) long; yellowish *key* with *narrow wing extending nearly to base of narrow body;* hanging in clusters; maturing in late summer and autumn.

Habitat: Moist alluvial soils along streams in floodplain forests.

Range: SE. Alberta east to Cape Breton Island; south to N. Florida, west to Texas; to 3000′ (914 m) in southern Appalachians.

The most widespread native ash, this species extends westward into the plains and nearly to the Rocky Mountains. A northeastern variation with twigs, leafstalks, and underleaf surfaces all densely covered with hairs has been called Red Ash. One of the most successful hardwoods in the Great Plains shelterbelts, hardy, fast-growing Green Ash is also planted on spoil banks after strip mining, as well as for shade.

---

346 **Pumpkin Ash**
"Red Ash"
Fraxinus profunda (Bush) Bush

Description: Large tree with enlarged and buttressed base and narrow crown of spreading branches.
Height: 80′ (24 m).
Diameter: 2′ (0.6 m).
Leaves: opposite; pinnately compound; 8–16″ (20–41 cm) long; with hairy axis. *Usually 7 or 9 leaflets* 3–7″ (7.5–18 cm) long, 2–3″ (5–7.5 cm) wide, sometimes larger; paired (except at end); *elliptical or narrowly ovate; without teeth* or slightly saw-toothed; slightly thickened. Dark green and becoming nearly hairless above, yellow-green and covered with *soft hairs* beneath.
Bark: gray; furrowed into forking ridges forming diamond pattern.
Twigs: light gray, stout, densely covered with hairs when young.
Flowers: ¼″ (6 mm) long; without corolla; in small clusters of many flowers each; before leaves in early spring. Male flowers yellowish and female flowers greenish, on separate trees. 
Fruit: 2–3″ (5–7.5 cm) long; yellowish key with *broad elliptical wing, extending nearly to base of thick body;* hanging in clusters; maturing in late summer and autumn.

Habitat: Wet soils of swamps and river valleys submerged part of year; in swamp forests.

Range: S. Maryland south to N. Florida, west to Louisiana, and north to S. Illinois and SW. Ohio; to 500′ (152 m).

This uncommon species is similar to Green Ash but has very large leaves and fruit. The scientific name, meaning "profound" or "deep," refers to the swampy habitat.

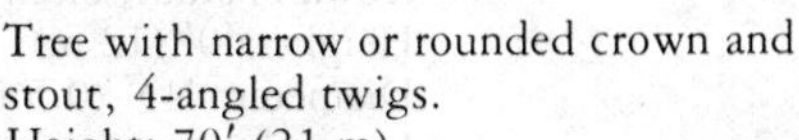

326 Blue Ash

Fraxinus quadrangulata Michx.

Description: Tree with narrow or rounded crown and stout, 4-angled twigs.
Height: 70′ (21 m).
Diameter: 1½′ (0.5 m).
Leaves: opposite; pinnately compound; 8–12″ (20–30 cm) long. *7–11 leaflets* 2½–4½″ (6–11 cm) long, 1–1½″ (2.5–4 cm) wide; paired (except at end); *lance-shaped* to ovate, *long-pointed; sharply saw-toothed;* slightly thickened. Yellow-green above, pale and often with tufts of hairs along midvein beneath; turning pale yellow in autumn.
Bark: gray; furrowed into scaly and shaggy plates.
Twigs: brown or gray; becoming hairless; stout, *4-angled* and slightly *winged.*
Flowers: less than ⅛″ (3 mm) long; *purplish, bisexual,* without calyx or corolla; in small clusters of many flowers each; before leaves in early spring.
Fruit: 1¼–2″ (3–5 cm) long; key with *broad oblong wings extending to base of flattened body;* hanging in clusters; maturing in autumn.

Habitat: Dry, rocky limestone slopes and moist soils of valleys; in hardwood forests.

Range: Ohio south to NW. Georgia, west to

NE. Oklahoma, north to extreme S. Wisconsin; local in extreme S. Ontario, S. Michigan, and W. West Virginia; at 400–2000′ (122–610 m).

Blue Ash is easily identifiable by its 4-angled twigs. Pioneers produced a blue dye for cloth by macerating the inner bark in water. The sap of the inner bark turns blue upon exposure to air.

351 Texas Ash

Fraxinus texensis (Gray) Sarg.

Description: Tree with short trunk and spreading crown of stout, often crooked branches.
Height: 40′ (12 m).
Diameter: 1½′ (0.5 m).
Leaves: opposite; pinnately compound; 5–8″ (13–20 cm) long. *3–7 leaflets* 1¼–3″ (3–7.5 cm) long, ¾–2″ (2–5 cm) wide; paired (except at end); *elliptical, rounded* or blunt at tip; *coarsely wavy-toothed* (except near base); *slightly thickened; slender-stalked.* Dark green above; *whitish beneath,* with network of veins, sometimes slightly hairy.
Bark: gray or reddish-brown; deeply furrowed into broad scaly ridges.
Twigs: greenish, slender, hairless or nearly so.
Flowers: ⅛″ (3 mm) long; purplish, without corolla; in small clusters of many flowers each; before leaves in early spring. Male and female flowers on separate trees.
Fruit: ⅝–1¼″ (1.5–3 cm) long; *key* with *narrow wing not extending down cylindrical body;* hanging in clusters; maturing in late spring.

Habitat: Dry rocky slopes and bluffs of canyons, especially limestone; in open forests.

Range: S. Oklahoma and Texas; at 400–1600′ (122–488 m).

Confined to Texas, except for a northern extension into the Arbuckle

Mountains of Oklahoma. This southwestern relative of White Ash has fewer and smaller leaflets and smaller fruit and is adapted to a warmer, less humid climate; some consider it a variety of that species.

103, 444 **Chinese Privet**
Ligustrum sinense Lour.

Description: Introduced shrub or small tree with several trunks, dense, much-branched crown, compact foliage, and abundant, showy white flowers.
Height: 20′ (6 m).
Diameter: 4″ (10 cm).
Leaves: deciduous or nearly evergreen; *opposite;* often in 2 rows; 1–2½″ (2.5–6 cm) long, ½–1″ (1.2–2.5 cm) wide. *Elliptical* or ovate; without teeth. *Dull green* above, light green and with fine hairs on midvein beneath.
Bark: gray; smooth.
Twigs: light brown, slender, covered with fine hairs.
Flowers: ¼″ (6 mm) long and wide; with *white bell-shaped corolla* of *4 short spreading lobes;* fragrant; in terminal clusters of many flowers each, 1½–4″ (4–10 cm) long, on short hairy branches; in late spring and early summer.
Fruit: ¼–5/16″ (6–8 mm) long; *berrylike, elliptical* or somewhat round, *black;* few seeds; maturing in autumn.

Habitat: Woodlands, thickets, and waste places; persisting along fencerows and at old homesites.

Range: Native of China. Widely naturalized in SE. United States from Virginia to Georgia, Texas, and Oklahoma.

The plants can be clipped into hedges of various shapes. The foliage is evergreen or nearly so southward and in mild winters. The fruit is considered poisonous.

68 Devilwood
"Wild-olive"
Osmanthus americanus (L.) Gray

Description: Evergreen shrub or small tree with narrow, oblong crown of paired, glossy, leathery leaves, and with dark blue fruit like small olives.
Height: 30′ (9 m), sometimes larger.
Diameter: 1′ (0.3 m).
Leaves: *opposite;* 3½–5″ (9–13 cm) long, ¾–1½″ (2–4 cm) wide. *Lance-shaped* to narrowly elliptical; *thick and leathery;* with *edges straight or turned under,* and obscure side veins. Shiny green above, dull and paler beneath. Stout leafstalks about ⅝″ (15 mm) long.
Bark: gray; thin, scaly, exposing dark red inner bark.
Twigs: red-brown, slender, slightly angled, becoming hairless.
Flowers: 3/16″ (5 mm) wide; with *bell-shaped 4-lobed whitish or yellowish corolla;* very fragrant; almost stalkless; in branching clusters of many flowers each, at leaf bases; in early spring. Male and female flowers (sometimes bisexual) usually on separate plants.

Fruit: ⅜–¾″ (10–19 mm) long; *elliptical, dark blue;* thin pulp; large stone; maturing in autumn and remaining attached in winter.

Habitat: Moist soils from river valleys to sandy uplands and dunes; in understory of forests.

Range: SE. Virginia to central Florida and west to SE. Louisiana; also Mexico; to 500′ (152 m).

Devilwood was so named because the fine-textured wood is difficult to split and work. The fruit resembles the cultivated Olive in the same family. The genus name, *Osmanthus,* from the Greek words for "odor" and "flower," refers to the fragrant blossoms.

BORAGE FAMILY
(Boraginaceae)

Herbs, shrubs, and trees; widespread, about 2000 species worldwide; 4 native and 1 naturalized tree and many native herb and shrub species in North America.
Leaves: usually alternate, simple, generally not toothed, sometimes rough with stiff hairs, without stipules.
Flowers: in clusters, often asymmetrical in a spiral; sometimes large and showy; usually bisexual; regular, with calyx of 5 sepals, and tubular corolla often in form of funnel or bell, 5 stamens inserted in tube and alternate, and 1 pistil with superior ovary, 2-celled and becoming 4-celled with 4 ovules, 1 style, and 1 stigma, sometimes with 2 or 4 lobes.
Fruit: a drupe of 1–4 seeds or 4 nutlets.

74, 562 **Anacua**
"Sugarberry" "Knockaway"
Ehretia anacua (Terán & Berland.) Johnst.

Description: Shrub or tree with 1 or more trunks and compact, rounded crown, evergreen or partly deciduous northward, with dull green foliage, abundant white flowers, and showy yellow-orange fruit.

Height: 40′ (12 m).
Diameter: 1′ (0.3 m).
Leaves: *evergreen except northward;* 1¼–3¼″ (3–8 cm) long, ¾–1½″ (2–4 cm) wide. *Elliptical* or ovate, abruptly short-pointed at tip, rounded at base; sometimes coarsely toothed beyond middle; *thick and stiff.* Dull green and *very rough* above, paler and hairy with coarse veins beneath.
Bark: reddish-brown to gray; furrowed into narrow ridges with peeling scales.
Twigs: brown to gray, slender, crooked, with stiff hairs when young

Flowers: ⁵⁄₁₆″ (8 mm) wide; *bell-shaped,* with 5 *rounded white corolla lobes;* fragrant; short-stalked; in clusters of many flowers each, 2–3″ (5–7.5 cm) long and wide, at end of twigs; branched and curved to one side; in early spring or occasionally in autumn.

Fruit: ⁵⁄₁₆″ (8 mm) in diameter; *round, yellow-orange;* thin, juicy, edible pulp; 2 stones, each 2-seeded; in clusters of several fruits; maturing in autumn or spring.

Habitat: Commonly a shrub forming thickets on dry hillsides or a forest tree in moist soils of river valleys.

Range: Central and S. Texas and E. Mexico; to 1000′ (305 m).

A popular ornamental in Texas, this species is hardy in dry areas and north to central Texas, where the plants may die back in cold winters. Wildlife consume the fruit, and the wood has served for fenceposts and tool handles. The name Anacua is from "Anachuite," a Mexican name for this and related species. That word is from two others of the Nahuatl language of the Aztecs meaning "paper" and "tree," perhaps referring to the scaly peeling bark. The English name "Knockaway" is a corruption from the same source.

VERBENA FAMILY
(Verbenaceae)

Herbs, shrubs, woody vines, and often large trees; about 3000 species, mostly in tropical and subtropical, also warm temperate regions; 3 native tree species and many native herb and shrub species in North America.
Leaves: opposite (sometimes whorled), generally simple, often deciduous without stipules.
Twigs: often 4-angled.
Flowers: mostly small, often colored, commonly in clusters, bisexual, usually irregular, usually with 5-lobed or 5-toothed calyx, corolla with short tube and 5 short, unequal, spreading lobes or in 2 lips, stamens usually 4 in pairs and sometimes 1 sterile stamen inserted in tube, and 1 pistil with superior ovary of 2 or 4 cells, each containing 1 ovule, long style, and 1–2 stigmas.
Fruit: generally a drupe with 2–4 nutlets or only 2–4 nutlets (capsule in Black-mangrove, *Avicennia*).

56, 435 **Black-mangrove**
"Blackwood" "Mangle"
Avicennia germinans (L.) L.

Description: Evergreen shrub or, in tropical regions, a tree with rounded crown of spreading branches.
Height: 40′ (12 m).
Diameter: 1′ (0.3 m).
Leaves: *evergreen; opposite;* 2–4″ (5–10 cm) long, ¾–1½″ (2–4 cm) wide. *Narrowly elliptical* or lance-shaped; without teeth; *slightly thickened.* Yellow-green and often shiny above, *gray-green* and covered with fine hairs beneath; both surfaces with scattered salt crystals.
Bark: dark gray or brown; smooth, becoming thick, fissured, and scaly.
Twigs: gray or brown; covered with

hairs when young; with *rings at enlarged nodes.*

Flowers: ⅜″ (10 mm) wide; with 4 *white corolla lobes,* rounded or notched; stalkless; crowded in upright clusters branched on 4-angled stalks at end of twigs; in summer or in tropics, nearly throughout year.

Fruit: 1–1¼″ (2.5–3 cm) long; *flat elliptical capsules* with tiny point; yellow-green; covered with fine hairs; maturing throughout year. 1 large dark green seed often germinating on tree, splitting capsule into 2 parts.

Habitat: Silt seashores, in salt and brackish water; in mangrove swamp forest.

Range: Coasts and islands from N. to S. Florida, west to S. Louisiana and S. Texas; at sea level; also Bermuda, West Indies, and from Mexico to Brazil and Peru.

Black-mangrove is the hardiest of the four species forming the mangrove swamp forests of southern Florida. It penetrates farthest inland into brackish water of rivers and farthest north along the Gulf Coast, where it becomes smaller and shrubby and is killed in cold winters. New seedlings, however, invade from seeds transported by currents and persist a few years. An important honey plant, it yields clear whitish honey of high quality. The other three native mangroves, outside the scope of this field guide, are tropical trees confined mostly to southern Florida but do extend northward along the coast to the central part of that state.

FIGWORT FAMILY
(Scrophulariaceae)

Herbs and small shrubs, sometimes shrubs and trees; worldwide, about 3000 species; numerous species of native herbs and a few shrubs in southern North America. No native trees; 1 naturalized tree species.
Leaves: generally deciduous, alternate or opposite, simple; toothed, lobed, or not toothed; without stipules.
Flowers: irregular, with deeply 4- to 5-lobed calyx, tubular corolla, often bell-shaped and 2-lipped, with 4–5 lobes, 4–5 stamens on corolla, and 1 pistil with superior 2-celled ovary containing many ovules, 1 style, and 2-lobed stigma.
Fruit: a many-seeded capsule.

98, 189, 461 **Royal Paulownia**
"Princess-tree" "Empress-tree"
Paulownia tomentosa (Thunb.) Steud.

Description: Naturalized tree with short trunk, broad, open crown of stout, spreading branches, *very large leaves,* and *showy purple flowers.*
Height: 50′ (15 m).
Diameter: 2′ (0.6 m).
Leaves: *opposite;* 6–16″ (15–41 cm) long, 4–8″ (10–20 cm) wide. *Broadly ovate;* long-pointed at tip; with *several veins from notched base;* sometimes slightly 3-toothed or 3-lobed. Dull light green and slightly hairy above, paler and densely covered with hairs beneath. Leafstalks 4–8″ (10–20 cm) long.
Bark: gray-brown; with network of irregular shallow fissures.
Twigs: light brown, stout; densely covered with soft hairs when young.
Flowers: 2″ (5 cm) long; with *bell-shaped* pale violet corolla ending in 5 rounded *unequal lobes;* fragrant; in upright

clusters, 6–12″ (15–30 cm) long, on stout hairy branches; in early spring from rounded brown hairy buds formed previous summer.

Fruit: 1–1½″ (2.5–4 cm) long; *egg-shaped capsule,* pointed, brown, thick-walled; *splitting into 2 parts;* many tiny winged seeds; maturing in autumn and remaining attached.

Habitat: Waste places, roadsides, and open areas.

Range: Native of China. Cultivated and naturalized from S. New York south to N. Florida, west to S. Texas, and north to Missouri.

This handsome, rapid-growing ornamental and shade tree resembles catalpas. Vigorous shoots, with enormous leaves 2′ (0.6 m) or more in length and width, can be produced by pruning back almost to the base. The soft, lightweight, whitish wood of this weed tree is exported to Japan for furniture and special uses, such as for sandals. Named for Anna Paulowna (1795–1865), of Russia, princess of the Netherlands and ancestor of the present queen, Juliana.

BIGNONIA FAMILY
(Bignoniaceae)

Mostly woody vines, also shrubs and often large trees, rarely herbs; about 700 species in tropical and warm temperate regions; 5 species of native trees and 2 species of native woody vines in North America.
Leaves: mostly opposite, sometimes alternate; palmately or pinnately compound or bipinnate, sometimes simple; without stipules.
Flowers: generally large, showy, in clusters; bisexual; slightly irregular, with tubular 5-toothed or 5-lobed calyx, large funnel- or bell-shaped tubular corolla, commonly yellow, pink, or whitish, with 5 unequal lobes sometimes in 2 lips, usually 4 large stamens in pairs and 1 sterile stamen inserted in tube, and on disk, 1 pistil with superior 2-celled ovary containing many ovules, long thin style, and 2 stigmas.
Fruit: usually a capsule, often long and podlike, splitting into 2 parts, with many winged seeds; or a berry.

94 Southern Catalpa
"Catawba" "Indian-bean"
Catalpa bignonioides Walt.

Description: Short-trunked tree with broad, rounded crown of spreading branches, large, heart-shaped leaves, large clusters of showy white flowers, and long, beanlike fruit.

Height: 50′ (15 m).
Diameter: 2′ (0.6 m).
Leaves: *3 at a node* (whorled) and opposite; 5–10″ (13–25 cm) long, 4–7″ (10–18 cm) wide. Ovate, *abruptly long-pointed* at tip, notched at base; *without teeth.* Dull green above, paler and covered with soft hairs beneath; turning blackish in autumn. With

unpleasant odor when crushed. Slender leafstalk 3½–6″ (9–15 cm) long.
Bark: brownish-gray; scaly.
Twigs: green, turning brown; stout, hairless or nearly so.
Flowers: 1½″ (4 cm) long and wide; with *bell-shaped corolla* of 5 *unequal rounded fringed lobes, white* with 2 orange stripes and many purple spots and stripes inside; slightly fragrant; in upright branched clusters to 10″ (25 cm) long and wide; in late spring.
Fruit: 6–12″ (15–30 cm) long, 5⁄16–⅜″ (8–10 mm) in diameter; narrow, cylindrical, *dark brown capsule; cigarlike,* thin-walled, splitting into 2 parts; many flat light brown seeds with 2 papery wings; maturing in autumn, remaining attached in winter.

Habitat: Moist soils in open areas such as roadsides and clearings.

Range: Original range uncertain; probably native in SW. Georgia, NW. Florida, Alabama, and Mississippi; widely naturalized from S. New England south to Florida, west to Texas, and north to Michigan; at 100–500′ (30–152 m).

Catalpa is the American Indian name, while the scientific name refers to a related vine with flowers of similar shape. Planted as a shade tree and an ornamental for the abundant showy flowers, cigarlike pods, and coarse foliage.

97, 441 **Northern Catalpa**
"Hardy Catalpa" "Indian-bean"
Catalpa speciosa Warder ex Engelm.

Description: Tree with rounded crown of spreading branches; large, heart-shaped leaves; large, showy flowers; and long, beanlike fruit.
Height: 50–80′ (15–24 m).
Diameter: 2½′ (0.8 m).
Leaves: *3 at a node* (whorled) and

opposite; 6–12″ (15–30 cm) long, 4–8″ (10–20 cm) wide. *Ovate, long-pointed,* straight to notched at base; *without teeth.* Dull green above, paler and covered with soft hairs beneath; turning blackish in autumn. Slender leafstalk 4–6″ (10–15 cm) long.

Bark: brownish-gray; smooth, becoming furrowed into scaly plates or ridges.

Twigs: green, turning brown; stout; becoming hairless.

Flowers: 2–2¼″ (5–6 cm) long and wide; with *bell-shaped corolla of 5 unequal rounded fringed lobes,* white with 2 orange stripes and purple spots and lines inside; in branched upright clusters, 5–8″ (13–20 cm) long and wide; in late spring.

Fruit: 8–18″ (20–46 cm) long, ½–⅝″ (12–15 mm) in diameter; narrow, cylindrical, *dark brown capsule; cigarlike,* thick-walled, splitting into 2 parts; many flat light brown seeds with 2 papery wings; maturing in autumn, remaining attached in winter.

Habitat: Moist valley soils by streams; naturalized in open areas such as roadsides and clearings.

Range: Original range uncertain; native apparently from SW. Indiana to NE. Arkansas; widely naturalized in SE. United States; at 200–500′ (61–152 m).

Northern Catalpa is the northernmost New World example of its tropical family and is hardier than Southern Catalpa, which blooms later and has slightly smaller flowers and narrower, thinner-walled capsules. Both are called "Cigartree" and "Indian-bean" because of the distinctive fruit.

MADDER FAMILY
(Rubiaceae)

One of the largest plant families, worldwide, mostly tropical and subtropical, with about 6000 species of trees, shrubs, and few herbs; 7 native tree species and many native herb and shrub species in North America.
Leaves: opposite, sometimes whorled; simple, not toothed, with paired stipules that form bud and leave ring scars at nodes.
Flowers: generally in clusters of many flowers each, small to large, bisexual, regular, with 4- to 5-lobed calyx often persisting, tubular corolla usually 4- to 5-lobed, generally colored and often showy, stamens 4–5 alternate and inserted in tube, and 1 pistil with inferior ovary, commonly 2-celled and containing many ovules (but sometimes as few as 1), 1 style, and 2 stigmas.
Fruit: a capsule or berry, sometimes a drupe.

73, 402 **Buttonbush**
"Honey-balls" "Globe-flowers"
Cephalanthus occidentalis L.

Description: Spreading, much-branched shrub or sometimes small tree with many branches (often crooked and leaning), irregular crown, balls of white flowers resembling pincushions, and buttonlike balls of fruit.

Height: 20′ (6 m).
Diameter: 4″ (10 cm).
Leaves: *opposite or 3 at a node* (whorled); 2½–6″ (6–15 cm) long, 1–3″ (2.5–7.5 cm) wide. *Ovate or elliptical,* pointed at tip, rounded at base; without teeth. Shiny green above, paler and sometimes hairy beneath; at southern limit nearly evergreen.
Bark: gray or brown; becoming deeply furrowed into rough scaly ridges.

Twigs: *mostly in 3's;* reddish-brown, stout, sometimes hairy, with rings at nodes.
Flowers: ⅝″ (15 mm) long; with narrow, tubular, white 4-lobed corolla and long threadlike style; fragrant; stalkless; crowded in upright long-stalked *white balls* of many flowers each, 1–1½″ (2.5–4 cm) in diameter; from late spring through summer.
Fruit: ¾–1″ (2–2.5 cm) in diameter; compact *rough brown balls* composed of many small, narrow, dry nutlets ¼″ (6 mm) long, each 2-seeded; maturing in autumn.

Habitat: Wet soils bordering streams and lakes.

Range: S. Quebec and SW. Nova Scotia, south to S. Florida, west to Texas, and north to SE. Minnesota; to 3000′ (914 m); in Arizona and California to 5000′ (1524 m); also Mexico, Central America, and Cuba.

The poisonous foliage of this abundant and widespread species is unpalatable to livestock. The bitter bark has served in home remedies, but its medicinal value is doubtful. Buttonbush is a handsome ornamental suited to wet soils and is also a honey plant. Ducks and other water birds and shorebirds consume the seeds.

81, 464 **Pinckneya**
"Fever-tree" "Georgia-bark"
Pinckneya pubens Michx.

Description: Shrub or small tree, usually with several trunks, a narrow, rounded crown, and greenish-yellow, tubular flowers, some with a beautiful, pink, petal-like lobe.
Height: 20′ (6 m).
Diameter: 6″ (15 cm).
Leaves: opposite; 4–8″ (10–20 cm) long, 2–4″ (5–10 cm) wide. *Elliptical,* short-pointed; *without teeth.* Dark green

and covered with fine hairs above, paler and with soft hairs beneath. Slender leafstalks about 1″ (2.5 cm) long.
Bark: light brown; scaly, bitter.
Twigs: slender, covered with gray hairs, turning light brown; with *rings at nodes*.
Flowers: about 1½″ (4 cm) long; *narrow*, with *tubular greenish-yellow corolla* with red spots inside and 5 *narrow lobes;* in open clusters of several flowers each, 6″ (15 cm) wide; at or near ends of leafy twigs; in late spring. On some flowers 1 calyx lobe expands greatly, becoming showy and *petal-like*, elliptical, 2–3″ (5–7.5 cm) long, and *cream to rose pink*.
Fruit: ¾″ (19 mm) in diameter; slightly hairy *rounded capsules; brown* with whitish dots; thin-walled; 2-celled and splitting in 2 parts; many flat, light brown, broad-winged seeds; maturing in autumn, remaining attached in winter.

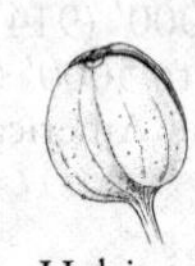

Habitat: Wet soils of swamps and stream borders; in understory of forests.

Range: Extreme S. South Carolina, Georgia, and N. Florida; to 300′ (91 m).

Among the most beautiful native trees and also grown as an ornamental, Pinckneya is the single, rare species of its genus. The very showy, pinkish, enlarged calyx lobes remain attached for several weeks. The bitter bark formerly served in home remedies for reducing fever.

HONEYSUCKLE FAMILY
(Caprifoliaceae)

About 400 species of shrubs, sometimes woody vines and small trees, rarely herbs; widespread, mostly in north temperate regions and in tropical mountains; 11 native tree species and many native shrub species in North America.
Leaves: opposite, usually simple (pinnately compound in *Sambucus*), stipules absent or minute.
Flowers: usually small, sometimes showy, often in clusters, bisexual, regular or irregular, with minute calyx of 5 (sometimes 4) teeth or lobes, tubular corolla of 5 (or 4) lobes, 5 (or 4) stamens alternate and inserted in tube, and 1 pistil with inferior ovary of 1–5 cells, each with 1 ovule, style, and stigma to 5 lobes.
Fruit: a berry or drupe.

336, 413, 539 **American Elder**
"Elderberry" "Common Elder"
Sambucus canandensis L.

Description: Large shrub or small tree with irregular crown of few, stout, spreading branches, clusters of white flowers, and many small black or purple berries.
Height: 16′ (5 m).
Diameter: 6″ (15 cm).
Leaves: opposite; *pinnately compound;* 5–9″ (13–23 cm) long; with yellow-green axis. *3–7 leaflets* 1½–4″ (4–10 cm) long, ¾–2″ (2–5 cm) wide; *paired (except at end); elliptical; sharply saw-toothed;* stalkless or nearly so. *Shiny green* above, dull light green and hairy along midvein beneath. Leaves sometimes partly bipinnate, with up to 13 leaflets.
Bark: light gray or brown with raised dots; smooth or becoming fissured and rough.
Twigs: light green, stout, angled, with

ringed nodes and *thick white pith.*
Flowers: ¼″ (6 mm) wide; with *white corolla* of 5 or 4 lobes; fragrant; many in upright *flat-topped,* much-branched clusters, 4–8″ (10–20 cm) wide; in late spring and early summer, shedding early.
Fruit: ¼″ (6 mm) in diameter; *black* or purplish-black *berry;* juicy and slightly sweet; 5 or fewer 1-seeded nutlets; maturing in late summer and autumn.

Habitat: Wet soils, especially in open areas near water at forest edges.

Range: SE. Manitoba east to Nova Scotia, south to S. Florida, and west to S. Texas; to 5000′ (1524 m).

This common, widespread shrub sprouts from roots. Elderberries are used for making jelly, preserves, pies, and wine. Birds and mammals of many species also feed on the berries. The bark, leaves, and flowers have served in home remedies. Whistles, popguns, and other toys can be made by removing the thick pith from the stems. Near the Gulf Coast the plants of the southern tree variety, Florida Elder (var. *laciniata* Gray), are evergreen or nearly so and bear flowers and fruit throughout the year.

223, 541, 542 **Arrowwood**
"Southern Arrowwood"
"Arrowwood Viburnum"
Viburnum dentatum L.

Description: Much-branched shrub with many shoots from base, or sometimes a small tree, with showy clusters of white flowers and blue-black fruit.

Height: 3–10′ (0.9–3 m), rarely 20′ (6 m).
Diameter: 3″ (7.5 cm).
Leaves: opposite; 1½–4″ (4–10 cm) long, 1–3½″ (2.5–9 cm) wide. *Ovate or rounded,* pointed at tip, blunt or

notched at base; with *many straight* sunken *side veins ending in large teeth;* leafstalks long, slender. Dull green and nearly hairless above, paler and *hairy beneath;* turning shiny red in autumn.
Bark: gray or reddish-brown; smooth.
Twigs: brown, slender, usually hairy, with ringed nodes.
Flowers: ¼" (6 mm) wide; with 5 *rounded white corolla lobes;* in branched, upright *long-stalked clusters* 2–3½" (5–9 cm) wide, of many flowers each; in spring and early summer.
Fruit: ¼–⅜" (6–10 mm) long; *rounded* or elliptical, blue or *blue-black,* juicy; large flattened stone; maturing in late summer and autumn.

Habitat: Moist to dry soils, especially sandy; forming thickets in open areas, at border and in understory of forest.

Range: Illinois east to Massachusetts, south to Florida, and west to E. Texas; to 4500' (1372 m) in southern Appalachians.

Arrowwood is a common, widespread shrub varying in leaf shape, size, and hairiness. In eastern Texas it sometimes becomes a small tree. Indians used the straight young stems as arrow shafts; hence the common name. The fruit is consumed by birds.

173, 414, 582 **Nannyberry**
"Blackhaw" "Sheepberry"
Viburnun lentago L.

Description: Shrub or small tree with short trunk, compact, rounded crown of drooping branches, small white flowers in clusters, and small bluish-black fruit.

Height: 20' (6 m).
Diameter: 6" (15 cm).
Leaves: opposite; 2½–4" (6–10 cm) long, 1½–2½" (4–6 cm) wide. *Elliptical,* long-pointed; *finely saw-toothed;* with prominent network of veins; broad, often hairy leafstalk. *Shiny*

green above, yellow-green with tiny black dots beneath; turning purplish-red and orange in autumn.
Bark: reddish-brown or gray; irregularly furrowed into scaly plates; with unpleasant skunklike odor.
Twigs: light green, slender, slightly hairy when young, ending in long-pointed hairy reddish bud.
Flowers: ¼″ (6 mm) wide; with 5 *rounded white corolla lobes;* slightly fragrant; in branched upright, stalkless clusters of many flowers each, 3–5″ (7.5–13 cm) wide; in late spring.
Fruit: ½″ (12 mm) long; *elliptical* or sometimes nearly round, slightly *flat, blue-black* with whitish bloom; sweet juicy pulp; somewhat flat stone; drooping on slender reddish stalks; maturing in autumn and remaining attached in winter.

Habitat: Moist soils of valleys and rocky uplands; at forest edges.

Range: SE. Saskatchewan east to New Brunswick and Maine, south to West Virginia, west to Nebraska and NE. Wyoming; local in SW. Virginia; to 2500′ (762 m); to 5000′ (1524 m) in Black Hills.

When cut, the plants sprout from roots, and old branches will often arch down and take root. Songbirds, gamebirds, and mammals eat the fruit in winter.

80, 410 Possumhaw Viburnum

"Possumhaw" "Swamphaw"
Viburnum nudum L.

Description: Shrub or small tree with spreading, open crown of irregular branches, many small, white or yellowish flowers, and small blue fruit.
Height: 16′ (5 m).
Diameter: 4″ (10 cm).
Leaves: opposite; 2–5″ (5–13 cm) long,

1–2½″ (2.5–6 cm) wide. *Elliptical* to narrowly elliptical; edges turned under and *without teeth* or slightly wavy; *slightly thick;* with *prominent* curved side *veins* raised beneath. *Shiny green* and becoming hairless above; paler, *rusty scurfy* (especially when young), and with tiny dots beneath; turning red in autumn. Short leafstalk covered with rust-colored hairs.
Bark: gray to brown; smooth.
Twigs: brown; slender; *rusty scurfy* when young; ending in long-pointed bud covered with rust-colored hairs.
Flowers: ½″ (12 mm) wide; with 5 rounded *creamy white* or yellowish *corolla lobes,* in upright short-stalked flat clusters, 2½–5″ (6–13 cm) wide; in spring.
Fruit: 5/16″ (8 mm) in diameter; nearly *round;* turning from pink to deep *blue* or blue-black with a bloom; bitter pulp; slightly flat stone; on slender stalks; maturing in autumn.

Habitat: Moist soil near streams and swamps and less frequently in upland slopes; in open forests, pinelands, and thickets.

Range: SW. Connecticut to central Florida, west to E. Texas, and north to central Arkansas and W. Kentucky; to 3000′ (914 m).

As the common name suggests, wildlife consume the fruit; deer also browse the foliage. The scientific name, meaning "naked," refers to the stalked, leafless flower clusters.

53 **Walter Viburnum**
"Blackhaw" "Small-leaf Viburnum"
Viburnum obovatum Walt.

Description: Evergreen shrub or small tree with a few trunks; broad, spreading crown; small, white flowers; and small, shiny black fruit.
Height: 20′ (6 m).

Diameter: 4″ (10 cm).
Leaves: *evergreen,* opposite; 1–2½″ (2.5–6 cm) long, ⅜–1¼″ (1–3 cm) wide; *spoon-shaped; few teeth* toward blunt tip *or not toothed; thick and leathery;* short-stalked base. *Shiny dark green* above, light green with reddish-brown gland-dots beneath.
Bark: blackish; thick, furrowed into many angular blocks.
Twigs: red-brown to gray; partly short, stiff side twigs or spurs, ending in narrow reddish-brown hairy bud.
Flowers: ¼″ (6 mm) wide; with 5 rounded *white corolla lobes;* in upright flat, almost stalkless clusters 1½–2½″ (4–6 cm) wide; in early spring with new leaves.

Fruit: 5⁄16″ (8 mm) long; *elliptical,* red turning *shiny black;* thin, almost tasteless pulp; slightly flat stone; on short stalks; maturing in summer or autumn and remaining attached.

Habitat: Moist valley soils, especially stream banks, flood plains, and swamps.

Range: E. South Carolina to central Florida; to 200′ (61 m).

Walter Viburnum honors Thomas Walter (1740–89), English-born planter of South Carolina, who described this species in his *Flora Caroliniana.* The Latin species name refers to the leaf shape. Birds consume quantities of the fruit.

---

203 Blackhaw
"Stagbush" "Sweethaw"
Viburnum prunifolium L.

Description: Shrub or small tree with short trunk, spreading, rounded or irregular crown, many showy, small, white flowers, and small, blue-black fruit.
Height: 20′ (6 m).
Diameter: 4″ (10 cm).
Leaves: opposite; 1½–3″ (4–7.5 cm)

long, ¾–2″ (2–5 cm) wide. *Elliptical; finely saw-toothed;* slightly thick; hairless or nearly so. *Shiny green* with network of sunken veins above, dull light green beneath; turning shiny red in autumn.
Bark: gray; rough, furrowed into rectangular plates.
Twigs: gray, slender, stiff, ending in flat, oblong, hairy brown bud.
Flowers: ¼″ (6 mm) wide; with 5 rounded *white corolla lobes;* in upright *flat,* stalkless clusters, 2–4″ (5–10 cm) wide; in spring.
Fruit: ½″ (12 mm) long; *elliptical,* slightly *flat,* dark *blue-black* with whitish bloom; thin, slightly sweetish edible pulp; somewhat flat stone; drooping on long slender reddish stalks; maturing in autumn, remaining attached into early winter.

Habitat: Moist soils, especially in valleys, and on slopes; in thickets and at borders of forests.

Range: SW. Connecticut south to Alabama, west to E. Kansas, and north to SE. Wisconsin and SW. Iowa; to 3000′ (914 m).

The fruit is consumed by songbirds, gamebirds, and mammals and can be made into preserves. The astringent bark was formerly used medicinally. The Latin species name refers to the leaves' resemblance to plum leaves.

204, 411 **Rusty Blackhaw**
"Bluehaw" "Rusty Nannyberry"
Viburnum rufidulum Raf.

Description: Large shrub or small tree with short trunk; spreading, irregular crown; many small white flowers; leaves covered with rust-colored hairs beneath; and small blue fruit.
Height: 20′ (6 m).
Diameter: 6″ (15 cm).
Leaves: opposite; 2–4″ (5–10 cm) long,

1–2½" (2.5–6 cm) wide. *Elliptical; finely saw-toothed;* slightly thick. *Shiny green* with network of sunken veins above, covered with *rust-colored hairs beneath,* especially along veins; turning shiny red in autumn. Short leafstalks covered with rust-colored hairs.
Bark: gray; rough, furrowed into rectangular plates.
Twigs: slender, stiff, covered with rust-colored hairs when young, turning gray; ending in oblong, flat, *reddish-brown hairy bud.*
Flowers: ¼" (6 mm) wide; with 5 rounded *white corolla lobes;* in upright *flat,* stalkless clusters, 3–6" (7.5–15 cm) wide, and covered with rust-colored hairs.
Fruit: ½" (12 mm) long; *elliptical;* slightly *flat; blue with whitish bloom;* slightly sweetish edible pulp; somewhat flat stone; drooping on slender reddish stalks; maturing in late summer and autumn, remaining attached into early winter.

Habitat: Uplands and less often in valleys; in forests and at edges of woods.

Range: SE. Virginia, south to N. Florida, west to central Texas and north to central Missouri; to 2500′ (762 m).

Rusty Blackhaw is distinguished from more northern Blackhaw primarily by the reddish-brown hairs on foliage and other parts, as well as by the slightly larger leaves and paler blue fruit. The Latin species name, meaning "reddish," also refers to the hairs. The two species intergrade where their ranges meet.

COMPOSITE FAMILY
(Compositae)

The second largest family of seed plants, nearly all herbs and a few shrubs; in the tropics, sometimes small to medium-sized trees. Worldwide, with about 13,000 species; 2 native tree, numerous herb, and some shrub species in North America.
Leaves: generally alternate, sometimes opposite; simple, thin, often toothed or lobed, without stipules.
Flowers: crowded in heads bordered by persisting green bracts; small, usually bisexual (sometimes male and female), with calyx reduced to hairs or scales and corolla tubular, regular with 5 teeth (disk flower) or irregular with flat ray (ray flower); 5 stamens inserted in tube and united by anthers, and 1 pistil.
Fruit: dry, seedlike (akene), often somewhat flat, with hairs or scales at tip.

211, 513 **Eastern Baccharis**
"Saltbrush" "Groundsel-tree"
Baccharis halimifolia L.

Description: Shrub or small, much-branched, spreading and rounded tree with fruit resembling silvery paintbrushes.

Height: 16′ (5 m).
Diameter: 4″ (10 cm).
Leaves: deciduous (in far Southeast, evergreen); 1–2½″ (2.5–6 cm) long, ⅜–1¼″ (1–3 cm) wide. *Elliptical* to obovate; usually a *few coarse teeth* toward short-pointed tip; tapering to short-stalked base; slightly thick; hairless or resinous. *Dull gray-green* above and beneath; falling late in autumn.
Bark: dark brown; thin, fissured into forking ridges.
Twigs: green, gray, or brown; slender, angled, hairless or nearly so.

Flowers: tiny; crowded in *greenish bell-shaped heads* less than ¼″ (6 mm) long; the heads in large upright branching clusters; in late summer and autumn. Male and female flowers on separate plants.

Fruit: ½″ (12 mm) long and wide; *brushlike whitish head* composed of many narrow seeds, each with tuft of *whitish hairs* or bristles to ⅜″ (10 mm) long; maturing in late autumn.

Habitat: Moist soils, including salt marshes, borders of streams, roadside ditches, open woods, and waste places.

Range: Massachusetts south to S. Florida, west to S. Texas, and north to SE. Oklahoma and Arkansas; to 800′ (244 m); also Bahamas and Cuba.

Apparently extending its natural range inland from the coastal plain, Eastern Baccharis is the only native eastern species of the Composite family reaching tree size. Baccharis is the ancient Greek name (derived from the god Bacchus) of a plant with fragrant roots. The Latin species name means "with the leaves of *Halimus,*" an old name for Saltbush, an unrelated shrub. Tolerant of saltwater spray, this handsome ornamental is one of the few eastern shrubs suitable for planting near the ocean.

Part III
Appendices

GLOSSARY

Achene A small, dry, seedlike fruit with a thin wall that does not open.

Acorn The hard-shelled, 1-seeded nut of an oak, with a pointed tip and a scaly cup at the base.

Aggregate fruit A fused cluster of several fruits, each one formed from an individual ovary, but all derived from a single flower, as in a magnolia.

Alternate Arranged singly along a twig or shoot, and not in whorls or opposite pairs.

Anther The terminal part of a stamen, containing pollen in 1 or more pollen-sacs.

Axis The central stalk of a compound leaf or flower cluster.

Bark The outer covering of the trunk and branches of a tree, usually corky, papery, or leathery.

Berry A fleshy fruit with more than 1 seed.

Bipinnate With leaflets arranged on side branches off a main axis; twice-pinnate; bipinnately compound.

Bisexual With male and female organs in the same flower.

Blade The broad, flat part of a leaf.

Bract A modified and often scalelike leaf, usually located at the base of a flower, a fruit, or a cluster of flowers or fruits.

Bud A young and undeveloped leaf, flower, or shoot, usually covered tightly with scales.

Calyx Collective term for the sepals of a flower.

Capsule A dry, thin-walled fruit containing 2 or more seeds and splitting along natural grooved lines at maturity.

Catkin A compact and often drooping cluster of reduced, stalkless, and usually unisexual flowers.

Cell A cavity in an ovary or fruit, containing ovules or seeds.

Compound leaf A leaf whose blade is divided into 3 or more smaller leaflets.

Cone A conical fruit consisting of seed-bearing, overlapping scales around a central axis.

Cone-scale One of the scales of a cone.

Conifer A cone-bearing tree of the Pine family, usually evergreen.

Corolla Collective term for the petals of a flower.

Crown The mass of branches, twigs, and leaves at the top of a tree, with particular reference to its shape.

Cultivated Planted and maintained by man.

Deciduous Shedding leaves seasonally, and leafless for part of the year.

Drupe A fleshy fruit with a central stonelike core containing 1 or more seeds.

Elliptical Elongately oval, about twice as long as wide, and broadest at the middle; like an ellipse.

Entire Smooth-edged, not lobed or toothed.

Escaped Spread from cultivation and now growing and reproducing without aid from man.

Family A group of related genera.

Filament The threadlike stalk of a stamen.

Fleshy fruit A fruit with juicy or mealy pulp.

Flower The reproductive structure of a tree or other plant, consisting of at least 1 pistil or stamen, and often including petals and sepals.

Follicle A dry, 1-celled fruit, splitting at maturity along a single grooved line.

Fruit The mature, fully developed ovary of a flower, containing 1 or more seeds.

Genus A group of closely related species. Plural, genera.

Gland-dot A tiny, dotlike gland or pore, usually secreting a fluid.

Habit The characteristic growth form or general shape of a plant.

Herb A plant with soft, not woody, stems, dying to the ground in winter.

Hybrid A plant or animal of mixed parentage, resulting from the interbreeding of 2 different species.

Intergradation The gradual merging of 2 or more distinct forms or kinds, through a series of intermediate forms.

Introduced Intentionally or accidentally established in an area by man, and not native; exotic or foreign.

Irregular flower A flower with petals of unequal size.

Keel A sharp ridge or rib, resembling the keel of a boat, found on some fruits and seeds.

Key A dry, 1-seeded fruit with a wing; a samara.

Lanceolate Shaped like a lance, several times longer than wide, pointed at the tip and broadest near the base.

Leader The highest, terminal shoot of a plant.

Leaflet One of the leaflike subdivisions of a compound leaf.

Linear Long, narrow, and parallel-sided.

Lobed With the edge of the leaf deeply but not completely divided.

Male cone The conical, pollen-bearing male element of a conifer.

Midvein The prominent, central vein or rib in the blade of a leaf; midrib.

Multiple fruit A fused cluster of several fruits, each one derived from a separate flower, as in a mulberry.

Native Occurring naturally in an area and not introduced by man; indigenous.

Naturalized Successfully established and reproducing naturally in an area where not native.

Needle The very long and narrow leaf of pines and related trees.

Node The point on a shoot where a leaf, flower, or bud is attached.

Nut A dry, 1-seeded fruit with a thick, hard shell that does not split along a natural grooved line.

Nutlet One of several small, nutlike parts of a compound fruit, as in a sycamore; the hard inner core of some fruits, containing a seed and surrounded by softer flesh, as in a hackberry.

Oblanceolate Reverse lanceolate; shaped like a lance, several times longer than wide, broadest near the tip and pointed at the base.

Oblong With nearly parallel edges.

Obovate Reverse ovate; oval, with the broader end at the tip.

Opposite Arranged along a twig or shoot in pairs, with 1 on each side, and not alternate or in whorls.

Ovary The enlarged base of a pistil, containing 1 or more ovules.

Ovate Oval, with the broader end at the base.

Ovule A small structure in the cell of an ovary, containing the egg and ripening into a seed.

Palmate With leaflets attached directly to the end of the leafstalk and not arranged in rows along an axis; palmately compound; digitate.

Palmate-veined With the principal veins arising from the end of the leafstalk and radiating toward the edge of the leaf, and not branching from a single midvein.

Parallel-veined	With the veins running more or less parallel toward the tip of the leaf.
Persistent	Remaining attached, and not falling off.
Petal	One of a series of flower parts lying within the sepals and next to the stamens and pistil, often large and brightly colored.
Pinnate	With leaflets arranged in 2 rows along an axis; pinnately compound.
Pinnate-veined	With the principal veins branching from a single midvein, and not arising from the end of the leafstalk and radiating toward the edge of the leaf.
Pistil	The female structure of a flower, consisting of stigma, style, and ovary.
Pith	The soft, spongy, innermost tissue in a stem.
Pod	A dry, 1-celled fruit, splitting along natural grooved lines, with thicker walls than a capsule.
Pollen	Minute grains containing the male germ cells and released by the stamens.
Pollen-sac	The part of an anther containing the pollen.
Pome	A fruit with fleshy outer tissue and a papery-walled, inner chamber containing the seeds.
Regular flower	A flower with petals all of equal size.
Resin	A plant secretion, often aromatic, that is insoluble in water but soluble in ether or alcohol.
Ring scar	A ringlike scar left on a twig after a leaf falls away.

Samara	A dry, 1-seeded fruit with a wing; a key.
Scale	One of the very short, pointed, and overlapping leaves of some conifers; the leaflike covering of a bud or of the cup of an acorn.
Scurfy	Covered with tiny, broad scales.
Seed	A fertilized and mature ovule, containing an embryonic plant.
Sepal	One of the outermost series of flower parts, arranged in a ring outside the petals, and usually green and leaflike.
Sheath	A tubular, surrounding structure; in some conifers, the papery tube enclosing the base of a bundle of needles.
Shoot	A young, actively growing twig or stem.
Shrub	A woody plant, smaller than a tree, with several stems or trunks arising from a single base; a bush.
Simple fruit	A fruit developed from a single ovary.
Simple leaf	A leaf with a single blade, not compound or composed of leaflets.
Sinus	The space between 2 lobes of a leaf.
Solitary	Borne singly, not in pairs or clusters.
Species	A kind or group of plants or animals, composed of populations of individuals that interbreed and produce similar offspring.
Spur	A short side twig, often bearing a cluster of leaves.
Stamen	One of the male structures of a flower, consisting of a threadlike filament and a pollen-bearing anther.

Stigma The tip of a pistil, usually enlarged, that receives the pollen.

Stipule A leaflike scale, often paired, at the base of a leafstalk in some trees.

Style The stalklike column of a pistil, arising from the ovary and ending in a stigma.

Toothed With an edge finely divided into short, toothlike projections.

Tree line The upper limit of tree growth at high latitudes or on mountains; timberline.

Trunk The major woody stem of a tree.

Tubular With the petals partly united to form a tube.

Unisexual With male or female organs, but not both, in a single flower.

Vein One of the riblike vessels in the blade of a leaf.

Whorled Arranged along a twig or shoot in groups of three or more at each node.

Wing A thin, flat, dry, shelflike projection on a fruit or seed, or along the side of a twig.

Wood The hard, fibrous inner tissue of the trunk and branches of a tree or shrub.

PICTURE CREDITS

All photographs were taken by Sonja Bullaty and Angelo Lomeo with the exception of those listed below.

The numbers in parentheses are plate numbers. Some photographers have pictures under agency names as well as their own. Agency names appear in boldface. Photographers hold the copyright to their works.

William Aplin (396, 461)
Les Blacklock (592)
Richard W. Brown (179 left)

Bruce Coleman, Inc.
Jane Burton (625)

Robert P. Carr (22 right, 374, 384, 476)
David Cavagnaro (16 left, 28 left and right, 29 left, 37 right, 38 left and right, 39 right, 159 left and right, 184 left, 200 left and right, 236 left, 265 right, 274 left and right, 295 left and right, 297 left, 299 left, 335 left, 428)
Scooter Cheatham (37 left, 40 left and right, 41 left and right, 45 right, 74 left and right, 96 left and right, 112 left and right, 120 left and right, 127 left and right, 144 left and right, 151 left and right, 171 left, 172 left and right, 183 right, 196 left and right, 197 left, 210 left and right, 216 left and right, 223 left and right, 227 left and right, 240 left and right, 248 left and right, 252 left and right, 272 left and right, 273 right, 296 left and right, 297 right, 298 left and right, 299 right, 300 left and right, 301 left and right, 302 left and right, 303 left and right, 307 left and right, 308 left and right, 318 left and right, 331 left and right, 340 right, 349 left and

right, 351 left and right, 354 left and right, 400, 406, 470, 471, 502, 504, 510, 553, 562, 563)
Stephen Collins (494, 542)
Steve Crouch (398, 505)
Kent and Donna Dannen (157 right, 357 left, 367, 567)
E. R. Degginger/Color-Pic, Inc. (160 right, 363 right, 472, 479, 594, 597)
Jack Dermid (2 left, 4 left, 51 left, 404, 429, 455, 588)
Wilbur H. Duncan (42 right, 48 right, 49 left and right, 109 right, 167 right, 269 right, 290 left, 323 left, 436, 491, 497, 558, 576)
Robert L. Eikum (51 right, 368)
William Fehrenbach (39 left, 294 left and right)
Fred Galle (30 left and right, 242 left, 290 right)
Lois Theodora Grady (6 right, 15 left, 194 left, 539, 545)
Farrell Grehan (366, 370, 379)
Raymond P. Guries (184 right, 414)
Pamela J. Harper (42 left, 43 left, 154 left, 157 left, 167 left, 178 left, 202 right, 236 right, 237 right, 238 right, 335 right, 365, 402, 410, 426, 433, 434, 442, 447, 464, 469, 503, 538, 557, 561, 565, 577, 580, 589, 591)
Hayes Regional Arboretum (146 left and right)
Grant Heilman Photography (399, 481, 530, 535, 564)
Walter H. Hodge (255 left, 256 left, 306 left, 415, 418, 420, 421, 446, 482, 547, 569, 571, 575)
Charles C. Johnson (6 left, 413, 438)
Bill Jordan (117 left and right)
J. James Kielbaso (7 left, 603)
John Kohout (334 left and right)
Donald J. Leopold (14 left, 175 left)
Clarence E. Lewis (282 right, 582, 593, 602)
Kenneth Lewis (604)
C. D. Luckhart (107 right)
John A. Lynch (180 left, 423, 583, 584, 587, 596, 598, 599, 600, 605, 606, 607, 608, 609, 610, 611, 612,

613, 616, 617, 618, 628)
Raymond A. Mendez (590)
Wendell D. Metzen (362 left)

Photo Researchers, Inc.
A–Z Collection (375, 378) A. W. Amber (386, 467, 570) Charles R. Belinky (549) Ken Brate (477, 520, 521) Kent and Donna Dannen (620) Michael P. Gadomski (484) P. W. Grace (478) W. Harlow (475, 512) Russ Kinne (480, 508, 581) Irwin L. Oakes (496) Richard Parker (519, 555) Alvin E. Staffan (372, 380, 381, 407, 492, 493, 498, 499, 500, 516, 522, 523, 525) Paul E. Taylor (495) Mary M. Thacher (541, 543) V. P. Weinland (191 right, 417, 513) Jeanne White (397)
Racz-Debreczy (486)
Betty Randall and Robert Potts (113 left)
Susan Rayfield (385, 387)
Clark Schaack (183 left)
Secrest Arboretum (395)
Terry L. Sharik (202 left)
B. M. Shaub (425)
Arlo I. Smith (451, 474)
John J. Smith (152 right, 182 right, 239 right, 405, 411, 412, 462)
Norman F. Smith (254 left)
Richard Spellenberg (100 left, 107 left, 340 left, 341 left and right, 391, 392, 546)
Alvin E. Staffan (147 left, 371, 394, 473, 483, 485, 501, 507, 514, 515, 517, 518, 525, 526, 528, 529, 531, 532, 533, 534, 536, 537, 540, 544, 548, 550, 551, 552, 554, 559, 560, 566, 568, 572, 573, 574, 614)
Rick Sullivan and Diana Rogers (13 left and right, 29 right, 45 left, 48 left, 56 left and right, 60 left and right, 61 left and right, 104 left and right, 109 left, 114 left and right, 145 left and right, 171 right, 197 right, 215 left and right, 217 left and right, 234 left and right, 249 left and right, 262 left and right, 269 left, 273 left,

337 left and right, 338 left and right, 435, 489, 527)

Gurdun Tarbox (408, 443)
Ray E. Umber (409, 437)
Larry West (152 left, 376, 458)
West Virginia University Herbarium (401)
Klinton M. Wigren (601)

INDEX

All species are listed by both the scientific and the accepted common names. Many species also have local names which are in quotation marks and refer the reader to the species description under the accepted name. Numbers in bold-face type refer to plate numbers. Numbers in italic type refer to page numbers. Circles preceding the accepted English names make it easy for you to keep a record of the trees you have seen.

A

B

P

THE AUTHOR

Elbert L. Little (1907–2004) was Chief Dendrologist (tree identification specialist) for the United States Forest Service for many years. After retiring, he served as a Research Associate in the Department of Botany, U.S. National Museum of Natural History, Smithsonian Institution. An international authority on tree classification, identification, and distribution, Dr. Little is the author of numerous technical and popular publications on the trees of the United States and tropical America, including *Checklist of United States Trees* (1979), the 6-volume *Atlas of United States Trees* (1971–81), *Common Trees of Puerto Rico and the Virgin Islands* (1964), and *Alaska Trees and Shrubs* (1972). He conducted field studies of New World trees from Alaska to Patagonia and, as a university visiting professor, taught tree identification to forestry students in Spanish and English. Dr. Little received MS and PhD degrees in botany from the University of Chicago in 1929; he was a Fellow of the Society of American Foresters and recipient of Distinguished Service Awards from both the U.S. Department of Agriculture and the American Forestry Association.

NATIONAL AUDUBON SOCIETY FIELD GUIDE SERIES

Also available in this unique all-color, all photographic format:

African Wildlife • Birds *(Eastern Region)* • Birds *(Western Region)* • Butterflies • Fishes • Fossils • Insects and Spiders • Mammals • Mushrooms • Night Sky • Reptiles and Amphibians • Rocks and Minerals • Seashells • Seashore Creatures • Trees *(Western Region)* • Tropical Marine Fishes • Weather • Wildflowers *(Eastern Region)* • Wildflowers *(Western Region)*

Prepared and produced by Chanticleer Press, Inc.

Founding Publisher: Paul Steiner
Publisher: Andrew Stewart

Staff for this book:

Editor-in-Chief: Gudrun Buettner
Executive Editor: Susan Costello
Managing Editor: Jane Opper
Guides Editor: Susan Rayfield
Project Editor: Olivia Buehl
Assistant Editor: Mary Beth Brewer
Natural Science Editor: John Farrand, Jr.
Production: Helga Lose
Art Director: Carol Nehring
Picture Library: Edward Douglas
Symbols and Range Maps: Paul Singer
Drawings: Bobbi Angell
Silhouettes: Dolores R. Santoliquido

Original series design by Massimo Vignelli

All editorial inquiries should be addressed to:
Field Guides
P.O. Box 479
Ascutney, VT 05030
editors@thefieldguideproject.com

To purchase this book or other National Audubon Society illustrated nature books, please contact:
Alfred A. Knopf
1745 Broadway
New York, NY 10019
(800) 733-3000
www.aaknopf.com